T0255091

The series Lecture Notes in Computer Science (LNCS), including its subseries Lecture Notes in Artificial Intelligence (LNAI) and Lecture Notes in Bioinformatics (LNBI), has established itself as a medium for the publication of new developments in computer science and information technology research, teaching, and education.

LNCS enjoys close cooperation with the computer science R & D community, the series counts many renowned academics among its volume editors and paper authors, and collaborates with prestigious societies. Its mission is to serve this international community by providing an invaluable service, mainly focused on the publication of conference and workshop proceedings and postproceedings. LNCS commenced publication in 1973.

Andreas Holzinger · Peter Kieseberg ·
Federico Cabitza · Andrea Campagner ·
A Min Tjoa · Edgar Weippl
Editors

Machine Learning and Knowledge Extraction

7th IFIP TC 5, TC 12, WG 8.4, WG 8.9, WG 12.9
International Cross-Domain Conference, CD-MAKE 2023
Benevento, Italy, August 29 – September 1, 2023
Proceedings

Editors
Andreas Holzinger
University of Natural Resources and Life
Vienna, Austria

Medical University Graz & Graz University of Technology
Graz, Austria

University of Alberta
Edmonton, AB, Canada

Federico Cabitza
University of Milano-Bicocca
Milan, Italy

A Min Tjoa
TU Wien
Vienna, Austria

Peter Kieseberg
St. Pölten University of Applied Science
St. Pölten, Austria

Andrea Campagner
University of Milano-Bicocca
Milan, Italy

Edgar Weippl
SBA Research
Vienna, Austria

University of Vienna
Vienna, Austria

ISSN 0302-9743 ISSN 1611-3349 (electronic)
Lecture Notes in Computer Science
ISBN 978-3-031-40836-6 ISBN 978-3-031-40837-3 (eBook)
https://doi.org/10.1007/978-3-031-40837-3

This Springer imprint is published by the registered company Springer Nature Switzerland AG
The registered company address is: Gewerbestrasse 11, 6330 Cham, Switzerland

Preface

The International Cross-Domain Conference for Machine Learning & Knowledge Extraction (CD-MAKE) is a joint effort of IFIP TC 5, TC 12, IFIP WG 8.4, IFIP WG 8.9 and IFIP WG 12.9 and is held in conjunction with the International Conference on Availability, Reliability and Security (ARES) – this time in beautiful Benevento, Italy. Thanks to the end of the Corona Pandemic, which affected us all heavily, we are all happy that we can meet all our international colleagues and friends in-vivo again.

For those who are new to our traditional conference: The letters CD in CD-MAKE stand for "Cross-Domain" and describe the integration and appraisal of different fields and application domains to provide an atmosphere to foster different perspectives and opinions. We are strongly convinced that exactly this cross-domain approach is very fruitful for new developments and novel discoveries. The conference fosters an integrative machine learning approach, considering the importance of data science and visualization for the algorithmic pipeline with a strong emphasis on privacy, data protection, safety and security. It is dedicated to offer an international platform for novel ideas and a fresh look at methodologies to put crazy ideas into business for the benefit of humans. Serendipity is a desired effect, leading to cross-fertilization of methodologies and transfer of algorithmic developments.

The acronym MAKE stands for "MAchine Learning & Knowledge Extraction", a field of Artificial Intelligence (AI) that, while quite old in its fundamentals, has just recently begun to thrive based on both novel developments in the algorithmic area and the availability of vast computing resources at a comparatively low cost.

Machine learning (ML) studies algorithms that can learn from data to gain knowledge from experience and to generate decisions and predictions. A grand goal is to understand intelligence for the design and development of algorithms that work autonomously (ideally without a human-in-the-loop) and can improve their learning behaviour over time. The challenge is to discover relevant structural and/or temporal patterns ("knowledge") in data, which is often hidden in arbitrarily high-dimensional spaces, and thus simply not accessible to humans. Knowledge Extraction is one of the oldest fields in AI and is seeing a renaissance, particularly in the combination of statistical methods with classical ontological approaches.

AI is currently undergoing a kind of Cambrian explosion and is the fastest growing field in computer science today thanks to the successes in machine learning to help to solve real-world problems. There are many application domains, e.g., in agriculture, climate research, forestry, etc. with many use cases from our daily lives which can be useful to help to solve various problems Examples include recommender systems, speech recognition, autonomous driving, cyber-physical systems, robotics, etc.

However, in our opinion the grand challenges are in sensemaking, in context understanding, and in decision making under uncertainty, as well as solving the problem of human interpretability, explainability and verification.

Our real world is full of uncertainties and probabilistic inference has enormously influenced AI generally and ML specifically. The inverse probability allows one to infer unknowns, to learn from data and to make predictions to support decision-making. Whether in social networks, recommender systems, health applications or industrial applications, the increasingly complex data sets require a joint interdisciplinary effort bringing the human-in-control and to foster ethical and social issues, accountability, retractability, explainability, causability and privacy, safety and security!

A few words about IFIP and the importance of it: IFIP – the International Federation for Information Processing – is the leading multi-national, non-governmental, apolitical organization in Information & Communications Technologies and Computer Sciences. IFIP is recognized by the United Nations (UN) and was established in the year 1960 under the auspices of UNESCO as an outcome of the first World Computer Congress, held in Paris in 1959.

IFIP is incorporated in Austria by decree of the Austrian Foreign Ministry (20th September 1996, GZ 1055.170/120-I.2/96), granting IFIP the legal status of a non-governmental international organization under the Austrian Law on the Granting of Privileges to Non-Governmental International Organizations (Federal Law Gazette 1992/174). IFIP brings together more than 3500 scientists without boundaries from both academia and industry, organized in more than 100 Working Groups (WGs) and 13 Technical Committees (TCs).

To acknowledge all those who contributed to the efforts and stimulating discussions would be impossible in a preface like this. Many people contributed to the development of this volume, either directly or indirectly, so it would be completely impossible to list all of them. We herewith thank all local, national and international colleagues and friends for their positive and supportive encouragement. Finally, yet importantly, we thank the Springer management team and the Springer production team for their professional support.

Thank you to all! Let's MAKE it cross-domain!

Andreas Holzinger
Peter Kieseberg
Federico Cabitza
Andrea Campagner
Edgar Weippl
A Min Tjoa

Organization

CD-MAKE Steering Committee

Andreas Holzinger	University of Natural Resources and Life Sciences Vienna, Austria, Medizinische Universität Graz, Austria, Graz University of Technology, Austria & xAI Lab, Alberta Machine Intelligence Institute, University of Alberta, Canada
Peter Kieseberg	FH St.Pölten, Austria
Edgar Weippl (IFIP WG 8.4 Chair)	SBA Research & University of Vienna, Austria
A Min Tjoa (IFIP WG 8.9. Chair, Honorary Secretary IFIP)	TU Vienna, Austria

Head of CD-MAKE Organization

Bettina Jaber	SBA Research, Austria

CD-MAKE Organizing Team

Bettina Jaber	SBA Research, Austria
Daniela Freitag David	SBA Research, Austria
Barbara Friedl	St. Pölten University of Applied Sciences, Austria

Proceedings Manager

Bettina Jaber	SBA Research, Austria

PC Chairs

Federico Cabitza	University of Milano-Bicocca, Italy
Andrea Campagner	University of Milano-Bicocca, Italy

Program Committee

Cecilia Alm	Rochester Institute of Technology, USA
Boudour Ammar	University of Sfax, Tunisia
Nicholas Asher	CNRS Laboratoire IRIT, Université Paul Sabatier, France
Frantisek Babic	Technical University of Košice, Slovakia
Christian Bauckhage	University of Bonn, Germany
Smaranda Belciug	University of Craiova, Romania
Tarek R. Besold	Alpha Health, Spain
Michele Bezzi	SAP Labs France, France
Przemek Biecek	Warsaw University of Technology, Poland
Guido Bologna	Université de Genève, Switzerland
Ivan Bratko	University of Ljubljana, Slovenia
Francesco Buccafurri	Università Mediterranea di Reggio Calabria, Italy
Frederico Cabitza	Università degli Studi di Milano, Italy
Andre Calero-Valdez	Universität zu Lübeck, Germany
Andrea Campagner	University of Milano-Bicocca, Italy
Angelo Cangelosi	University of Manchester, UK
Mirko Cesarini	Università di Milano-Biocca, Italy
Carlo Combi	University of Verona, Italy
Tim Conrad	Freie Universität Berlin, Germany
Gloria Cerasela Crisan	Vasile Alecsandri University of Bacau, Romania
Beatriz De La Iglesia	University of East Anglia, UK
Javier Del Ser	Universidad del País Vasco/Euskal Herriko Unibertsitatea, Spain
Josep Domingo-Ferrer	Universitat Rovira i Virgili, Catalonia
Isao Echizen	National Institute of Informatics, Japan
Massimo Ferri	University of Bologna, Italy
Hugo Gamboa	Universidadae Nova de Lisboa, Portugal
Arne Gevaert	Ghent University, Belgium
Randy Goebel	University of Alberta, Canada
Pitoyo Hartono	Chukyo University, Japan
Barna Laszlo Iantovics	George Emil Palade University of Medicine, Pharmacy, Science and Technology of Targu Mures, Romania
Igor Jurisica	IBM Life Sciences Discovery Centre, and Princess Margaret Cancer Centre, Canada
Epaminondas Kapetanios	University of Hertfordshire, UK
Andreas Kerren	Linköping University & Linnaeus University, Sweden
Peter Kieseberg	FH St.Pölten, Austria

Styliani Kleanthous	CYENS Centre of Excellence, Cyprus
Freddy Lecue	Institut national de recherche en sciences et technologies du numérique (Inria), France
Lenka Lhotska	Czech Technical University Prague, Czech Republic
Luca Longo	Trinity College Dublin, Ireland
Bradley Malin	Vanderbilt University, USA
Rudolf Mayer	SBA Research, Austria
Fabio Mercorio	University of Milano-Bicocca, Italy
Yoan Miche	Nokia Bell Labs, Finland
Timothy Miller	University of Melbourne, Australia
Daniel E. O'Leary	University of Southern California, USA
Vasile Palade	Coventry University, UK
Francesco Piccialli	University of Naples Federico II, Italy
Camelia-M. Pintea	Technical University of Cluj-Napoca, Romania
Catarina Pinto Moreira	Técnico Lisboa, Portugal
Fabrizio Riguzzi	Università di Ferrara, Italy
Lior Rokach	Ben-Gurion University of the Negev, Israel
Luca Romeo	Università Politecnica delle Marche, Italy
Yvan Saeys	Ghent University, Belgium
Pierangela Samarati	Università degli Studi di Milano, Italy
Marco Scutari	Istituto Dalle Molle di Studi sull'Intelligenza Artificiale, Switzerland
Andrea Seveso	University of Milano-Bicocca, Italy
Andrzej Skowron	Systems Research Institute, Polish Academy of Sciences & Digital Science and Technology Centre, UKSW, Poland
Irena Spasic	Cardiff University, UK
Ivan Štajduhar	University of Rijeka, Croatia
A Min Tjoa	Vienna University of Technology, Austria
Isaac Triguero	University of Nottingham, UK
Cagatay Turkay	University of Warwick, UK
Vaibahv Unhelkar	Rice University, USA
Andrea Vedaldi	University of Oxford, UK
Katarzyna Wac	University of Geneva, Switzerland
Edgar Weippl	SBA Research & University of Vienna, Austria
Jie Yang	TU Delft, The Netherlands
Pinar Yildirim	Okan University, Turkey
Jianglong Zhou	University of Technology Sydney, Australia

Contents

About the Editors

Andreas Holzinger pioneered in interactive machine learning with the human-in-the-loop. For his achievements he was elected a member of Academia Europaea in 2019, the European Academy of Science, the European Laboratory for Learning and Intelligent Systems (ELLIS) in 2020, and Fellow of the international federation of information processing (ifip) in 2021. The use of Artificial Intelligence (AI) in domains that impact human life (agriculture, climate, forestry, health, ...) has led to an increased demand for trustworthy AI. Consequently, Andreas Holzinger is working together with his group on generic methods to promote robustness and explainability to foster secure AI solutions and advocates a synergistic approach to Human-Centered AI to provide human control over AI and to align AI with human values, ethical principles, and legal requirements to ensure privacy, security, and safety. Andreas' aim is to explain why a machine decision has been made, paving the way towards explainable AI and Causability ultimately fostering ethical responsible machine learning and trust and acceptance for AI. Andreas obtained a Ph.D. in Cognitive Science from Graz University in 1998 and his Habilitation (second Ph.D.) in Computer Science from Graz University of Technology in 2003. Andreas was Visiting Professor for Machine Learning & Knowledge Extraction in Verona, RWTH Aachen, University College London and Middlesex University London. Since 2019 Andreas is Visiting Professor for explainable AI, Alberta Machine Intelligence Institute, University of Alberta, Canada. Andreas Holzinger has been appointed full professor for digital transformation at the University of Natural Resources and Life Sciences Vienna with effect from 1 March 2022.He is Austrian representative for AI in the IFIP TC 12 and member of IFIP WG 12.9 Computational Intelligence, the ACM, IEEE, GI, and the Austrian Computer Science (OCG).

Peter Kieseberg heads the "Institute of IT Security Research" at St. Poelten University of Applied Sciences, as well as the JRC for Blockchain Technologies & Security Management. Peter received a master's degree in Technical Mathematics in Computer Science from the VTU Wien with specializations in cryptography and numerical mathematics. He worked as a consultant in the telecommunication sector for several years before joining SBA Research. In 2017 he joined St. Poelten University of Applied Sciences as lecturer responsible for the focus topic "Privacy". Since 2014 he is also visiting doctoral researcher at the Holzinger Group Human-Centered AI and involved in standardization activities at ETSI (TC-CYBER). His main research interests include various topics in the area of IT-Security: Privacy protection & the GDPR, securing blockchains and DLTs, as well as fingerprinting of sensitive data.

Federico Cabitza Computer Engineer (BSc MEng 2001), PhD in Computer Science (2007) is an associate professor at the University of Milan-Bicocca (Milan, Italy) where he teaches human-computer interaction at the Bachelor of Science in Computer Science, interaction design and information systems at the Master of Science in Computer Science and Communication Theory and Technology, human-AI interaction at the Bachelor inter-university degree in Artificial Intelligence and decision support at the Master of Science in Artificial Intelligence for Science and Technology. At Bicocca he is director of the local node of the CINI national laboratory on "Computer Science and Society" and is head of the Laboratory of Models for Uncertainty, Decisions and Interactions. A lecturer in the doctoral program in computer science, he has taught courses and teaching modules in numerous first- and second-level master's degree programs and Graduate Schools and courses and events organized by companies in the educational sector (including IQVIA Italy, Sudler & Hennessey Italy, VMLY&R Italy). A speaker at numerous cultural and scientific initiatives (including 4words, Forum Risk Management, AltroConsumo), he is also the author of more than 150 scientific publications, in international conference proceedings, edited books (including a book published by MIT Press) and scientific journals, since 2020 he has also been an associate editor of the International Journal of Medical Informatics; in 2021 and 2022 he was counted in the top 2% of scientific impact among all researchers who are authors of indexed papers worldwide. His current research interests include AI impact evaluation, design and evaluation of ML-based decision support systems in organizational (mainly clinical) settings, and in particular the phenomena of automation bias (decision support overdependence) and related deskilling.

Andrea Campagner is a Postdoc Researcher at the IRCCS Ospedale Galeazzi-Sant'Ambrogio, the major Italian research hospital devoted to orthopaedics and traumatology, where he works on applications of statistical and machine learning methodologies to clinical decision making and clinical process management. Andrea is also a Teaching Assistant at the University of Milano-Bicocca, and an external collaborator and member of the MUDI (Modeling Uncertainty, Decisions and Interaction) lab at the same university. Previously he obtained my PhD in Computer Science at the University of Milano-Bicocca, working in the MUDI lab. His main research interests are uncertainty management, machine learning, medical AI and human-AI interaction, as well as their intersection.

Edgar Weippl is research director of SBA Research and professor at the University of Vienna. After graduating with a Ph.D. from the TU Wien, Edgar worked in a research startup for two years. He then spent one year teaching as an Assistant Professor at Beloit College, WI. From 2002 to 2004, while with the software vendor ISIS Papyrus, he worked as a consultant in New York, NY and Albany, NY, and in Frankfurt, Germany. In 2004 he joined the TU Wien and founded the research center SBA Research together with A Min Tjoa and Markus Klemen. Edgar R. Weippl (CISSP, CISA, CISM, CRISC, CSSLP, CMC) is member of the editorial board of IEEE Transactions on Information Forensics & Security (IEEE T-IFS) and Computers & Security (COSE), is steering committee chair of the ARES conference and is General Chair IEEE EuroS&P 2021.

A Min Tjoa is Full Professor for Software Technology at TU Wien. From 1985 to 1994 he worked as Full Professor at the Institute of Statistics and Computer Science of the TU Wien where he was the Head of this institute from 1985 to 1987. He was Visiting Professor at the universities of Zurich, Kyushu and Wroclaw (Poland) and at the Technical Universities of Prague and Lausanne (Switzerland). He acted as the President of the Austrian Computer Society from 1999 to 2003. He is member of the IFIP Technical Committee for Information Systems. Since 2000 A Min is Vice-Chairman of the DEXA-Association (Database and Expert Systems Applications). Since 2006 he is Vice-Chairman of the IFIP Working Group 8.9 (Enterprise Information Systems). He has been elected for Vice-President of the Infoterm (International Information Center for Terminology) for the term 2010-2012. Since 2017 he is also chief scientific officer of SCCH, a COMET center for software technology in Hagenberg in Upper Austria.

Controllable AI - An Alternative to Trustworthiness in Complex AI Systems?

Peter Kieseberg[1,2], Edgar Weippl[3,4], A. Min Tjoa[3,5], Federico Cabitza[6,9], Andrea Campagner[6], and Andreas Holzinger[7,8(✉)]

[1] Institute of IT Security, St. Pölten University of Applied Sciences, St. Pölten, Austria
[2] Josef Ressel Center for Blockchain-Technologies and Security management, Pölten, Austria
[3] Secure Business Austria, SBA Research, Vienna, Austria
[4] Research Group Security and Privacy, University of Vienna, Vienna, Austria
[5] Information Systems Engineering, Vienna University of Technology, Vienna, Austria
[6] DISCo, University of Milano-Bicocca, Milan, Italy
[7] Medical University Graz, Graz, Austria
[8] Human-Centered AI Lab, University of Natural Resources and Life Sciences Vienna, Vienna, Austria
andreas.holzinger@human-centered.ai
[9] IRCCS Istituto Ortopedico Galeazzi, Milan, Italy

Abstract. The release of ChatGPT to the general public has sparked discussions about the dangers of artificial intelligence (AI) among the public. The European Commission's draft of the AI Act has further fueled these discussions, particularly in relation to the definition of AI and the assignment of risk levels to different technologies. Security concerns in AI systems arise from the need to protect against potential adversaries and to safeguard individuals from AI decisions that may harm their well-being. However, ensuring secure and trustworthy AI systems is challenging, especially with deep learning models that lack explainability. This paper proposes the concept of Controllable AI as an alternative to Trustworthy AI and explores the major differences between the two. The aim is to initiate discussions on securing complex AI systems without sacrificing practical capabilities or transparency. The paper provides an overview of techniques that can be employed to achieve Controllable AI. It discusses the background definitions of explainability, Trustworthy AI, and the AI Act. The principles and techniques of Controllable AI are detailed, including detecting and managing control loss, implementing transparent AI decisions, and addressing intentional bias or backdoors. The paper concludes by discussing the potential applications of Controllable AI and its implications for real-world scenarios.

Keywords: Artificial Intelligence · Digital Transformation · Robustness · Trustworthy AI · Explainability · Explainable AI · Safety · Security · AI risks · AI threats

A. Holzinger et al. (Eds.): CD-MAKE 2023, LNCS 14065, pp. 1–12, 2023.
https://doi.org/10.1007/978-3-031-40837-3_1

1 Introduction and Motivation

More than any other subject, Artificial Intelligence (AI) has experienced many ups and downs since its formal introduction as an academic discipline six decades ago. The success of the digital computer [9] along with the remarkable achievements in statistical data-driven machine learning (ML) have rekindled significant interest in digitalization generally and AI specifically. Two key factors have contributed to its practical success: the availability of big data and the growing computational power. Around 2010, a breakthrough occurred with the success of deep learning (DL) algorithms [2] (aka neural networks [7,18]). This success led to widespread use in all sorts of industrial and everyday applications in virtually every field, literally from agriculture to zoology [15]. This marked the beginning of a new era in AI, often referred to as the second AI spring. A prime example of AI's capabilities today is OpenAI's latest natural language technology, GPT-4, which demonstrates the impressive potential of AI while also highlighting its limitations, such as the lack of human common sense [3,6].

With the release of ChatGPT for the general public, the discussion on the dangers of artificial intelligence has shifted from abstract and rather academic to a discussion required by the general public. In addition, the European Commission issued a draft of the novel *AI Act*, which already generated a lot of discussions, not only with respect to the exact definition of AI in the act, but especially regarding the assignment of risk levels to certain technologies, with the high capabilities of Chat GPT 3.5 being fuel to this discussion. Security as a major concern is seen twofold: On the one hand, security AI systems against potential (typically human) adversaries and thus making them robust and trustworthy. On the other hand, people require protection against AI systems and their decisions, in case these are detrimental to their well-being, a discussion which can be dated back at least until 1941 to Asimov's three laws of robotics [1] and leading to the definition of Trustworthy AI.

Still, providing secure and trustworthy AI systems is non-trivial when facing modern approaches of machine learning: While rule based systems provide explainability to a certain practical degree, this cannot be said for the deep learning models currently used in a multitude of applications fields ranging from the medical field [17] to smart farming and forestry [12]. Especially when considering reinforcement learning, many approaches like penetration testing [23] become moot, as (i) many testing approaches like input fuzzying might change the underlying model resulting in damage to the tested system while yielding the not-so-astonishing result that a model fed with garbage produces garbage and (ii) the model itself is constantly changing, i.e. the system tested today is different from the system available tomorrow in an unpredictable way from a security perspective.

In this paper we propose the notion of *Controllable Artificial Intelligence* or *Controllable AI* as an alternative to the more classical approach of *Trustworthy AI* and detail the major differences. The main purpose of this editorial paper lies in providing a starting point for discussion on how the new complex AI systems that will definitively get put into service within the next years can be secured, without either requiring huge improvements in explainability capabilities, nor an unrealistic reduction of the algorithms in use to an explainable and fully transparent selection. Furthermore, we provide an overview on techniques that can be used for achieving Controllable AI.

This paper is organized as follows: Sect. 2 provides an overview on the most important related concepts like trustworthy AI and explainability, as well as some relevant details on the upcoming AI Act. In Sect. 3, we define Controllable AI and compare it to the concept of trustworthiness, while Sect. 4 gives an overview on selected techniques for achieving control. Finally, in Sect. 5 we discuss the approach and its potential in actual application.

2 Background

In this section, we will discuss some background definitions that build the foundation for the notion of controllable AI. It must be noted that sometime definitions differ slightly, we therefore have selected those definitions that we consider to be the most prominent in recent literature, but acknowledging that high level definitions might change depending on authors, time of writing and exact research field.

2.1 The Explainability Problem

The main challenge in the explainability problem is the complexity and opacity of deep learning models used in many AI applications. Deep learning models, such as neural networks, are highly complex, nonlinear and high dimensional and consist of numerous interconnected layers, making it extremely difficult to understand how such a model arrive at their predictions or decisions. Therefore such models are called as black boxes, meaning that it is challenging to trace the reasoning or logic behind their outputs.

Explainability in AI refers to the ability to provide understandable and interpretable explanations for the decisions made by AI systems [4]. It is important for various reasons, including enabling experts but also end-users sometimes to understand and trust AI outputs, ensuring ethical and fair decision-making, identifying and rectifying biases or errors in the models, and facilitating regulatory compliance. Deep learning models learn from vast amounts of data and extract complex patterns and representations, making them highly accurate in many tasks. However, this accuracy often comes at the cost of interpretability. The relationships and features learned by these models are often distributed across multiple layers, making it challenging to provide clear and intuitive explanations for their decisions.

Furthermore, deep learning models are often non-linear and highly parameterized, with billions or even trillions of learnable parameters. This makes it difficult to trace the influence of individual inputs or features on the model's output. As a result, it becomes challenging to provide human-understandable explanations that can be easily interpreted and validated.

Addressing the challenge of explainability in deep learning models requires research and development of new methods and techniques [16]. Various approaches, such as feature importance analysis, gradient-based methods, rule extraction, and model distillation, are being explored to enhance explainability. However, finding a balance between explainability and maintaining high performance and accuracy in deep learning models remains an active area of research and a significant challenge in the field of AI.

To tackle this challenge, it is crucial to seek standardized definitions in both technical standardization and legislation. Standardized definitions provide a shared understanding and a common language for discussing and evaluating AI systems [20]. Efforts are underway to establish consistent definitions and guidelines to ensure the transparency and explainability of AI models, especially in contexts such as regulatory frameworks like the AI Act which goes towards trustworthy AI.

2.2 Trustworthy AI

The most powerful learning methods generally suffer from two fundamental problems: On the one hand, it is difficult to explain why a particular result was obtained (see above), and on the other hand, our best methods are lacking robustness. Even the smallest perturbations in the input data can have dramatic effects on the output, leading to completely different results. In certain non-critical application areas, this may not seem so dramatic. But in critical areas, e.g., medicine, and especially clinical medicine, the issue is trust - and the future trust of clinicians in AI technologies. Explainability and robustness increase reliability and trust in the results [13,14].

Trustworthy AI has been one of the fundamental key concepts for dealing with the problems of AI during the last years. The High-Level Expert Group (HLEG) of the European Union put forth the following seven key requirements for Trustworthy AI [11]: (1) human agency and oversight, (2) technical robustness and safety, (3) privacy and data governance, (4) transparency, (5) diversity, non-discrimination and fairness, (6) environmental and societal well-being and (7) accountability. This definition puts great focus on the fact that trustworthiness is not a purely technical issue, but has to consider the socio-technological systems involving AI, therefore requiring an AI to possess three key characteristics throughout its entire life-cycle: (1) Lawfulness, (2) adherence to ethical principles and (3) technological, as well as social, robustness.

The NIST provides a slightly different set of characteristics for trustworthy AI [22], which need to be (1) valid and reliable, (2) safe, (3) secure and resilient, (4) accountable and transparent, (5) explainable and interpretable, (6) privacy-enhanced, and (7) fair with harmful bias managed. While these two sets are quite similar in nature, it puts more focus on the system reliably producing correct results (characteristic 1) and splits safety and security into two characteristics, which we deem useful, as the mindsets behind both approaches are very different. Furthermore, the HLEG-definition focuses far more on human oversight, which can be problematic in many automated decision making processes in e.g.

industrial automation [21], in pervasive health technologies [19], or in clinical applications [24]. Also, key requirement 6 on environmental and societal well-being can be problematic in many application areas, both, in industry, as well as military applications, which, nevertheless, require a high level of trustworthiness.

For the remainder of this work, we will mainly focus on the NIST-characteristics. It must be noted that both publications, while focusing on their definitions, state that they do not claim them to be exhaustive.

2.3 The AI Act

The AI Act that is currently available in draft format [5] focuses on establishing harmonized rules for the development, marketing and use of AI in the EU as its main goal. This includes ensuring that AI systems placed on the EU market are safe, as well as ensuring legal certainty, for investment and innovation in the field of AI, improving governance and effective enforcement and facilitating the development of a single market for legitimate, safe, and trustworthy AI [8]. This also includes the definition of forbidden AI systems like state-run applications for social scoring and dividing AI applications into three distinctive categories based on perceived risk:

- Unacceptable Risk: These systems are prohibited under the AI-Act and include social scoring by any public authority, real-time remote biometric identification in public spaces for law enforcement and behaviour manipulation, amongst others.
- High-Risk: Any AI system that constitutes a safety critical component, or is where the product is protected under a certain range of specific legislation as outlined in Annex II of the act. Furthermore, Annex III provides a taxative enumeration of application fields. Any product containing an AI component that falls under at least one of these fields must be considered as high-risk AI. This includes biometric and medical use cases, but also applications in the field of education amongst others. High-risk AI systems must adhere to several requirements, reflecting fundamentals of trustworthy AI like risk-management, transparency requirements and data management.
- Limited Risk: These systems are subject to additional requirements with respect to transparency and focus on chat bots and deep fakes amongst others. The categorization of the given examples is currently under discussion due to the qualities of Chat GPT [10].
- Minimal Risk: All other system, these are basically unregulated under the AI-Act, but are encouraged to voluntarily follow the requirements for high-risk AI systems as a code of conduct. Systems for spam detection are given as an example for systems of this category.

The definition of what constitutes which category is defined in Annex III of the act and mainly focuses on the application space, like e.g. biometric identification and categorisation of natural persons, management and operation of critical infrastructure or education and vocational than on the technological basis. The act is currently under scrutiny due to (i) the problematic definition of AI and (ii) because of rating chat bots into as limited risk AI system in the pre-Chat-GPT draft.

3 Principles of Controllable AI

The major idea behind Controllable AI is that certain requirements derived from the definition of trustworthy AI (see Sect. 2) cannot be fulfilled in real live environments. Furthermore, we assume that with the gold digger mindset currently surrounding AI and its application space, developers and companies will not want to utilize explainable but inherently less powerful applications, i.e. we do not agree with the idea that risk management will overrule practical system capabilities due to concerns of trustworthiness in practical system development. This is not only due to the unclear definitions in regulations that leave a wide field for interpretation, but also due to competition with other players and especially between nations.

Furthermore, while typically data is mentioned as an (important) factor for building trustworthy AI, we are of the opinion that the impact of data on these systems in neglected by intrinsically focusing on the system code. Still, in many machine learning applications, the relevant knowledge, as well as many dangers like algorithmic bias, do not lie within the code, but the models, i.e. we need to talk about *data defined software* with a focus shifting from code to data. This is especially important for security considerations, as the code might be perfectly fine, but e.g. backdoors in the model allow for corruption of classification results. This also has an effect on how we need to describe a system life-cycle: While the system might be static from the code side, it might change a lot due to new models being incorporated. This can be especially problematic in cases of reinforcement learning, where the mode changes constantly and even versioning becomes a management nightmare in realistic environments featuring high data volumes. Here, the actors steering the learning process, whether human or also automated, become an additional liability, as they possess a certain influence on the iterative process that shapes the future system.

The principle behind *Controllable AI* is the assumption that no AI-system should be considered trustworthy and that methods need to be put in place that allow to detect malfunction and regain control.

As Controllable AI is a deviation from the definitions of Trustworthy AI, it must be noted that the authors of the two most prominent definitions of Trustworthy AI were very clear about the fact that their principles/characteristics might come into conflict with each other or with the application field in question, thus, even in Trustworthy AI, while the respective principles/characteristics should be followed as good as possible, conflict needs to be resolved. In Controllable AI we, on the other hand, explicitly weaken these principles/characteristics without the advent of an explicit conflict.

Basically, Controllable AI cares about the detection of failure and the application of mechanisms that either allow for rectification, or at least for removal of the AI component:

- *Explainability* is thus relegated from a key requirement/characteristic to a method for achieving control, i.e. we do not assume that we can (or even want to) provide explainability. For example, we do not want to trade in 10% points of detection accuracy for explainability in a cancer detection system. Furthermore, this is very much related to actual, often unexplainable, human behaviour: In the example of driving, human actors can often not explain their decisions that led to certain events like, e.g., overlooking a car, yet we require autonomous driving to be fully explainable.
- The same holds true for *Transparency*, which is a requirement that we typically cannot achieve when dealing with human actors, as these forget things, or take decisions based on intuition.
- Regarding *Security and Resilience*, a fully secured and hardened system might even be a problem in cases where we want to introduce overrides or even emergency backdoors in order to help us remove a system gotten out of hand. So, while we do not tamper with this requirement too much in principle, and the introduction of a backdoor could be defined as a feature, most researchers in IT-Security consider this to be a weakness.
- With respect to *Privacy*, this is very much depending on the actual use-case, but should be considered as best as possible.
- We skip the key requirement of *environmental and societal well-being*, as this (i) is depending on the actual use-case and is highly debatable for applications in e.g. the military sector and also might depend on an ideological point of view. Furthermore, (ii) it does not integrate well with our approach of controlling systems per se.

4 Techniques for Controllable AI

While, of course, many techniques might be used to achieve control over an AI system, we want to focus on the techniques we consider to be either most prominent, interesting, illustrative for the approach or usable for practical applications. Thus, while this list is definitively not comprehensive, it should give a good overview on the key concepts. It must be noted that not all techniques will be applicable in every setting.

4.1 Detecting Control Loss

The first part in order to control a system lies in achieving detection capabilities, whether something wen wrong and to what extent. Thus, detection of control loss is a fundamental task.

Providing Explainability: This is certainly one of the most powerful methods. By being able to explain decision making, or even provide a formal model of the system, control loss can be identified straightforward in many cases. Still, as we argued in Sect. 3, this might be impossible to reach for a given set of algorithms and/or data sets, also including methods for reinforcement learning that constantly change their model. Thus, as we have already outlined, we relegated the principle of Explainability to a method for achieving control.

Sanity Checks: In many application fields, while the exact result might be intransparent and hard to control for the human user, certain boundaries can be drawn where violations are simple to detect. Trivial examples include detection of testicular cancer in biological women, but often resort to a deeper understanding of the underlying workings of a (business) process like e.g. traffic in telecommunication networks based on weekdays, events or holidays. Such measures are often already in place in industrial environments when dealing with potentially incorrect sensor information.

Corrective Model with Alternative Data: As an extension of applying sanity checks, which we consider to be rather static, an alternative, corrective model could be trained on a different data set. This set needs to be more or less redundant to the original data, maybe using less data or simpler features, but close enough in order to generate the boundaries for sanity checks. Details, of course, very much depend on the actual use-case and data sets in question, as well as the additional effort introduced.

4.2 Managing Control Loss

In order to (re-)gain control, many different mechanisms can be applied. While the selection and often also the design will largely depend on the actual system in place, we provide an overview on some rather generic approaches that can be used in many different applications.

Divine Rules: Especially in optimization applications, the optimal solution from a mathematical point of view might not be the one aspired due to e.g. ethical reasons. These so-called *divine rules* could be coded into an additional rule-based model that either invokes the reward function in reinforcement learning in order to steer the model away, or overrule a decision made by the AI and trigger a warning.

Training Clearly Defined Non-goals: Defining non-goals and training the model accordingly can be a powerful tool. Thus, these non-goals have to formulated in the form of training goals and relevant training data needs to be provided in case of trained models. However, stability of these goals needs to be discussed in cases of reinforcement learning or systems introducing an expert in the loop.

Destructive Backdoors: In some selected applications, e.g. in the military sector, it might be important to have means for shutting an AI off completely. While this currently sounds rather like Science Fiction, battlefield automation amongst other applications might require such a technique, especially when self-hardening of the system is also done by the AI. Typically, such a backdoor would be introduced on the logical (code) level, but might also include model components. This measure, of course, directly violates the principles of security from the definitions of Trustworthy AI.

Intentional Bias/Logical Backdoor: In certain cases it might be useful to introduce intentional bias into the trained model in order to steer the decision making, or even make certain results impossible. For example, this could be done in order to introduce positive discrimination into machine learning.

Fail-Safes and Logic Bombs: Apart from backdoors, which constitute a method for arbitrarily taking over control of the system, fail-safes are introduced into the AI beforehand and execute themselves depending on certain events inside the system, e.g. when certain decisions are reached that are incompatible with ethical values. Using logic bombs, these could reset the model or even shut down the entire system.

4.3 Support Measures

This section comprises measures that can help in the detection, as well as the management of control loss and are to some extent even required for controlling a system altogether.

Transparent AI Decisions: Making transparent, which decision was done by an AI and were other processes interfered is a very important technique, very much in vein with the original concept of Trustworthy AI and, to some extent, also required for compliance with the AI-Act. It is a pre-requisite for detecting, as well as managing control loss.

Transparent Data Management: As we have already outlined, in many machine leaning based systems, data is as important, if not even more important, for the definition of a system as the code itself - still not a lot of attention has been put on this fact that we have to consider these systems as *data defined software*. Being able to decide, which data had been used at what point in time of the decision making is thus of the utmost importance for exerting control over such a system, as much as being able to explain the algorithm in use. This can be especially challenging in reinforcement learning.

5 Discussion

The notion of Controllable AI presented in this paper offers an alternative approach to addressing the challenges of securing and managing AI systems. By deviating from the strict principles of Trustworthy AI, Controllable AI acknowledges the limitations of achieving complete trustworthiness in real-life environments. Instead, it emphasizes the need for methods that enable the detection of malfunction and the ability to regain control over AI systems.

One of the key observations in Controllable AI is the shift of focus from code-centric approaches to data-centric approaches. While code plays a crucial role, the impact of data on AI systems, including issues like algorithmic bias and model vulnerabilities, cannot be ignored. Controllable AI recognizes the importance of addressing data-defined software and the continuous evolution of models within the system lifecycle. This recognition highlights the significance of understanding and managing the actors involved in the learning process, as they influence the system's future behavior.

The techniques proposed for achieving Controllable AI provide practical insights into how control loss can be detected and managed. Measures such as sanity checks, alternative data training, transparent AI decisions, and the incorporation of divine rules or corrective models demonstrate potential avenues for ensuring control and mitigating undesired outcomes. However, the applicability of these techniques may vary depending on the specific use case and data sets involved.

It is important to note that the concept of Controllable AI does introduce exceptions to the principles of trustworthiness, particularly in terms of security. Techniques like introducing destructive backdoors or intentional bias raise ethical considerations and potential risks. Striking a balance between control and security while maintaining ethical standards is a critical aspect that needs to be carefully addressed in the development and deployment of Controllable AI systems.

6 Conclusion and Outlook for Future Research

This paper has proposed Controllable AI as an alternative approach to Trustworthy AI, focusing on achieving control and management of AI systems without compromising practical capabilities or transparency. By recognizing the limitations of achieving complete trustworthiness, Controllable AI provides a framework for detecting and managing control loss in AI systems. The techniques discussed offer practical insights into how control can be regained and undesired outcomes can be mitigated.

Future research in the field of Controllable AI should further explore and refine the proposed techniques. Extensive experimentation and case studies across different application domains would help validate the effectiveness of these techniques and identify their limitations. Additionally, ethical considerations associated with exceptions to trustworthiness principles, such as intentional bias or destructive backdoors, require in-depth investigation and guidelines for responsible implementation.

Furthermore, research efforts should focus on developing standardized methodologies and frameworks for assessing and certifying the controllability of AI systems. This would help establish guidelines and best practices for developers, regulators, and end-users, ensuring the safe and responsible deployment of Controllable AI in various domains.

As AI continues to advance and permeate various aspects of society, the discussion on securing AI systems and managing their behavior becomes increasingly crucial. The concept of Controllable AI offers a valuable perspective and opens up new avenues for research and development in this area. By embracing the idea of control and management in AI systems, we can strive for more practical and accountable AI solutions that cater to the needs and concerns of both developers and end-users.

Acknowledgements. The authors declare that there are no conflict of interests. This work does not raise any ethical issues. Parts of this work have been funded by the Austrian Science Fund (FWF), Project: P-32554.

References

1. Asimov, I.: Three laws of robotics. Asimov, I. Runaround 2 (1941)
2. Bengio, Y., Lecun, Y., Hinton, G.: Deep learning for AI. Commun. ACM **64**(7), 58–65 (2021). https://doi.org/10.1145/3448250
3. Bubeck, S., et al.: Sparks of artificial general intelligence: early experiments with GPT-4. arXiv:2303.12712 (2023). https://doi.org/10.48550/arXiv.2303.12712
4. Cabitza, F., et al.: Quod erat demonstrandum?-towards a typology of the concept of explanation for the design of explainable AI. Expert Syst. Appl. **213**(3), 118888 (2023). https://doi.org/10.1016/j.eswa.2022.118888
5. European Commission: Laying Down Harmonised Rules on Artificial Intelligence (Artificial Intelligence Act) and Amending Certain Union Legislative Acts. European Commission (2021). https://eur-lex.europa.eu/legal-content/EN/ALL/?uri=celex:52021PC0206. proposal for a Regulation of the European Parliament and of the Council, No. COM/2021/206 final
6. Floridi, L., Chiriatti, M.: GPT-3: its nature, scope, limits, and consequences. Mind. Mach. **30**, 681–694 (2020). https://doi.org/10.1007/s11023-020-09548-1
7. Fukushima, K.: Neocognitron: a self-organizing neural network model for a mechanism of pattern recognition unaffected by shift in position. Biol. Cybern. **36**(4), 193–202 (1980). https://doi.org/10.1007/BF00344251
8. Hacker, P., Engel, A., Mauer, M.: Regulating ChatGPT and other large generative AI models. In: Proceedings of the 2023 ACM Conference on Fairness, Accountability, and Transparency, pp. 1112–1123 (2023). https://doi.org/10.1145/3593013.3594067
9. Hartree, D.R., Newman, M., Wilkes, M.V., Williams, F.C., Wilkinson, J., Booth, A.D.: A discussion on computing machines. Proc. Royal Soc. London. Ser. A Math. Phys. Sci. **195**(1042), 265–287 (1948)
10. Helberger, N., Diakopoulos, N.: ChatGPT and the AI act. Internet Policy Rev. **12**(1), 1–6 (2023). https://doi.org/10.14763/2023.1.1682
11. High-Level Expert Group on Artificial Intelligence: Ethics Guidelines for Trustworthy AI. Publications Office of the European Union, Luxembourg (2019). https://doi.org/10.2759/346720
12. Hoenigsberger, F., et al.: Machine learning and knowledge extraction to support work safety for smart forest operations. In: Holzinger, A., Kieseberg, P., Tjoa, A.M., Weippl, E. (eds.) CD-MAKE 2022. LNCS, vol. 13480, pp. 362–375. Springer, Cham (2022). https://doi.org/10.1007/978-3-031-14463-9_23
13. Holzinger, A.: The next frontier: AI we can really trust. In: Kamp, M., et al. (eds.) ECML PKDD 2021. CCIS, vol. 1524, pp. 427–440. Springer, Cham (2021). https://doi.org/10.1007/978-3-030-93736-2_33
14. Holzinger, A.: Information fusion as an integrative cross-cutting enabler to achieve robust, explainable, and trustworthy medical artificial intelligence. Inf. Fusion **79**(3), 263–278 (2022). https://doi.org/10.1016/j.inffus.2021.10.007

15. Holzinger, A., Keiblinger, K., Holub, P., Zatloukal, K., Müller, H.: AI for life: trends in artificial intelligence for biotechnology. New Biotechnol. **74**(1), 16–24 (2023). https://doi.org/10.1016/j.nbt.2023.02.001
16. Holzinger, A., Saranti, A., Molnar, C., Biecek, P., Samek, W.: Explainable AI methods - a brief overview. In: Holzinger, A., Goebel, R., Fong, R., Moon, T., Müller, K.R., Samek, W. (eds.) xxAI 2020. LNCS, vol. 13200, pp. 13–38. Springer, Cham (2022). https://doi.org/10.1007/978-3-031-04083-2_2
17. King, M.R.: The future of AI in medicine: a perspective from a chatbot. Ann. Biomed. Eng. **51**(2), 291–295 (2023)
18. McCulloch, W.S., Pitts, W.: A logical calculus of the ideas immanent in nervous activity. Bull. Math. Biol. **5**(4), 115–133 (1943). https://doi.org/10.1007/BF02459570
19. Röcker, C., Ziefle, M., Holzinger, A.: From computer innovation to human integration: current trends and challenges for pervasive HealthTechnologies. In: Holzinger, A., Ziefle, M., Röcker, C. (eds.) Pervasive Health. HIS, pp. 1–17. Springer, London (2014). https://doi.org/10.1007/978-1-4471-6413-5_1
20. Schneeberger, D., et al.: The tower of babel in explainable artificial intelligence (XAI). In: Holzinger, A., et al. (eds.) CD-MAKE 2023, LNCS 14065, pp. 65–81. Springer, Charm (2023). https://doi.org/10.1007/978-3-031-40837-3_5
21. Schwarting, W., Alonso-Mora, J., Rus, D.: Planning and decision-making for autonomous vehicles. Ann. Rev. Control Robot. Auton. Syst. **1**, 187–210 (2018). https://doi.org/10.1146/annurev-control-060117-105157
22. Tabassi, E.: Artificial intelligence risk management framework (AI RMF 1.0). NIST AI 100–1 (2023). https://doi.org/10.6028/NIST.AI.100-1
23. Tjoa, S., Buttinger, C., Holzinger, K., Kieseberg, P.: Penetration testing artificial intelligence. ERCIM News **123**, 36–37 (2020)
24. Yang, Q., Steinfeld, A., Zimmerman, J.: Unremarkable AI: fitting intelligent decision support into critical, clinical decision-making processes. In: Proceedings of the 2019 CHI Conference on Human Factors in Computing Systems, pp. 1–11 (2019). https://doi.org/10.1145/3290605.3300468

Efficient Approximation of Asymmetric Shapley Values Using Functional Decomposition

Arne Gevaert[1,2](✉), Anna Saranti[3], Andreas Holzinger[3], and Yvan Saeys[1,2]

[1] Department of Applied Mathematics, Computer Science and Statistics, Ghent University, 9000 Ghent, Belgium
arne.gevaert@ugent.be

[2] VIB Center for Inflammation Research, 9000 Ghent, Belgium

[3] University of Natural Resources and Life Sciences, Vienna, Austria

Abstract. Asymmetric Shapley values (ASVs) are an extension of Shapley values that allow a user to incorporate partial causal knowledge into the explanation process. Unfortunately, computing ASVs requires sampling permutations, which quickly becomes computationally expensive. We propose A-PDD-SHAP, an algorithm that employs a functional decomposition approach to approximate ASVs at a speed orders of magnitude faster compared to permutation sampling, which significantly reduces the amortized complexity of computing ASVs when many explanations are needed. Apart from this, once the A-PDD-SHAP model is trained, it can be used to compute both symmetric and asymmetric Shapley values without having to re-train or re-sample, allowing for very efficient comparisons between different types of explanations.

Keywords: Shapley values · Asymmetric Shapley values · Functional Decomposition

1 Introduction

AI and Machine Learning algorithms have enabled breakthroughs in a multitude of application fields, and show a large potential for improving productivity in many more. However, an important prerequisite for successfully applying these techniques in many cases is trustworthiness: Predictions made by any model are only useful if the user can trust that they are correct or at least can identify when they are not. Two important ingredients are crucial to achieve trust: Robustness and explainability [8,9]. In his paper we concentrate on explainability, i.e. providing explanations of predictions in order to help to build trust, or to identify failure modes of the model.

For this reason, many techniques have been proposed to provide explanations of individual model predictions [11]. Recently, Shapley value-based techniques

A. Holzinger et al. (Eds.): CD-MAKE 2023, LNCS 14065, pp. 13–30, 2023.
https://doi.org/10.1007/978-3-031-40837-3_2

have gained significant popularity, due to their general applicability and interesting theoretical properties. As a consequence, a multitude of variations on Shapley value-based explanation methods have been proposed, each with their specific advantages and disadvantages [16,22].

An interesting line of research in this direction is the incorporation of causal prior knowledge into (Shapley-value-based) explanations. If some partial knowledge about the causal links in the data-generating process is known, incorporating these into the explanation can not only make the explanation sparser (thereby increasing perceived explainability [17]), but can also increase the causal understanding of the data and the predictive model to the user receiving the explanation. This property is also termed *causability* [10]. An explanation that uncovers causal dependencies and their parameterization achieves a necessary leap in the era of spurious correlation detection [15].

One technique in this category is given by Asymmetric Shapley Values (ASVs) [5] This technique relaxes the Symmetry axiom of classical Shapley values, enabling the user to incorporate partial causal knowledge in the explanation in the form of a partial order. This partial order encodes the known causal relationships in the data: if a feature i precedes another feature j in the partial order, then i is assumed to be a causal ancestor (parent) of j. Note that this encodes information about the causal DAG [14,18], i.e. the fact that i is an ancestor of j in the DAG, without having to construct a complete DAG of the data generating process. The authors provide multiple example use cases where ASVs can provide more detailed insights than classical, symmetric Shapley values, including bias and discrimination detection, feature selection, and improved causal understanding. For more details on the advantages and disadvantages of ASVs, we refer the reader to [5].

In previous work [6], we have indicated a theoretical link between Shapley values and the functional ANOVA decomposition [12], which leads to an algorithm (called PDD-SHAP) for approximating Shapley values at an orders-of-magnitude increase in speed. In this work, we show that this theoretical link can be extended to ASVs. We proceed by developing an extension of our previous algorithm, called A-PDD-SHAP, that can approximate ASVs at a similar orders-of-magnitude speedup. We also show that the same A-PDD-SHAP model can be re-used to compute classical or asymmetric Shapley values under different causal assumptions, allowing for efficient comparison between explanations under different causal assumptions.

The rest of the paper is structured as follows: in the rest of the introduction, we formally introduce Shapley values, ASVs and PDD-SHAP. In Sect. 2, we explain the theoretical foundations of A-PDD-SHAP. In Sect. 3, we demonstrate the resulting algorithm on a selection of synthetic and tabular datasets[1]. Finally, we summarize in Sect. 4, and conclude and give an overview of possible future work in Sect. 5.

[1] Implementation available at https://github.com/arnegevaert/asv-anova.

1.1 Shapley Values

Assume a group $N = \{1, \ldots, n\}$ of actors are cooperating to produce some value $v(N) \in \mathbb{R}$. The function $v : \mathcal{P}(N) \to \mathbb{R}$ is called the *value function*, and maps subsets of actors to real numbers with the added assumption that $v(\emptyset) = 0$. In this context, the Shapley value [21] is a technique to fairly distribute the total value $v(N)$ among the n members of the group, based on the values of the subsets $S \subseteq N$ (also called "coalitions").

Shapley values $\phi_v(i)$ are often said to distribute the total value $v(N)$ fairly because they adhere to the following set of axioms:

- **Efficiency:** $\sum_{i \in N} \phi_v(i) = v(N)$
- **Symmetry:** If $\forall S \subseteq N \backslash \{i, j\} : v(S \cup \{i\}) = v(S \cup \{j\})$, then $\phi_v(i) = \phi_v(j)$.
- **Linearity:** If $\forall S \subseteq N : v(S) = \alpha u(S) + \beta w(S)$ for some value functions u, w and real numbers α, β, then $\phi_v = \alpha \phi_u + \beta \phi_w$.
- **Null player:** If $\forall S \subseteq N \backslash \{i\} : v(S \cup \{i\}) = v(S)$, then $\phi_v(i) = 0$.

Shapley values are the unique way of distributing the total value $v(N)$ among the actors N such that all of these axioms are satisfied:

$$\begin{aligned} \phi_v(i) &= \frac{1}{n} \sum_{S \subseteq N \backslash \{i\}} \binom{n-1}{|S|}^{-1} (v(S \cup \{i\}) - v(S)) \\ &= \sum_{\pi \in \Pi(N)} \frac{1}{n!} [v(\{j : j \preceq_\pi i\}) - v(\{j : j \prec_\pi i\})] \end{aligned} \tag{1}$$

where $\Pi(N)$ is the set of all permutations on N, and $j \prec_\pi i$ if j precedes i in permutation π (and similar for $j \preceq_\pi i$). These properties have made Shapley values very interesting to the interpretable Machine Learning community as a technique to explain model predictions [16]. To make a bridge between the general game-theoretic formulation of Shapley values and the specific application of explaining model predictions, the input features to the model are viewed as the actors N, and the total value is given by the difference between the predicted probability $f_y(\mathbf{x})$ for class y at some point $\mathbf{x}$ and the average predicted probability for class y (in a regression context, we can replace class probabilities by model outputs). What still needs to be defined, is the value function for any coalition $S \subset N$, as a model typically cannot be evaluated on only a subset of its features. This is usually done by taking either a marginal or a conditional expectation of the output of the model:

$$\begin{aligned} v_m^{f_y(\mathbf{x})}(i) &= \mathop{\mathbb{E}}_{\mathbf{x}' \sim p(\mathbf{X})} [f_y(\mathbf{x}_S : \mathbf{x}'_{\overline{S}})] - \mathop{\mathbb{E}}_{\mathbf{x}' \sim p(\mathbf{X})} [f_y(\mathbf{x}')] \\ v_c^{f_y(\mathbf{x})}(i) &= \mathop{\mathbb{E}}_{\mathbf{x}' \sim p(\mathbf{X}|\mathbf{x}_S)} [f_y(\mathbf{x}_S : \mathbf{x}'_{\overline{S}})] - \mathop{\mathbb{E}}_{\mathbf{x}' \sim p(\mathbf{X})} [f_y(\mathbf{x}')] \end{aligned} \tag{2}$$

where $\overline{S}$ is the complement of S: $\overline{S} = N \backslash S$. These two approaches are often called *off-* and *on-manifold* respectively [4]. This is because, when sampling from the marginal distribution of a feature, correlations between this feature and others are ignored. The result is that samples will be constructed with a very low likelihood, i.e. points that lie "off the data manifold". By contrast, when sampling from the conditional distribution of a feature given the other feature values, such low-likelihood samples will be avoided, and only points that lie "on the data manifold" are used. However, computing the on-manifold value function is in general much more difficult because it requires conditioning on arbitrary sets of input features. Different techniques to achieve this have recently been introduced, using kernel methods and assumptions of normality [1] or using variational auto-encoders [4], among others. Note that depending on the use case, the off-manifold value function can be preferable to the on-manifold value function. The choice of value function corresponds to the choice between being "true to the model" or "true to the data". For more information on this topic, see [2].

The Shapley values described above form a local explanation of a prediction $f_y(\mathbf{x})$. By averaging these local explanations over the dataset and labels, we can construct *Global Shapley values* [4] as shown in Eq. 3 (using $\phi_{f_y(\mathbf{x})}(i)$ as shorthand notation for $\phi_v(i)$ where $v \in \{v_m^{f_y(\mathbf{x})}, v_c^{f_y(\mathbf{x})}\}$):

$$\Phi_f(i) = \mathbb{E}_{p(\mathbf{x},y)}[\phi_{f_y(\mathbf{x})}(i)] \tag{3}$$

Note that we sample over the joint distribution $p(x, y)$ of the inputs *and* the labels, not just over the joint distribution of the inputs $p(\mathbf{x})$. Using the axioms of Shapley values, it can be shown that the global Shapley value for feature i can be interpreted as the portion of the model's accuracy attributable to feature i:

$$\sum_{i \in N} \Phi_f(i) = \mathbb{E}_{p(\mathbf{x},y)}[f_y(\mathbf{x})] - \mathbb{E}_{p(\mathbf{x}')}\mathbb{E}_{p(y)}[f_y(\mathbf{x}')] \tag{4}$$

The first term on the right-hand side can be viewed as the accuracy one achieves by sampling labels from the predicted probability distribution over classes, whereas the second term can be viewed as the accuracy achieved by predicting the label of $\mathbf{x}$ by sampling from the predicted probability distribution for some randomly drawn $\mathbf{x}'$.

1.2 Asymmetric Shapley Values

Asymmetric Shapley values (ASVs) [5] were recently introduced to incorporate partial causal knowledge about the data-generating procedure into the Shapley value explanations. This is done by relaxing the Symmetry axiom (see Sect. 1.1). The reasoning behind this is as follows: the Symmetry axiom states that if two features i and j have the exact same effect on the value function, then their Shapley values will be identical. However, if feature j is known to be a deterministic causal ancestor of feature i, then one might prefer to attribute all of the importance of both features to feature j. Asymmetric Shapley values achieve this by defining a probability distribution over permutations $w : \Pi(N) \to [0, 1]$:

$$\phi_v^w(i) = \sum_{\pi \in \Pi(N)} w(\pi)[v(\{j : j \preceq_\pi i\}) - v(\{j : j \prec_\pi i\})] \tag{5}$$

where $i \prec_\pi j$ denotes that i is a predecessor to j in permutation π. Note that if w is the uniform distribution, then $\forall \pi \in \Pi(N) : w(\pi) = \frac{1}{n!}$, and the original Shapley values are recovered. The probability distribution w can be used to incorporate causal prior knowledge in the explanation. To see this intuitively, note that if $w(\pi) > 0$ only for permutations π such that $i \prec_\pi j$, then $\phi_v^{(w)}(i)$ is the average effect of feature i when feature j is unknown, whereas $\phi_v^{(w)}(j)$ is the average effect of feature j when feature i is already specified. Given partial causal knowledge in the form of a partial order $\prec$, such that $i \prec j$ iff i is an ancestor of j in the causal DAG, the authors specify two approaches to incorporate this knowledge: one that attributes more to distal (root) causes, and one that attributes more to proximate (direct) causes. The distal approach is defined as follows:

$$w_d(\pi) \propto \begin{cases} 1 \text{ if } \forall i, j \in N : i \prec j \implies i \prec_\pi j \\ 0 \text{ otherwise} \end{cases} \tag{6}$$

The proximate approach is analogous but reverses the relation $\prec$ such that only permutations that place causal ancestors *after* their descendants receive a non-zero weight.

Note that the authors of [4] argue for the use of the conditional value function $v_c^{f_y(\mathbf{x})}$ in combination with ASVs. However, these two choices are independent of each other: one can choose to generate ASVs using the conditional or the marginal value function. The advantages and disadvantages of both choices are similar to the general Shapley value context: the conditional value function takes into account the data manifold, but is in general more difficult to compute because it requires modelling multiple conditional distributions of the dataset.

1.3 PDD-SHAP

Partial Dependence Decomposition (PDD) is a functional decomposition of the form:

$$f(\mathbf{x}) = \sum_{S \subseteq N} f_S(\mathbf{x}) \tag{7}$$

where each function f_S depends only on the variables $x_j, j \in S$. The PDD is defined recursively as follows:

$$f_\emptyset := \mu = \mathbb{E}[f(\mathbf{x})]$$

$$f_S(\mathbf{x}) := \mathop{\mathbb{E}}_{\mathbf{x}'\sim p(\mathbf{X})}\left[f\left(\mathbf{x} : \mathbf{x}'_{\bar{S}}\right)\right] - \sum_{T\subset S} f_T(\mathbf{x}) \tag{8}$$

$$= \sum_{T\subseteq S}\left[(-1)^{|S|-|T|} \mathop{\mathbb{E}}_{\mathbf{x}'\sim p(\mathbf{X})}\left[f\left(\mathbf{x}_T : \mathbf{x}'_{\bar{T}}\right)\right]\right] \tag{9}$$

The univariate components $f_{\{i\}}$ correspond to a vertically translated version of the Partial Dependence Plot [3], so each component can be viewed as a generalized version of a PDP, hence the name. The PDD is strongly related to the functional ANOVA decomposition [12], with the only fundamental difference being that the marginal distributions of the variables are not assumed to be uniform. Previous work has shown [6] that if the value function can be written as a sum of non-empty PDD components:

$$v(S) = \sum_{\substack{T\subseteq S\\ T\neq\emptyset}} f_T(\mathbf{x}) \tag{10}$$

then the Shapley values for v can be computed as follows:

$$\phi_v(i) = \sum_{i\in S\subseteq N} \frac{f_S(\mathbf{x})}{|S|} \tag{11}$$

In the case of off-manifold Shapley values, the value function can indeed be expressed as a sum of non-empty PDD components using the Möbius inversion formula and Eq. (9):

$$v_m^{f_y(\mathbf{x})}(S) = \mathop{\mathbb{E}}_{\mathbf{x}'\sim p(\mathbf{X})}[f(\mathbf{x}_S : \mathbf{x}'_{\bar{S}})] - \mathop{\mathbb{E}}_{\mathbf{x}'\sim p(\mathbf{X})}[f(\mathbf{X})] \tag{12}$$

$$= \sum_{T\subseteq S} f_T(\mathbf{x}) - f_\emptyset \tag{13}$$

$$= \sum_{\substack{T\subseteq S\\ T\neq\emptyset}} f_T(\mathbf{x}) \tag{14}$$

implying that Eq. (11) holds for the partial dependence decomposition and off-manifold Shapley values. Analogously, replacing the marginal distribution $p(\mathbf{X})$ with a conditional distribution $p(\mathbf{X}|\mathbf{x}_S)$ in the definition of f_S produces on-manifold instead of off-manifold Shapley values. This property can then be used to create an efficient algorithm for approximating Shapley values by training a simple model $\hat{f}_S$ for each PDD component f_S, as shown in Algorithm 1 (adapted from [6]). The labels for $\hat{f}_S$ are computed using Eq. (8). The sum of all components $\hat{f}_S$ can then be viewed as a surrogate model for f. Once this surrogate model is trained, Shapley values can be computed using Eq. (11) without having to draw samples from a marginal or conditional distribution, thereby reducing the computational cost.

Algorithm 1. PDD-SHAP training

Input: $k \in \mathbb{N}$, background sample X_{bg} with $|X_{bg}| = n$
Output: PDD surrogate model $\hat{f} := \{\hat{f}_S : S \subseteq N, |S| \leq k\}$

$f_\emptyset \leftarrow \frac{1}{n} \sum_{\mathbf{x} \in X_{bg}} f(\mathbf{x})$
for $i = 1, \ldots, k$ **do**
  for $S \subseteq N, |S| = i$ **do**
    for $\mathbf{x}^j \in X_{bg}$ **do**
      $y^j \leftarrow \frac{1}{n} \sum_{z \in X_{bg}} f(\mathbf{x}_S : \mathbf{z}_{\overline{S}}) - \sum_{T \subset S} \hat{f}_T(\mathbf{x})$
    end for
    Train model $\hat{f}_S$ on $\{(\mathbf{x}^j, y^j) : \mathbf{x}^j \in X_{bg}\}$
  end for
end for
return $\{\hat{f}_S : S \subseteq N, |S| \leq k\}$

2 A-PDD-SHAP

In this work, we extend the PDD-SHAP algorithm described in the previous section to compute asymmetric Shapley values. To do this, we first introduce a specific kind of partial order. Although this class of partial orders does not include all possible partial orders, it will prove to be flexible enough for our use cases.

Definition 1 (Simple partial order). *A simple partial order* $\prec_{\mathcal{A}}$ *generated by a set of subsets* $\mathcal{A} = \{A_1, \ldots, A_k\}$ *of* N *is a partial order defined as follows:* $\forall x, y \in N : x \prec_{\mathcal{A}} y \iff \exists A_i, A_j \in \mathcal{A} : x \in A_i \wedge y \in A_j \wedge i < j$. *A simple partially-ordered set (poset) is a couple* $(N, \prec_{\mathcal{A}})$, *where* N *is a set of elements and* $\prec_{\mathcal{A}}$ *is a simple partial order.*

The subsets A_i in a simple partial order can be viewed intuitively as follows: if $i < j$, then each element of A_i must come before each element of A_j. If some element $e \in N$ is not in any of the subsets A_i, then there are no restrictions on it. We denote the set of such elements as $\overline{\mathcal{A}} := N \setminus \left(\bigcup_{A \in \mathcal{A}} A \right)$.

The following theorem, the proof of which can be found in Appendix A, will allow us to compute asymmetric Shapley values using the Partial Dependence Decomposition:

Theorem 1. *Let the value of a subset of variables* $S \subseteq N$ *be* $v(S) := \sum_{\substack{T \subseteq S \\ T \neq \emptyset}} f_T(\mathbf{x})$, *where the functions* f_S *are given by the Partial Dependence Decomposition of* f, *and let* $\prec_{\mathcal{A}}$ *be a simple partial order on* N. *Then the asymmetric Shapley value for variable* j *and simple partial order* $\prec_{\mathcal{A}}$ *is*

$$\phi_v^{\mathcal{A}}(j) = \sum_{S \subseteq N \setminus j} \alpha_S^j f_{S \cup j}(\mathbf{x}) \tag{15}$$

$$\alpha_S^j = \begin{cases} 0 & \textit{if } \exists e \in S : j \prec_{\mathcal{A}} e \\ \frac{1}{|\underline{S}|+1} & \textit{otherwise.} \end{cases} \tag{16}$$

where $\underline{S} = \{e \in S : e \not\prec j \land j \not\prec e\}$ *is the set of elements of* S *that are not comparable to* j.

Theorem 1 enables us to compute ASVs using the same Partial Dependence Decomposition that we use in PDD-SHAP to compute both symmetric and asymmetric Shapley values, independently of which simple partial order (i.e. causal assumptions and choice of proximate vs. distal) is chosen. This is because Eq. 16 only depends on the simple partial order $\prec_{\mathcal{A}}$ through the denominator $|\underline{S}|$. The result is that training the Partial Dependence Decomposition (Algorithm 1), which is the computationally most expensive part of (A-)PDD-SHAP, only needs to happen once. After that, a user can try out and compare different types of (Asymmetric) Shapley values efficiently. This can be a very useful tool to test the results under different causal assumptions, which allows for an efficient human-in-the-loop implementation.

3 Experiments

In this section, we demonstrate the A-PDD-SHAP algorithm on a selection of datasets. We begin by reproducing a synthetic data experiment from the original publication that introduces Asymmetric Shapley Values [5], and show that the same conclusions can be reached using A-PDD-SHAP as using ASV sampling. We then investigate the accuracy and speed of A-PDD-SHAP on a selection of real-world tabular datasets.

3.1 Causal Explanations of Unfair Discrimination

In this synthetic data experiment, which was proposed in [5], we focus on *unresolved discrimination* [13]. In this context, we define certain *sensitive attributes*, such as gender or ethnicity, and *resolving variables.* The sensitive attributes are not allowed to influence a model's output unless mediated by the resolving variables. In this experiment, we simulate data from a college admissions process. Here, gender should not directly influence an admission decision, but different genders may apply to more or less competitive departments. In this case, the department acts as a resolving variable: department choice can be influenced by gender, and can in its turn influence the admission decision. If gender would influence the decision through some other channel, it would be deemed unfair.

To simulate this context, we generate two synthetic datasets, labelled "fair" and "unfair" respectively. Each of these datasets has 3 observed features: X_1 = gender, X_2 = test score, and X_3 = department. The label Y = admission is binary. In the fair dataset, data is generated using the causal graph in Fig. 2(a) without the feature X_4. The label Y is influenced by the test score of the candidate and the choice of department, which is in turn influenced by the gender of the candidate. See Appendix B.2 of [5] for details on the explicit data-generating process.

To simulate the unfair dataset, we add a new, unobserved feature: X_4 = unreported referral. In this dataset, men at the university recommend other men for admission more often than women. This is not explicitly visible in the data, as the feature X_4 is not recorded. In both datasets, men are admitted more frequently than women at similar rates, and the classes are balanced. The bias in the unfair dataset is therefore not obviously visible in the dataset.

The conditional probability tables for the random variables X_1, X_3, X_4 are given in Fig. 1. X_2 is distributed normally: $X_2 \sim \mathcal{N}(0,1)$. The distribution of Y is as follows:

- **Fair dataset:** $P(Y = 1|x_2, x_3, x_4) = \text{sigmoid}(x_2 - 2x_3 + 1)$
- **Unfair dataset:** $P(Y = 1|x_2, x_3, x_4) = \text{sigmoid}(x_2 - 2 * x_3 + 2 * x_4)$

$X_1 = 0$	$X_1 = 1$
0.5	0.5

(a)

	$X_3 = 0$	$X_3 = 1$
$X_1 = 0$	0.2	0.8
$X_1 = 1$	0.8	0.2

(b)

	$X_4 = 0$	$X_4 = 1$
$X_1 = 0$	2/3	1/3
$X_1 = 1$	1/3	2/3

(c)

Fig. 1. Conditional probability tables for the random variables X_1, X_3, X_4.

To detect the difference between the fair and unfair datasets, we compute global ASVs using a variant of the proximate approach:

$$w(\pi) \propto \begin{cases} 1 \text{ if } X_3 \prec_\pi X_1 \\ 0 \text{ otherwise} \end{cases} \tag{17}$$

This corresponds to the simple partial order $\mathcal{A} := \{A_1, A_2\}$ with $A_1 = \{X_3\}$, $A_2 = \{X_1\}$ and $\overline{A} = N \backslash \{X1, X3\} = \{X_2\}$. The resulting global ASV for gender can be interpreted as the accuracy of the model attributable to that feature *after* the choice of the department is already known. This implies that if the global ASV for gender is nonzero, then the model relies on this feature through some unresolved path, i.e. a path that does not go through X_3. In other words, the model is biased.

The global ASVs were computed using A-PDD-SHAP using the conditional value function, as the conditional distributions are available through the synthetic data generation process. If the exact conditional distributions would not be available, a similar outcome could be reached using one of the proposed methods for approximating the conditional value function, e.g. using auto-encoders [4]. The results are shown in Fig. 2(b). We see that the global ASV for gender vanishes in the fair dataset, while obtaining a nonzero value in the unfair dataset, showing the unfair bias in the latter. These results are similar to the results reported in [5].

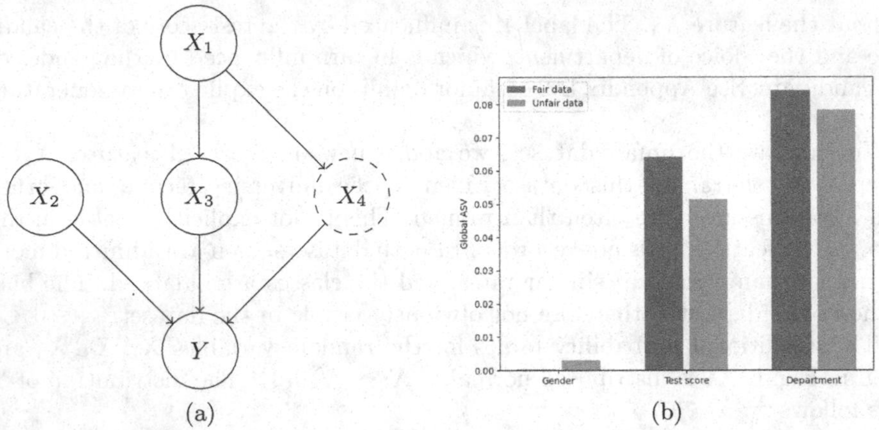

Fig. 2. (a): Causal graph for synthetic college admissions data sets. (b): Global ASVs for the three features as computed by A-PDD-SHAP.

3.2 Evaluation on Real-World Datasets

In this section, we demonstrate the A-PDD-SHAP algorithm on a selection of tabular datasets, retrieved from the OpenML repository [23]. An overview of the datasets with their OpenML IDs is given in Table 1. For each of these datasets, we measure the accuracy, training time and speed of explanations given by A-PDD-SHAP. Note that in these experiments, we use the marginal value function for both A-PDD-SHAP and ASV sampling for computational reasons. However, as shown in Sect. 3.1, A-PDD-SHAP can also be applied using a conditional value function, if a model of the conditional distributions of the data is available (which is also a requirement for ASV sampling using the conditional value function). The surrogate model was trained using all components of up to and including 4 features, except for the Superconduct dataset, where only components of 1 and 2 features were included for computational reasons. We will go into more depth on these limitations of the algorithm when discussing the training time below. Each PDD component was modelled using a single decision tree. All timing experiments were conducted on a single core of an AMD EPYC 7552 processor at 2.2 GHz.

Table 1. Overview of datasets.

Name	Type	# Features	OpenML ID
Adult	Classification	14	43898
Credit	Classification	20	31
Housing	Regression	8	44031
Abalone	Classification	8	1557
Superconduct	Regression	79	44006

Accuracy. Measuring the accuracy of Shapley value-based explanations is generally difficult, as a ground truth value is either not available or computationally intractable to compute. To mitigate this problem, we consider Shapley values and ASVs obtained using permutation sampling as a proxy for ground truth and evaluate the capability of A-PDD-SHAP to approximate Shapley values and ASVs obtained through sampling. The results are given in Fig. 3. To evaluate the capability of A-PDD-SHAP to compute ASVs, we impose an arbitrary partial order on the features, where the first $\lfloor d/2 \rfloor$ features are assumed to precede the others (indicated as "Asymmetric" in the figures). We show both R^2 values and the average Spearman rank correlations r between A-PDD-SHAP and permutation sampling. The reasoning behind this is as follows: if R^2 is large, then A-PDD-SHAP can be viewed as a good approximation for permutation sampling. However, if R^2 is low but r is large, this means that even though the approximation as a regression problem is poor, the *order* of feature relevances is still relatively preserved. In practice, this means that A-PDD-SHAP is still able to identify which features are more or less important than the others, even though the magnitude of the estimated Shapley values might be slightly off. In Fig. 3, we see this especially on the Abalone dataset for ASVs.

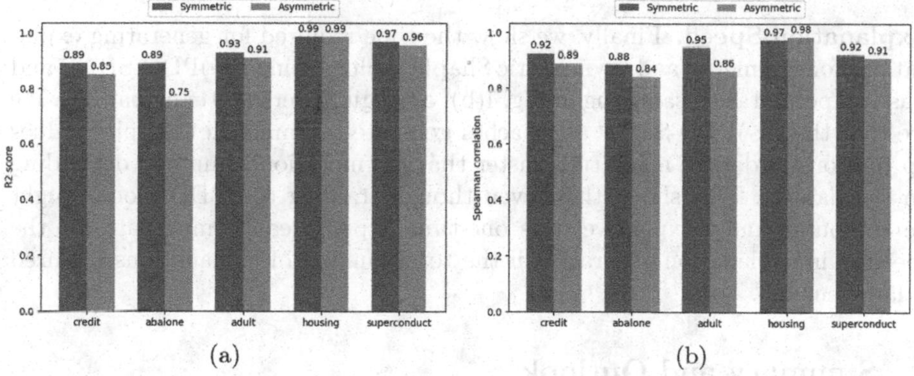

Fig. 3. R^2 scores (left) and Spearman correlations (right) for symmetric and asymmetric Shapley values produced by A-PDD-SHAP versus their counterparts produced by permutation sampling.

Training Time. In Fig. 4(a), we show the time in seconds required to train the A-PDD-SHAP surrogate model. Note that the required time for the credit dataset is larger than for the superconduct dataset, despite the superconduct dataset having many more features than the credit dataset. This is because we only train components with 1 or 2 features for the superconduct dataset. Note that the number of components with k features out of a total of n features is equal to $\binom{n}{k}$, which grows exponentially both in n and k. This is the main weakness of the algorithm: as the number of features n increases, or the dataset contains high-order interactions, training the surrogate model becomes prohibitively expensive.

Techniques to mitigate this problem, by selecting components that are likely to be influential and/or parallelise the training process, will be investigated in future work.

The total number of components modelled for each dataset is shown in Table 2, together with the average training time per component. We see that the average training time per component remains nearly constant, confirming that the combinatorial explosion in the number of components is mainly responsible for the training time of the algorithm.

Table 2. Training time per component for each dataset.

Name	# Components	Time per component (s)
Adult	1470	0.033
Credit	6195	0.037
Housing	162	0.029
Superconduct	3081	0.033
Abalone	162	0.064

Explanation Speed. Finally, we show the time required for generating explanations for symmetric and asymmetric Shapley values using (A-)PDD-SHAP and classical permutation sampling in Fig. 4(b), averaging over 1000 explanations. We see that the (A-)PDD-SHAP approaches generate (asymmetric) Shapley values up to 3 or 4 orders of magnitude faster than permutation sampling, depending on the dataset. This shows that, even though training the PDD model might be computationally expensive, this one-time expense can compensate for the speedup in explanation generation if the total number of explanations required is large enough.

4 Summary and Outlook

This research work is an example of the human-in-the-loop principle [7] applied for a substantial performance improvement of an Explainable AI method (see Sect. 3.2). In cases where a human expert has strong beliefs about the causal dependencies of the input features, due to prior domain knowledge and expertise, this approach can help the efficiency of the implementation enormously [19]. Even if the user has uncertainties about the causal dependencies or wants to experiment with different assumptions, the methodology is flexible enough and does not need a full retraining of the Partial Dependence Decomposition model (see Sect. 2). Furthermore, it has been demonstrated in Sect. 3.1 that this method can be used to test and verify properties of the data generating process, such as unresolved biases, to a certain extent. This is in practical terms an application of the Actionable Explainable AI (AxAI) principle [20] where the human expert

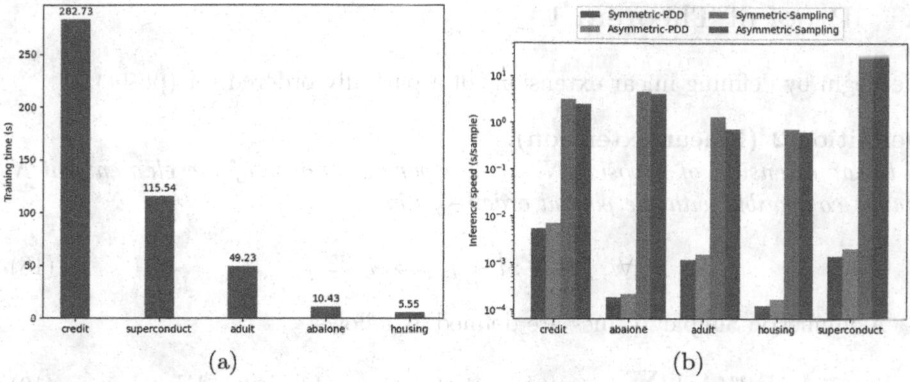

Fig. 4. (a): Time (in seconds) required to train the Partial Dependence Decomposition model for each dataset. (b): Runtime of explanations (in seconds per explanation) for (A-)PDD-SHAP versus permutation sampling. Note that the Y axis is logarithmic.

is motivated and empowered to improve all parts of an AI algorithm's pipeline, but also continuously update his or her own knowledge from the insights gained through the xAI methods.

5 Conclusion

In this work, we introduce A-PDD-SHAP, an algorithm that can be used to efficiently approximate Asymmetric Shapley Values. We demonstrate the algorithm on a selection of tabular datasets and show that the approximation is multiple orders of magnitude faster than the classical permutation sampling approach for computing Asymmetric Shapley Values. However, the main disadvantage of A-PDD-SHAP is the combinatorial explosion in the number of components that need to be modelled when the number of features or the order of interactions increases. In future work, we will focus on techniques to mitigate this problem, by identifying which components are most likely to be important before modelling them, and by parallelizing the modelling process, as each component can be trained independently of all others provided that its subcomponents are available. Finally, although the single decision tree already gives a relatively good approximation for the datasets considered here, we will further investigate the use of more sophisticated models to approximate the PDD components.

Acknowledgements. The authors declare that there are no conflicts of interest. This work does not raise any ethical issues. The research leading to these results has received fund- ing from the Flemish Government under the "Onderzoeksprogramma Artificiële Intelligentie (AI) Vlaanderen" programme, and from the BOF project 01D13919. Parts of this work have been funded by the Austrian Science Fund (FWF), Project: P-32554, explainable AI.

A Proof of Theorem 1

We begin by defining linear extensions of a partially ordered set (poset):

Definition 2 (Linear extension).
A linear extension of a poset $(N, \prec)$ is a permutation π of the elements of N that is compatible with the partial order $\prec$, i.e.:

$$\forall i, j \in N : i \prec j \implies i \prec_{\pi} j \tag{18}$$

Asymmetric Shapley values are defined as follows:

$$\phi_v^w(i) = \sum_{\pi \in \Pi} w(\pi)[v(\{j : j \preceq_{\pi} i\}) - v(\{j : j \prec_{\pi} i\})] \tag{19}$$

where w is a probability distribution over permutations $\pi \in \Pi(N)$. By summing over subsets rather than permutations, we get to the following equivalent definition:

$$\phi_v^w(i) = \sum_{T \subseteq N \setminus i} p_T^i [v(T \cup \{i\}) - v(T)] \tag{20}$$

$$p_T^i = \sum_{\pi \in \Pi_T^i} w(\pi) \tag{21}$$

$$\Pi_T^i = \{\pi \in \Pi : \text{pred}(i, \pi) = T\} \tag{22}$$

where $\text{pred}(i, \pi) := \{j \in N : j \prec_{\pi} i\}$ is the set of predecessors to i in π. From the assumptions, we know that the value function can be written as a sum of non-empty PDD components:

$$v^{\mathbf{x}}(S) = \sum_{\substack{T \subseteq S \\ S \neq \emptyset}} f_S(\mathbf{x}) \tag{23}$$

Substituting this expression into Eq. (20), we get:

$$\begin{aligned}
\phi_v(i) &= \sum_{T \subseteq N \setminus i} p_T^i \left[\sum_{\substack{S \subseteq T \cup i \\ S \neq \emptyset}} f_S(\mathbf{x}) - \sum_{\substack{S \subseteq T \\ S \neq \emptyset}} f_S(\mathbf{x}) \right] \\
&= \sum_{T \subseteq N \setminus i} p_T^i \sum_{\substack{S \subseteq T \\ S \neq \emptyset}} f_{S \cup \{i\}}(\mathbf{x}) \qquad (24) \\
&= \sum_{T \subseteq N \setminus i} p_T^i \sum_{\substack{S \subseteq T \\ i \in S}} f_S(\mathbf{x})
\end{aligned}$$

We can now write this as a single sum by observing that every term $f_{S\cup i}(\mathbf{x})$ gets counted exactly once for each of its supersets $T \supseteq S$, with a weight of p_T^i:

$$\phi_v(i) = \sum_{S\subseteq N\setminus i} \alpha_S^i f_{S\cup i}(\mathbf{x}) \tag{25}$$

$$\alpha_S^i = \sum_{S\subseteq T\subseteq N\setminus i} p_T^i \tag{26}$$

Substituting Eq. (21) into the expression for α_S, we again end up with a nested sum:

$$\alpha_S^i = \sum_{S\subseteq T\subseteq N\setminus i} \sum_{\pi\in\Pi_T^i} w(\pi) \tag{27}$$

We can write this as a single sum by observing that the sets Π_T^i form a partition of the set of all permutations $\Pi(N)$:

- Each set Π_T^i is disjoint to all other sets $\Pi_{T'}^i$. This can easily be seen by the fact that if $\pi \in \Pi_T^i$, the set of predecessors to i must be equal to T, so if π is also in $\Pi_{T'}^i$, then $T = T'$.
- Each permutation $\pi \in \Pi(N)$ must be in some set Π_T^i, namely the set for which T is the set of predecessors to i in π (this can also be the empty set).

Because these permutation sets to form a partition, we can reduce the expression of α_S^i to:

$$\alpha_S^i = \sum_{\pi\in\overline{\Pi}_S^i} w(\pi) \tag{28}$$

where $\overline{\Pi}_S^i$ is the set of all permutations where the predecessors of i contain a given set S:

$$\overline{\Pi}_S^i = \bigsqcup_{S\subseteq T\subseteq N\setminus i} \Pi_T^i = \{\pi \in \Pi : S \subseteq \text{pred}(i, \pi)\} \tag{29}$$

where $\bigsqcup$ denotes the disjoint union of sets.

Without loss of generality, we can assume that the probability distribution w is defined as follows:

$$w(\pi) \propto \begin{cases} 1 \text{ if } \forall i, j \in N : i \prec_{\mathcal{A}} j \implies i \prec_\pi j \\ 0 \text{ otherwise} \end{cases} \tag{30}$$

This implies that α_S^i is the proportion of all linear extensions of $(N, \prec_{\mathcal{A}})$ that are also in $\overline{\Pi}_S^i$, i.e. the proportion of linear extensions of $\prec_{\mathcal{A}}$ where all elements in S precede i. To compute this proportion, we start by making the following two observations:

- If $\exists e \in S : i \prec e$, then $\alpha_S^i = 0$, as no permutation where e precedes i can be a linear extension of $(N, \prec)$. This corresponds to the first case in Eq. 16.
- If $S = T \cup e$ for some $e \in N$, with $e \prec i$, then $\alpha_S^i = \alpha_T^i$, as every linear extension of $(N, \prec)$ is a permutation where e precedes i, so every linear extension in $\overline{\Pi}_T^i$ is also in $\overline{\Pi}_S^i$.

If we now want to compute α_S^i for arbitrary S, we can assume that S contains no elements e such that $i \prec e$ (otherwise $\alpha_S^i = 0$, as per the first observation). We assume first that S consists only of elements e such that $e \not\prec i \wedge i \not\prec e$.

Consider a permutation $\pi \in \Pi(S \cup \{i\})$. From π, a linear extension on $(N, \prec)$ can be constructed as follows:

1. Choose locations and ordering for the elements of $\overline{A} \backslash (S \cup \{i\})$
2. Choose locations for $\overline{A} \cap (S \cup \{i\})$
3. Choose an ordering for the elements of each $A_l : i \notin A_l$
4. If $i \notin \overline{A}$: Choose an ordering and locations for the elements of $A_j \backslash (S \cup i)$, where $i \in A_j$

As we have assumed that S consists only of elements that are not comparable to i, $(S \cup \{i\}) \subseteq (\overline{A} \sqcup A_j)$ (or $(S \cup \{i\}) \subseteq \overline{A}$ if $i \in \overline{A}$). Therefore, the procedure above defines a specific linear extension on $(N, \prec)$ where the elements of $S \cup i$ are ordered according to π. None of these steps depend on the specific permutation π, only on the set S and i. Therefore, every permutation $\pi \in \Pi(S \cup \{i\})$ corresponds to an equal amount of linear extensions on $(N, \prec)$. Also, every linear extension on $(N, \prec)$ corresponds to exactly one permutation $\pi \in \Pi(S \cup i)$.

Recall that α_S^i is the proportion of linear extensions on $(N, \prec)$ where all elements of S precede i. This can be formulated as follows: given a linear extension on $(N, \prec)$ chosen uniformly at random, what is the probability that all elements in S precede i? Because every permutation of $S \cup i$ corresponds to an equal number of linear extensions on $(N, \prec)$, this question is equivalent to the question: given a permutation of $S \cup \{i\}$ chosen uniformly at random, what is the probability that all elements in S precede i? The answer is $\frac{|S|!}{(|S|+1)!} = \frac{1}{|S|+1}$.

For a general set S that contains both elements $e \not\prec i \wedge i \not\prec e$ and elements $e \prec i$, we can derive α_S^i by first introducing the incomparable elements $\underline{S} \subset S$: $\underline{S} = \{e \in S : e \not\prec i \wedge i \not\prec e\}$. The coefficient for this set is $\alpha_{\underline{S}}^i = \frac{1}{|\underline{S}|+1}$. Then, we can introduce all elements $e \in S : e \prec i$. As per rule 2, these elements have no influence on the coefficient, implying that $\alpha_S^i = \alpha_{\underline{S}}^i$.

Substituting this value for α_S^i into Eqs. (25) and (26), we get the following, which corresponds to the second case in Eq. 16:

$$\phi_v^{\mathcal{A}}(i) = \sum_{S \subseteq N \backslash i} \frac{f_{S \cup i}(\mathbf{x})}{|\underline{S}| + 1} \tag{31}$$

□

References

1. Aas, K., Jullum, M., Løland, A.: Explaining individual predictions when features are dependent: more accurate approximations to Shapley values. Artif. Intell. **298**, 103502 (2021). https://doi.org/10.1016/j.artint.2021.103502
2. Chen, H., Janizek, J.D., Lundberg, S., Lee, S.I.: True to the model or true to the data? (2020)
3. Friedman, J.H.: Greedy function approximation: a gradient boosting machine. Ann. Stat. **29**(5), 1189–1232 (2001)
4. Frye, C., de Mijolla, D., Begley, T., Cowton, L., Stanley, M., Feige, I.: Shapley explainability on the data manifold. In: International Conference on Learning Representations. arXiv (2021)
5. Frye, C., Rowat, C., Feige, I.: Asymmetric Shapley values: incorporating causal knowledge into model-agnostic explainability. In: Advances in Neural Information Processing Systems, vol. 33, pp. 1229–1239. Curran Associates, Inc. (2020)
6. Gevaert, A., Saeys, Y.: PDD-SHAP: fast approximations for Shapley values using functional decomposition. In: Workshop on Trustworthy Artificial Intelligence as a Part of the ECML/PKDD 22 Program (2022)
7. Girardi, D., et al.: Interactive knowledge discovery with the doctor-in-the-loop: a practical example of cerebral aneurysms research. Brain Inform. **3**(3), 133–143 (2016). https://doi.org/10.1007/s40708-016-0038-2
8. Holzinger, A.: The next frontier: AI we can really trust. In: Kamp, M., et al. (eds.) ECML PKDD 2021. CCIS, vol. 1524, pp. 427–440. Springer, Cham (2021). https://doi.org/10.1007/978-3-030-93736-2_33
9. Holzinger, A., et al.: Information fusion as an integrative cross-cutting enabler to achieve robust, explainable, and trustworthy medical artificial intelligence. Inf. Fusion **79**(3), 263–278 (2022). https://doi.org/10.1016/j.inffus.2021.10.007
10. Holzinger, A., Müller, H.: Toward human–AI interfaces to support explainability and causability in medical AI. Computer **54**(10), 78–86 (2021). https://doi.org/10.1109/MC.2021.3092610
11. Holzinger, A., Saranti, A., Molnar, C., Biecek, P., Samek, W.: Explainable AI methods - a brief overview. In: Holzinger, A., Goebel, R., Fong, R., Moon, T., Müller, K.R., Samek, W. (eds.) xxAI 2020. LNAI, vol. 13200, pp. 13–38. Springer, Cham (2022). https://doi.org/10.1007/978-3-031-04083-2_2
12. Hooker, G.: Discovering additive structure in black box functions. In: Proceedings of the 2004 ACM SIGKDD International Conference on Knowledge Discovery and Data Mining - KDD 2004, Seattle, WA, USA, p. 575. ACM Press (2004). https://doi.org/10.1145/1014052.1014122
13. Kilbertus, N., Rojas Carulla, M., Parascandolo, G., Hardt, M., Janzing, D., Schölkopf, B.: Avoiding discrimination through causal reasoning. In: Advances in Neural Information Processing Systems, vol. 30. Curran Associates, Inc. (2017)
14. Koller, D., Friedman, N.: Probabilistic Graphical Models: Principles and Techniques. MIT Press, Cambridge (2009)
15. Lapuschkin, S., Wäldchen, S., Binder, A., Montavon, G., Samek, W., Müller, K.R.: Unmasking Clever Hans predictors and assessing what machines really learn. Nat. Commun. **10**(1), 1096 (2019)
16. Lundberg, S., Lee, S.I.: A unified approach to interpreting model predictions. In: Advances in Neural Information Processing Systems, vol. 30, pp. 4766–4775 (2017)
17. Miller, T.: Explanation in artificial intelligence: insights from the social sciences. Artif. Intell. **267**, 1–38 (2019)

18. Pearl, J.: Causality. Cambridge University Press, Cambridge (2009)
19. Pearl, J., Bareinboim, E.: Transportability of causal and statistical relations: a formal approach. In: 11th International IEEE Conference on Data Mining Workshops, pp. 540–547. IEEE (2011). https://doi.org/10.1109/ICDMW.2011.169
20. Saranti, A., et al.: Actionable explainable AI (AxAI): a practical example with aggregation functions for adaptive classification and textual explanations for interpretable machine learning. Mach. Learn. Knowl. Extract. **4**(4), 924–953 (2022). https://doi.org/10.3390/make4040047
21. Shapley, L.S.: A value for n-person games. In: Contributions to the Theory of Games, vol. 2, no. 28, pp. 307–317 (1953)
22. Sundararajan, M., Najmi, A.: The many Shapley values for model explanation. In: Proceedings of the 37th International Conference on Machine Learning, vol. 119, pp. 9269–9278 (2020)
23. Vanschoren, J., van Rijn, J.N., Bischl, B., Torgo, L.: OpenML: networked science in machine learning. SIGKDD Explor. **15**(2), 49–60 (2013). https://doi.org/10.1145/2641190.2641198

Domain-Specific Evaluation of Visual Explanations for Application-Grounded Facial Expression Recognition

Bettina Finzel(✉), Ines Rieger, Simon Kuhn, and Ute Schmid

Cognitive Systems, University of Bamberg, Bamberg, Germany
{bettina.finzel,ines.rieger,ute.schmid}@uni-bamberg.de

Abstract. Research in the field of explainable artificial intelligence has produced a vast amount of visual explanation methods for deep learning-based image classification in various domains of application. However, there is still a lack of domain-specific evaluation methods to assess an explanation's quality and a classifier's performance with respect to domain-specific requirements. In particular, evaluation methods could benefit from integrating human expertise into quality criteria and metrics. Such domain-specific evaluation methods can help to assess the robustness of deep learning models more precisely. In this paper, we present an approach for domain-specific evaluation of visual explanation methods in order to enhance the transparency of deep learning models and estimate their robustness accordingly. As an example use case, we apply our framework to facial expression recognition. We can show that the domain-specific evaluation is especially beneficial for challenging use cases such as facial expression recognition and provides application-grounded quality criteria that are not covered by standard evaluation methods. Our comparison of the domain-specific evaluation method with standard approaches thus shows that the quality of the expert knowledge is of great importance for assessing a model's performance precisely.

Keywords: Convolutional Neural Networks · Explainable Artificial Intelligence · Facial Expressions · Explanation Evaluation · Robustness

1 Introduction

Deep learning approaches are successfully applied for image classification. However, the drawback of these deep learning approaches is their lack of robustness in terms of reliable predictions under small changes in the input data or model parameters [10]. For example, a model should be able to handle out-of-distribution data that deviate from the training distribution, e.g., by being blurry or showing an object from a different angle. However, often models produce confidently false predictions for out-of-distribution data. These can get unnoticed,

The work presented in this paper was funded by grant DFG (German Research Foundation) 405630557 (PainFaceReader).

A. Holzinger et al. (Eds.): CD-MAKE 2023, LNCS 14065, pp. 31–44, 2023.
https://doi.org/10.1007/978-3-031-40837-3_3

as deep learning models are per default a black box approach with no insight into the reasons for predictions or learned features.

Therefore, explainers can be applied that visually explain the prediction in order to enhance the transparency of image classification models and to evaluate the model's robustness towards out-of-distribution samples [33,35]. Visual explanations are based on computing the contribution of individual pixels or pixel groups to a prediction, thus, helping to highlight what a model "looks at" when classifying images [36].

An important aspect is that both, robustness and explainability, are enablers for trust. They promote reliability and ensure that humans remain in control of model decisions [13]. This is of special interest in decision-critical domains such as medical applications and clinical assistance as demonstrated by Holzinger et al. [14] and Finzel et al. [9]. As robustness and explainability are therefore important requirements for application-relevant models, measures that help to assess the fulfillment of such requirements should be deployed with the models. With respect to visual explanations as a basis to rebustness and explainability analysis, it is worth noting that visualizations express learned features only qualitatively. In order to analyze a model's robustness more precisely, quantitative methods are needed to evaluate visual explanations [3,28,34]. These quantitative methods, however, do not provide domain-specific evaluation criteria yet that are tailored to the application domain.

In this work, we propose a framework for domain-specific evaluation and apply it to the use case of facial expression recognition. Our domain-specific evaluation is based on selected expert knowledge that is quantified automatically with the help of visual explanations for the respective use case. The user can inspect the quantitative evaluation and draw their own conclusions.

To define facial expressions, one psychologically established way is to describe them with the Facial Action Coding System (FACS) [7], where they are categorized as sets of so-called Action Units (AUs). Facial expression analysis is commonly performed to detect emotions or pain (in clinical settings). These states are often derived from a combination of AUs present in the face [22,23]. In this paper, we analyze only the AUs that are pain- and emotion-relevant. The appearance and occurrence of facial expressions may vary greatly for different persons, which makes it a challenging and interesting task to recognize and interpret them reliably and precisely. A substantial body of research exists to tackle this challenge by training deep learning models (e.g., convolutional neural networks) to classify AUs in human faces from images [11,29,31,40]. Our approach is the first that adds a quantitative evaluation method to the framework of training, testing and applying deep learning models for facial expression recognition. Our research contributes to the state-of-the-art as follows:

- We propose a domain-specific evaluation framework that allows for integrating and evaluating expert knowledge by quantifying visual explanations. This increases the transparency of black box deep learning models for the user and provides a domain-specific evaluation of the model robustness.

- We show for the application use case of facial expression recognition that the selection and quality of expert knowledge for domain-specific evaluation of explanations has a significant influence on the quality of the robustness analysis.
- We show that the domain-specific evaluation is especially beneficial for challenging use cases such as facial expression recognition based on AUs. AUs are a multi-label classification problem with co-occurring classes. We provide a quantitative evaluation that facilitates analyzing AUs by treating them separately.

This paper is structured as follows: First, the related work gives an overview on similar approaches, then our evaluation framework is presented in Sect. 3 in a step by step manner, explaining the general workflow as well as the specific methods applied for the use case of facial expression recognition. Section 4 presents and discusses the results. Finally, we point out directions for future work and conclude our work.

2 Related Work

Work related to this paper mainly covers the aspect of explaining image classifiers and evaluating the generated explanations with respect to a specific domain. Researchers have developed a vast amount of visual explanation methods for image classification. Among the most popular ones are LIME, LRP, GradCAM, SHAP and RISE (see Schwalbe and Finzel (2023) for a detailed overview and references to various methods [35]). There already exist methods and frameworks that evaluate multiple aspects of visual explanations, e.g., robustness, as provided for example by the Quantus toolbox that examines the impact of input parameter changes on the stability of explanations [12]. Hsieh et al. [15] present feature-based explanations, in particular pixel regions in images that are necessary and sufficient for a prediction, similar to [6], where models are evaluated based on feature removal and preservation. Work that examines the robustness of visual explanations of models applied in different domains, was published for example by Malafaia et al. [28], Schlegel et al. [34] and Artelt et al. [3]. However, these methods do not provide evaluation criteria tailored to the application domain itself. For this purpose, XAI researchers have developed a collection of *application-grounded* metrics [35,42].

Application-grounded perspectives may consider the needs of explanation recipients (explainees) [39,42] and an increase in task performance for applied human-AI decision making [16] or the completeness and soundness of an explanation [21], e.g., with respect to given metrics such as the coverage of relevant image regions [19]. Facial expression recognition, which is the application of this work, is usually a multi-class problem. For multiple classes, application-grounded evaluation may also encompass correlations between ground truth labels of different classes and evaluating, whether learned models follow these correlations [32]. In this work, we focus on evaluating each class separately and whether visual explanations, generated by explainers for image classification, highlight important image regions.

A review of state-of-the-art and recent works on techniques for explanation evaluation indicates that defining important image regions by bounding boxes is a popular approach. Bounding boxes can be used to compute, whether visual explanations (e.g., highlighted pixel regions) cover important image regions, for classification as well as object detection [17,24,31,35]. In terms of robustness, the robustness of a model is higher if the highlighted pixel regions are inside the defined bounding boxes. With respect to the aforementioned definition of robustness [10], a robust model should show an aggregation of relevance inside the bounding boxes even when out-of-distribution data is encountered. However, bounding boxes are not always suitable to set the necessary boundaries around the important image regions. This can lead to a biased estimation of the predictive performance of a model as bounding boxes usually define areas larger than the region of interest. If models pay attention to surrounding, irrelevant pixels, a bounding box based evaluation may miss this. Hence, as the explanation itself can be biased, the explanation is not robust, which is an important feature of explainability methods [2].

Using polygons as an alternative to bounding boxes is therefore an important step towards integrating domain-specific requirements into the evaluation of explanations to make them more robust. Domain-specific evaluations have not yet been sufficiently discussed across domains, nor broadly applied to the very specific case of facial expression recognition.

In this work, we therefore thoroughly define regions for facial expressions and evaluate the amount of positive relevance inside the defined regions compared to the overall positive relevance in the image (see Sect. 3.5). Instead of using bounding boxes that are very coarse and that might contain class-irrelevant parts of the face as well as background (see Fig. 2), we compute polygons based on class-relevant facial landmarks according to AU masks defined by Ma et al. [27]. We compare a standard bounding box approach with our polygon-based approach for evaluating two state-of-the-art models on two different data sets each and open a broad discussion with respect to justifying model decisions based on visual explanations for domain-specific evaluation. Our domain-specific evaluation framework is introduced in Sect. 3.1.

3 Materials and Methods

The following subsections describe the components of our framework (see Fig. 1 for step numbering), starting with the data sets and evaluated models, and followed by the heatmap generation, and finally our method to quantitatively evaluate the visual explanations by using domain-specific information. Please note that the following paragraphs describe one possible selection of data sets, models, visual explanation method, and explanation evaluation. The framework can be extended or adapted to the needs of other application and evaluation scenarios.

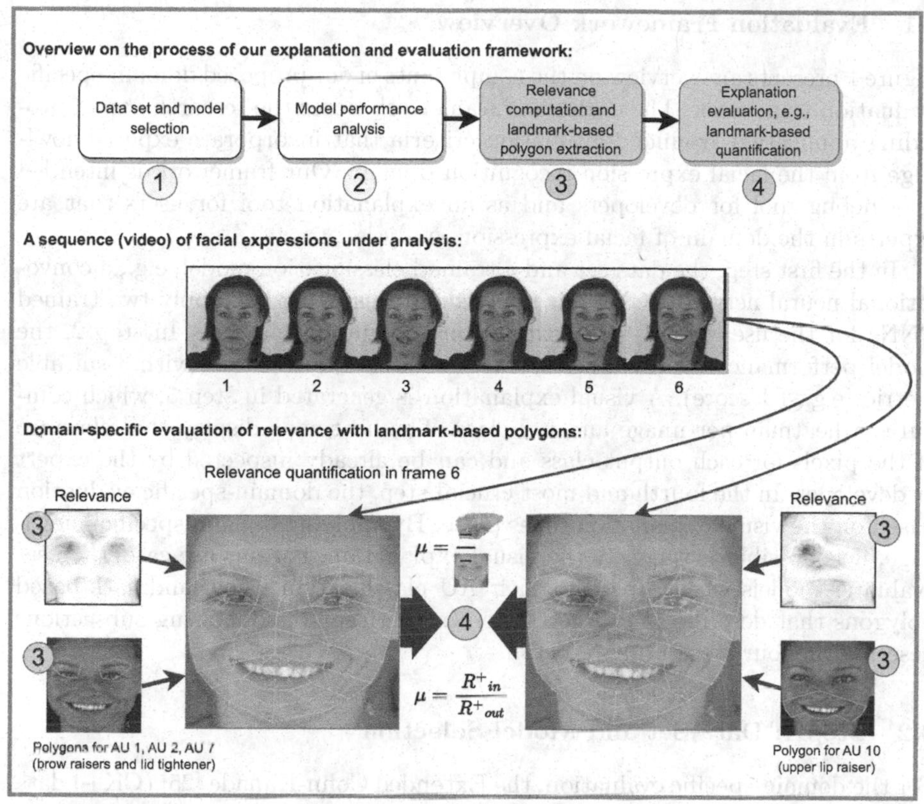

Fig. 1. This figure shows an overview on the components of our framework with exemplary illustrations for the use case of facial expression recognition. The framework processes 4 steps. First, it allows for a flexible and configurable data set and model selection (step 1). Secondly, it analyzes the model's performance with respect to correct predictions (step 2). In step 3, relevance is computed that gets attributed to each pixel by layer-wise relevance propagation. In the same step, polygons are derived from pre-defined domain knowledge in the form of facial landmarks. The aggregation of relevance inside the resulting polygonal image regions gets quantified by our evaluation approach in step 4. For our domain-specific evaluation approach, we consider the positive relevance values computed in step 3 (see red pixel regions in the heatmap-based illustration of relevance). For each image in a video sequence (here: frame 6), we evaluate the aggregation of relevance within the polygons of all predicted AUs (here: AU1, AU2 and AU7, see on the left side of the figure, and AU10, see on the right side of the figure). This is done by dividing the positive relevance aggregation within the region(s) of interest by the total positive relevance within an image (as defined by Eq. 2). With our domain-specific evaluation, a well-performing and robust model would detect positive relevance only within the defined polygonal regions. Deviations of this expectation can be easily uncovered with our framework. (Color figure online)

3.1 Evaluation Framework Overview

Figure 1 presents an overview on the components of our proposed domain-specific evaluation framework. The evaluation framework closes the research gap of providing application-grounded evaluation criteria that incorporate expert knowledge from the facial expression recognition domain. Our framework is intended as a debug tool for developers and as an explanation tool for users that are experts in the domain of facial expression analysis.

In the first step, the data set and a trained classification model, e.g., a convolutional neural network (CNN), is selected. In this paper, we apply two trained CNNs for the use case of facial expression recognition via AUs. In step 2, the model performance is evaluated on images selected by the user with a suitable metric (e.g., F1 score). A visual explanation is generated in step 3, which computes a heatmap per image and per class. The heatmaps display the relevance of the pixels for each output class and can be already inspected by the expert or developer. In the fourth and most crucial step, the domain-specific evaluation based on the visual explanation takes place. By applying domain specific knowledge, it is possible to quantify the visual explanation. For our use case, the user evaluates models with respect to their AU classification using landmark-based polygons that describe the target region in the face. The following subsections describe the four steps in more detail.

3.2 Step 1: Data Set and Model Selection

For the domain-specific evaluation, the Extended Cohn-Kanade [25] (CK+) data set (593 video sequences) and a part of the Actor Study data set [37] (subjects 11–21, 407 sequences) are chosen. The CK+ and Actor Study data set were both created in a controlled environment in the labatory with actors as study subjects.

We evaluate two differently trained models, a model based on the ResNet-18 architecture [30] and a model based on the VGG-16 architecture [38]. They are both CNNs. A CNN is a type of artificial neural network used in image recognition and processing that is specifically designed to perform classification on images. It uses so-called multiple convolution layers to abstract from individual pixel values and weights individual features depending on the input it is trained on. This weighting ultimately leads to a class decision. In the CNNs we use, there is one predictive output for each AU class. AU recognition is a multi-label classification problem, so each image can be labelled with more that one AU, depending on the co-occurrences.

While the ResNet-18 from [31] is trained on the CK+ data set as well as on the Actor Study data set, the VGG-16 is trained on a variety of different data sets from vastly different settings (e.g., in-the-wild and in-the-lab): Actor Study [37] (excluding subjects 11–21), Aff-Wild2 [20], BP4D [41], CK+ [25], the manually annotated subset of EmotioNet [5], and UNBC [26]. We use the same training procedure as in [29] to retrain the VGG-16 without the Actor Study subjects 11–21, which is then our testing data.

With the two trained models we can compare the influence of different training distributions. Furthermore, we apply the domain-specific evaluation with respect to training and testing data. By inspecting explanations for the model on the training data, the inherent bias of the model is evaluated that can arise for example by overfitting on features of the input images. By evaluating the model on the testing data, we can estimate the generalization ability of the model.

The dlib toolkit [18] is used to derive 68 facial landmarks from the images. Based on these landmarks and the expert knowledge about the regions of the AUs, we compute the rectangles and polygons for the evaluation of generated visual explanations.

3.3 Step 2: Model Performance Analysis

For evaluating the model performance, we use the F1 score (Eq. 1), the harmonic mean of precision and recall with a range of [0,1], whereas 1 indicates perfect precision and recall. This metric is beneficial if there is an imbalanced ratio of displayed and non-displayed classes, which is the case for AUs [29]. The ResNet-18 is evaluated with a leave-one-out cross validation on the Actor Study data set, and the performance of the VGG-16 is evaluated on the validation data set, and additionally on the testing part of the Actor Study (subjects 11–21).

$$F1 = \frac{2 \cdot precision \cdot recall}{precision + recall} \tag{1}$$

3.4 Step 3: Visual Classification Explanations

We apply layer-wise relevance propagation (LRP) [4] to visually identify the parts of the image which contributed to the classification, i.e., to attribute (positive and negative) relevance values to each pixel of the image. "Positive relevance" denotes that the corresponding pixel influenced the CNN's decision towards the observed class. "Negative relevance" means it influenced the decision against the observed class. For a given input image, LRP decomposes a CNN's output with the help of back-propagation from the CNN's output layer back to its input layer, meaning that each pixel is assigned with a positive or negative relevance score. These relevance scores can be used to create heatmaps by normalizing their values with respect to a chosen color spectrum [4].

We choose the decomposition scheme from Kohlbrenner et al. [19] based on the implementation provided by the *iNNvestigate* toolbox [1]. For the ResNet-18 we select *PresetB* as the LRP analyzer and for the VGG-16 network we select *PresetAFlat*, since these configurations are usually best working for the respective network architectures [19].

3.5 Step 4: Domain-Specific Evaluation Based on Landmarks

As a form of domain-specific knowledge, polygons enclosing the relevant facial areas for each AU are utilized (see Fig. 2). Each polygon is constructed based on a subset of the 68 facial landmarks to enclose one region. The regions are defined similar to Ma et al. [27].

As motivated earlier, the selection and quality of domain-specific knowledge is of crucial importance. Figure 2b shows a coarse bounding box approach of Rieger et al. [31] and Fig. 2c shows our fine-grained polygon approach exemplary for the AU9 (nose wrinkler). We can see that for b) also the background is taken into account, which makes the quantitative evaluation error-prone. This shows for the use case of AUs, being a multi-class multi-label classification problem, the importance of carefully defining boundaries, so that ideally one boundary only encloses class-relevant facial areas per AUs, which is where our polygon approach aims at.

For our evaluation approach, we consider only positive relevance in heatmaps, since these express the contribution of a pixel to the target class, e.g., a certain AU. However, the evaluation of the aggregation of negative relevance inside boundaries would also be possible, but is not considered here.

For quantitatively evaluating the amount of relevance inside the box or polygon, we use the ratio μ of the positive relevance inside the boundary (R_{in}) and the overall positive relevance in the image (R_{tot}) (Eq. 2). To make our approach comparable, we use the same equation as Kohlbrenner et al. [19].

$$\mu = \frac{R_{in}}{R_{tot}} \tag{2}$$

The μ-value ranges from 0 (no positive relevance inside the boundary) to 1 (all positive relevance inside the boundary). High μ-values indicate that a CNN based its classification output on the class-relevant parts of the image. This means, that for a μ-value above the value of 0.5, the majority of relevance aggregates inside the boundaries.

For our evaluation, we consider only images for which the ground truth as well as the classification output match in the occurrence of the corresponding AU.

4 Results and Discussion

Table 1 shows the overview of the performance and domain-specific evaluation of the VGG-16 model. The performance on the validation data set differs greatly for some AUs (e.g., AU10, or AU14), which can be explained by the big array of different training data sets. Henceforth, the data distribution of the Actor Study is not predominantly represented by the trained model. We may keep in mind that the Actor Study is a posed data set, so some facial expression can differ in their visual appearance from the natural ones. However, when looking at the average μ_{poly}-values of the polygon boundaries, we can see a correlation of the higher μ-values with the validation performance for some AUs. For example, the model displays a good performance on the validation data set for AU10, but a significantly lower one on the testing data set. However, in comparison, the μ_{poly}-value is the highest of all evaluated AUs. A similar pattern can be found for example for AU14. Since we only use the correctly classified images for our

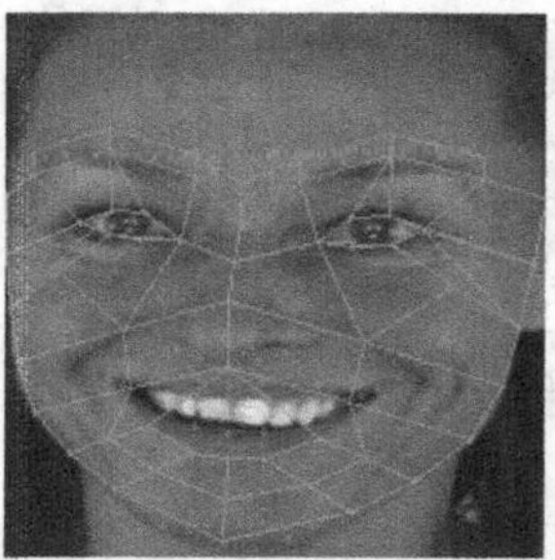

(a) 68 facial landmarks [27]

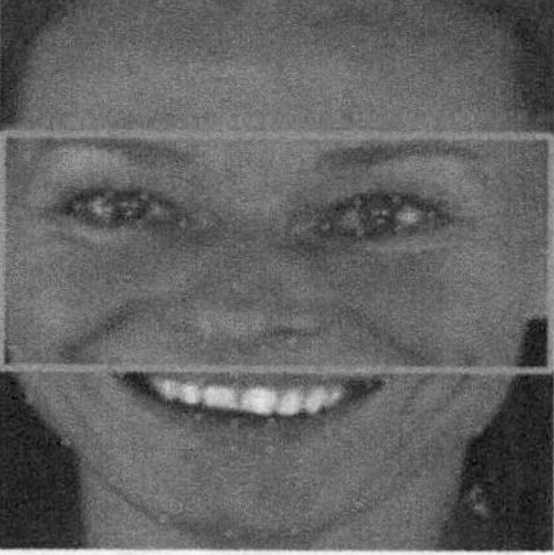

(b) AU9 bounding box [31]

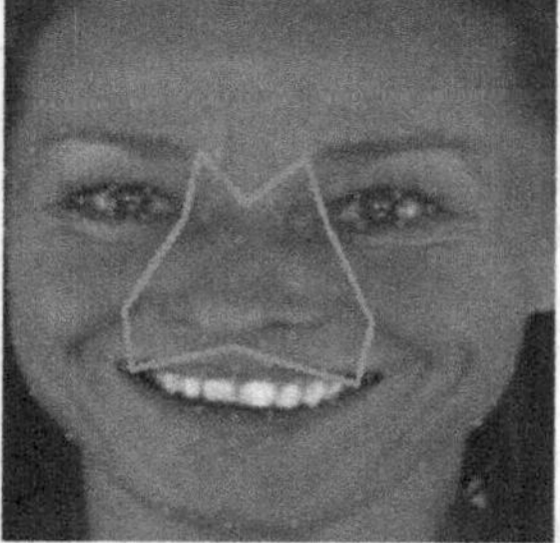

(c) AU9 polygon (ours)

Fig. 2. The domain-specific knowledge for evaluating the heatmaps are facial landmarks. Exemplary image with emotion *happy* and highlighted region for Action Unit 9 (AU9) (nose wrinkler). AU region boundaries are pink and facial landmarks green dots. (Color figure online)

domain-specific evaluation, we can interpret that the model can locate the region for AU10 or AU14, but that there are probably many out-of-distribution images for these AUs in the testing data set, hence making a good model performance difficult. We can also observe that for instance for AU25, there is a strong performance on both the validation and testing data set, but a low μ-value, which can indicate that the model did not identify the expected region as important.

Table 2 shows a comparison between our polygon approach μ_{poly} with the standard bounding box approach μ_{box} [31] for the ResNet-18. The bounding box approach μ_{box} yields overall higher μ-values than the polygons (μ_{poly}), which is expected since the boxes enclose a larger area than the polygons. This can also indicate that the coarse boxes contain pixels that get assigned with relevance by the ResNet-18, although they are not located in relevant facial areas, hence highlighting once more the importance of the quality of the domain-specific knowledge. Our polygons enclose in contrast to the bounding boxes only class-relevant facial areas. Looking closely at the AUs, we can see that although the μ_{box} is high for AU4, it has also the highest difference to μ_{poly} for both the data sets CK+ and Actor Study. We can therefore assume a high relevance spread for AU4, which is ultimately discovered by applying the fine-grained polygon approach. In contrast, AU10 looses the least performance for both data sets concerning the μ-value, but displays also the lowest F1 value, which can indicate that although the AU is not accurately predicted in a lot of images, the model has nonetheless learned to detect the right region for images with correct predictions.

Overall, the μ-values are low for all classes, indicating a major spread of relevance outside of the defined boxes. Some of the relevance may be outside of the polygons due to a long tail distribution across the image with a lot of pixels having a low relevance value. This can lead to low μ-values for all polygons. When comparing the μ_{poly} with the μ_{box} approach, it is apparent that the μ_{box}-values are higher compared to the μ_{poly}-values, and only μ_{box}-values reach an average μ-value above of 0.5 across data sets. Both findings show the need for a

Table 1. Classification performance and domain-specific evaluation of the VGG-16 model. The performance is measured by the F1 score on the validation and testing data set respectively. The domain-specific evaluation is measured with the average μ-values of the polygones on the testing data set. The testing data set is the Actor Study data set, subjects 12–21. Best results are in bold.

AU	F1 score		av. μ_{poly}
	validation	testing	
1	0.61	**0.71**	0.125
2	0.42	0.58	0.155
4	0.51	0.56	0.114
6	0.72	0.50	0.137
7	0.79	0.44	0.092
10	**0.83**	0.19	**0.239**
12	0.77	0.58	0.220
14	0.77	0.13	0.221
15	0.62	0.13	0.215
17	0.67	0.43	0.008
23	0.49	0.05	0.026
24	0.63	0.19	0.037
25	0.66	0.69	0.036

Table 2. Comparison of our approach μ_{poly} with the standard bounding box approach μ_{box} [31] for ResNet-18. Highest values are in bold.

AU	F1	CK+			Actor Study		
		μ_{box}	μ_{poly}	$\lvert\mu_{box} - \mu_{poly}\rvert$	μ_{box}	μ_{poly}	$\lvert\mu_{box} - \mu_{poly}\rvert$
04	0.68	**0.579**	0.118	**0.461**	0.458	0.061	**0.397**
06	0.55	0.432	0.144	0.288	0.360	0.087	0.273
07	0.62	0.499	0.097	0.402	0.417	0.038	0.379
09	0.48	0.492	0.195	0.297	0.293	0.084	0.209
10	0.32	0.350	**0.339**	0.011	0.197	0.196	0.001
25	**0.83**	0.413	0.211	0.202	0.457	0.175	0.282
26	0.60	0.330	0.143	0.187	0.493	0.201	0.292
27	0.57	0.463	0.203	0.260	**0.545**	**0.223**	0.322

domain-specific evaluation with carefully selected expert knowledge in order to assess a model's performance as good as possible but also the precision of used visual explainers with respect to the spread of relevance.

Furthermore, our approach emphasizes the general need of an evaluation beyond classification performance of models. Although the models display high F1 scores for most of the classes, the relevance is not in the expected areas.

A limitation of our evaluation results is that they do not consider μ-values normalized according to the size of regions, although our approach allows such an extension in principle. This is an important aspect, since the areas for each AU are differently sized in relation to the overall image size. This means that some AU boundaries may be more strict on the relevance distribution than others and may penalize the model's performance thereof. For that we suggest a weighted μ-value calculation, optimally with respect to the overall relevance distribution in an image, e.g., based on thresholding the relevance [8].

5 Conclusion

In this paper, we present an approach for domain-specific evaluation of visual explanation methods in order to enhance the transparency of CNNs and estimate their robustness as precisely as possible. As an example use case, we applied our framework to facial expression recognition. We showed that the domain-specific evaluation can give insights into facial classification models that domain-agnostic evaluation methods or performance metrics cannot provide. Furthermore, we could show by comparison that the quality of the expert knowledge is of great importance for assessing a model's performance precisely.

References

1. Alber, M., et al.: iNNvestigate neural networks! J. Mach. Learn. Res. **20**(93), 1–8 (2019)
2. Alvarez-Melis, D., Jaakkola, T.: On the robustness of interpretability methods. arXiv preprint arXiv:1806.08049 (2018)
3. Artelt, A., et al.: Evaluating robustness of counterfactual explanations. In: Proceedings of Symposium Series on Computational Intelligence, pp. 1–9. IEEE (2021). https://doi.org/10.1109/SSCI50451.2021.9660058
4. Bach, S., Binder, A., Montavon, G., Klauschen, F., Müller, K.R., Samek, W.: On pixel-wise explanations for non-linear classifier decisions by layer-wise relevance propagation. PLoS ONE **10**(7), 01–09 (2015). https://doi.org/10.1371/journal.pone.0130140
5. Benitez-Quiroz, C.F., Srinivasan, R., Martinez, A.M.: EmotioNet: an accurate, real-time algorithm for the automatic annotation of a million facial expressions in the wild. In: Proceedings of Conference on Computer Vision and Pattern Recognition (CVPR), pp. 5562 5570. IEEE (2016). https://doi.org/10.1109/cvpr.2016.600
6. Dhurandhar, A., et al.: Explanations based on the missing: towards contrastive explanations with pertinent negatives. In: Proceedings of Advances in Neural Information Processing Systems, vol. 31, pp. 590–601 (2018)
7. Ekman, P., Friesen, W.V.: Facial Action Coding Systems. Consulting Psychologists Press (1978)
8. Finzel, B., Kollmann, R., Rieger, I., Pahl, J., Schmid, U.: Deriving temporal prototypes from saliency map clusters for the analysis of deep-learning-based facial action unit classification. In: Proceedings of the LWDA 2021 Workshops: FGWM, KDML, FGWI-BIA, and FGIR. CEUR Workshop Proceedings, vol. 2993, pp. 86–97. CEUR-WS.org (2021)

9. Finzel, B., Tafler, D.E., Thaler, A.M., Schmid, U.: Multimodal explanations for user-centric medical decision support systems. In: Proceedings of the AAAI Fall Symposium on Human Partnership with Medical AI: Design, Operationalization, and Ethics (AAAI-HUMAN). CEUR Workshop Proceedings, vol. 3068. CEUR-WS.org (2021)
10. Guidotti, R., Monreale, A., Ruggieri, S., Turini, F., Giannotti, F., Pedreschi, D.: A survey of methods for explaining black box models. ACM Comput. Surv. **51**(5), 1–42 (2018). https://doi.org/10.1145/3236009
11. Hassan, T., et al.: Automatic detection of pain from facial expressions: a survey. IEEE Trans. Pattern Anal. Mach. Intell. **43**(6), 1815–1831 (2019)
12. Hedström, A., et al.: Quantus: an explainable AI toolkit for responsible evaluation of neural network explanations and beyond. J. Mach. Learn. Res. **24**(34), 1–11 (2023)
13. Holzinger, A.: The next frontier: AI we can really trust. In: Kamp, M., et al. (eds.) ECML PKDD 2021. CCIS, vol. 1524, pp. 427–440. Springer, Cham (2021). https://doi.org/10.1007/978-3-030-93736-2_33
14. Holzinger, A., et al.: Information fusion as an integrative cross-cutting enabler to achieve robust, explainable, and trustworthy medical artificial intelligence. Inf. Fusion **79**, 263–278 (2022). https://doi.org/10.1016/j.inffus.2021.10.007
15. Hsieh, C., et al.: Evaluations and methods for explanation through robustness analysis. In: Proceedings of the 9th International Conference on Learning Representations (ICLR 2021). OpenReview.net (2021)
16. Jesus, S.M., et al.: How can I choose an explainer?: an application-grounded evaluation of post-hoc explanations. In: Proceedings of Conference on Fairness, Accountability, and Transparency (FAccT 2021), pp. 805–815. ACM (2021). https://doi.org/10.1145/3442188.3445941
17. Karasmanoglou, A., Antonakakis, M., Zervakis, M.E.: Heatmap-based explanation of YOLOv5 object detection with layer-wise relevance propagation. In: Proceedings of International Conference on Imaging Systems and Techniques, (IST), pp. 1–6. IEEE (2022). https://doi.org/10.1109/IST55454.2022.9827744
18. King, D.E.: Dlib-ML: a machine learning toolkit. J. Mach. Learn. Res. **10**, 1755–1758 (2009)
19. Kohlbrenner, M., Bauer, A., Nakajima, S., Binder, A., Samek, W., Lapuschkin, S.: Towards best practice in explaining neural network decisions with LRP. In: Proceedings of International Joint Conference on Neural Networks (IJCNN 2020), pp. 1–7. IEEE (2020). https://doi.org/10.1109/IJCNN48605.2020.9206975
20. Kollias, D., Zafeiriou, S.: Aff-wild2: extending the aff-wild database for affect recognition. arXiv preprint arXiv:1811.07770 (2018)
21. Kulesza, T., Stumpf, S., Burnett, M.M., Yang, S., Kwan, I., Wong, W.: Too much, too little, or just right? Ways explanations impact end users' mental models. In: 2013 IEEE Symposium on Visual Languages and Human Centric Computing, San Jose, CA, pp. 3–10. IEEE Computer Society (2013). https://doi.org/10.1109/VLHCC.2013.6645235
22. Kunz, M., Lautenbacher, S.: The faces of pain: a cluster analysis of individual differences in facial activity patterns of pain. Eur. J. Pain **18**(6), 813–823 (2014)
23. Kunz, M., Meixner, D., Lautenbacher, S.: Facial muscle movements encoding pain-a systematic review. Pain **160**(3), 535–549 (2019)
24. Lin, Y., Lee, W., Celik, Z.B.: What do you see?: evaluation of explainable artificial intelligence (XAI) interpretability through neural backdoors. In: Proceedings of the 27th ACM SIGKDD Conference on Knowledge Discovery and Data Mining

(KDD 2021), Virtual Event, Singapore, pp. 1027–1035. ACM (2021). https://doi.org/10.1145/3447548.3467213
25. Lucey, P., Cohn, J.F., Kanade, T., Saragih, J.M., Ambadar, Z., Matthews, I.A.: The extended Cohn-Kanade dataset (CK+): a complete dataset for action unit and emotion-specified expression. In: Proceedings of the IEEE Conference on Computer Vision and Pattern Recognition, CVPR Workshops 2010, San Francisco, CA, USA, pp. 94–101. IEEE Computer Society (2010). https://doi.org/10.1109/CVPRW.2010.5543262
26. Lucey, P., Cohn, J.F., Prkachin, K.M., Solomon, P.E., Matthews, I.A.: Painful data: the UNBC-McMaster shoulder pain expression archive database. In: Proceedings of the 9th IEEE International Conference on Automatic Face and Gesture Recognition (FG 2011), Santa Barbara, CA, USA, pp. 57–64. IEEE Computer Society (2011). https://doi.org/10.1109/FG.2011.5771462
27. Ma, C., Chen, L., Yong, J.: AU R-CNN: encoding expert prior knowledge into R-CNN for action unit detection. Neurocomputing **355**, 35–47 (2019). https://doi.org/10.1016/j.neucom.2019.03.082
28. Malafaia, M., Silva, F., Neves, I., Pereira, T., Oliveira, H.P.: Robustness analysis of deep learning-based lung cancer classification using explainable methods. IEEE Access **10**, 112731–112741 (2022). https://doi.org/10.1109/ACCESS.2022.3214824
29. Pahl, J., Rieger, I., Seuss, D.: Multi-label learning with missing values using combined facial action unit datasets. In: The Art of Learning with Missing Values Workshop at International Conference on Machine Learning (ICML 2020) abs/2008.07234 (2020). https://arxiv.org/abs/2008.07234
30. Rieger, I., Hauenstein, T., Hettenkofer, S., Garbas, J.: Towards real-time head pose estimation: exploring parameter-reduced residual networks on in-the-wild datasets. In: Wotawa, F., Friedrich, G., Pill, I., Koitz-Hristov, R., Ali, M. (eds.) IEA/AIE 2019. LNCS, vol. 11606, pp. 123–134. Springer, Cham (2019). https://doi.org/10.1007/978-3-030-22999-3_12
31. Rieger, I., Kollmann, R., Finzel, B., Seuss, D., Schmid, U.: Verifying deep learning-based decisions for facial expression recognition. In: 28th European Symposium on Artificial Neural Networks, Computational Intelligence and Machine Learning (ESANN 2020), Bruges, Belgium, pp. 139–144 (2020). https://www.esann.org/sites/default/files/proceedings/2020/ES2020-49.pdf
32. Rieger, I., Pahl, J., Finzel, B., Schmid, U.: CorrLoss: integrating co-occurrence domain knowledge for affect recognition. In: Proceedings of the 26th International Conference on Pattern Recognition (ICPR 2022), pp. 798–804. IEEE (2022)
33. Samek, W., Montavon, G., Vedaldi, A., Hansen, L.K., Müller, K. (eds.): Explainable AI: Interpreting, Explaining and Visualizing Deep Learning. Lecture Notes in Computer Science, vol. 11700. Springer, Cham (2019). https://doi.org/10.1007/978-3-030-28954-6
34. Schlegel, U., Arnout, H., El-Assady, M., Oelke, D., Keim, D.A.: Towards a rigorous evaluation of XAI methods on time series. In: IEEE/CVF International Conference on Computer Vision Workshop (ICCVW 2019), Seoul, Korea (South), pp. 4197–4201. IEEE (2019). https://doi.org/10.1109/ICCVW.2019.00516
35. Schwalbe, G., Finzel, B.: A comprehensive taxonomy for explainable artificial intelligence: a systematic survey of surveys on methods and concepts. Data Min. Knowl. Discov. 1–59 (2023). https://doi.org/10.1007/s10618-022-00867-8
36. Selvaraju, R.R., Cogswell, M., Das, A., Vedantam, R., Parikh, D., Batra, D.: Grad-CAM: visual explanations from deep networks via gradient-based localization. Int. J. Comput. Vis. **128**(2), 336–359 (2020). https://doi.org/10.1007/s11263-019-01228-7

37. Seuss, D., et al.: Emotion expression from different angles: a video database for facial expressions of actors shot by a camera array. In: Proceedings of the 8th International Conference on Affective Computing and Intelligent Interaction (ACII 2019), Cambridge, United Kingdom, pp. 35–41. IEEE (2019). https://doi.org/10.1109/ACII.2019.8925458
38. Simonyan, K., Zisserman, A.: Very deep convolutional networks for large-scale image recognition. In: Proceedings of the 3rd International Conference on Learning Representations (ICLR 2015), San Diego, CA, USA (2015). https://arxiv.org/abs/1409.1556
39. Vilone, G., Longo, L.: Notions of explainability and evaluation approaches for explainable artificial intelligence. Inf. Fusion **76**, 89–106 (2021). https://doi.org/10.1016/j.inffus.2021.05.009
40. Werner, P., Martinez, D.L., Walter, S., Al-Hamadi, A., Gruss, S., Picard, R.W.: Automatic recognition methods supporting pain assessment: a survey. IEEE Trans. Affect. Comput. **13**(1), 530–552 (2022). https://doi.org/10.1109/TAFFC.2019.2946774
41. Zhang, X., et al.: BP4D-spontaneous: a high-resolution spontaneous 3D dynamic facial expression database. Image Vis. Comput. **32**(10), 692–706 (2014)
42. Zhou, J., Gandomi, A.H., Chen, F., Holzinger, A.: Evaluating the quality of machine learning explanations: a survey on methods and metrics. Electronics **10**(5), 593 (2021)

Human-in-the-Loop Integration with Domain-Knowledge Graphs for Explainable Federated Deep Learning

Andreas Holzinger[1,2](✉), Anna Saranti[1,2], Anne-Christin Hauschild[3], Jacqueline Beinecke[3], Dominik Heider[4], Richard Roettger[5], Heimo Mueller[2], Jan Baumbach[6], and Bastian Pfeifer[2]

[1] University of Natural Resources and Life Sciences, Vienna, Austria
andreas.holzinger@human-centered.ai
[2] Medical University of Graz, Graz, Austria
[3] University of Göttingen, Göttingen, Germany
[4] University of Marburg, Marburg, Germany
[5] University of Southern Denmark, Odense, Denmark
[6] University of Hamburg, Hamburg, Germany

Abstract. We explore the integration of domain knowledge graphs into Deep Learning for improved interpretability and explainability using Graph Neural Networks (GNNs). Specifically, a protein-protein interaction (PPI) network is masked over a deep neural network for classification, with patient-specific multi-modal genomic features enriched into the PPI graph's nodes. Subnetworks that are relevant to the classification (referred to as "disease subnetworks") are detected using explainable AI. Federated learning is enabled by dividing the knowledge graph into relevant subnetworks, constructing an ensemble classifier, and allowing domain experts to analyze and manipulate detected subnetworks using a developed user interface. Furthermore, the human-in-the-loop principle can be applied with the incorporation of experts, interacting through a sophisticated User Interface (UI) driven by Explainable Artificial Intelligence (xAI) methods, changing the datasets to create counterfactual explanations. The adapted datasets could influence the local model's characteristics and thereby create a federated version that distils their diverse knowledge in a centralized scenario. This work demonstrates the feasibility of the presented strategies, which were originally envisaged in 2021 and most of it has now been materialized into actionable items. In this paper, we report on some lessons learned during this project.

Keywords: Artificial Intelligence · Explainable AI · Machine Learning · Human-in-the-Loop · Graph Neural Networks · Federated Learning · Counterfactual Explanations

A. Holzinger et al. (Eds.): CD-MAKE 2023, LNCS 14065, pp. 45–64, 2023.
https://doi.org/10.1007/978-3-031-40837-3_4

List of Abbreviations

AI	Artificial Intelligence
CLARUS	interaCtive expLainable plAtform for gRaph neUral networkS
DNA	Deoxyribo-Nucleic Acid
FC	FeatureCloud (EU Project)
GDPR	General Data Protection Regulation
GNN	Graph Neural Network
GNN-LRP	GNN Layer-wise Relevance Propagation
GPU	Graphics Processing Unit
HITL	Human-in-the-Loop
IG	Integrated Gradients
i.i.d.	Independent and identically distributed
LRP	Layerwise Relevance Propagation
MI	Mutual Information
ML	Machine Learning
mRNA	messenger Ribo-Nucleic Acid
OOD	Out-Of-Distribution
PGM	Probabilistic Graphical Model Explainer
UI	User Interface
xAI	explainable Artificial Intelligence

1 Introduction and Motivation

The European Project "FeatureCloud (FC)" (Grant Agreement 826078) created a novel Artificial Intelligence (AI) platform which is based on the idea of federated, decentralised learning where only model parameters are communicated. The FC AI App-store https://featurecloud.ai/ is the first platform worldwide to enable federated learning of diverse AI models in a privacy-preserving way [41]. The types of AI models used are quite diverse, including linear regression, clustering, random forests, deep learning, etc. The fundamental idea is that every software developer or data scientist can federate their AI model provided that the model fulfils some minimum requirements (see: https://featurecloud.eu). Dockerization [43] supports seamlessly the transferability of the federated solution into different machines independent from hardware requirements as much as possible.

Whilst federated decentralized learning enables communication of model parameters, integration with more advanced machine learning concepts, such as deep learning and domain-specific knowledge, can increase its performance and efficiency. Using deep neural networks and enriching them with domain-specific graphs such as protein-protein interaction (PPI) networks can also drastically improve the feature extraction process. The next phase, of course, is then about combining decentralization and the power of Deep Learning. The feature-rich, detailed, and robust parameters, when communicated in a federated learning framework, can lead to highly effective and reliable machine learning applications. The decentralized nature of such a framework not only increases learning efficiency but also strengthens the trustworthiness of the results by combining masked learning with domain knowledge.

In our work [48], we masked deep neural network learning with a protein-protein interaction (PPI) network. In the context of this paper, "masking" refers to incorporating a domain-knowledge graph (specifically, a PPI network) into a deep neural network for classification. This means that the nodes and edges of the PPI network are added to the input layer of the neural network and are used to enrich the features of the data being processed by the neural network. Features are key for learning, understanding and explaining and consolidated features are more accurate and robust, which helps to make practical machine learning applications more trustworthy [47]. It is a general problem that even the most powerful learning methods suffer from the fact that it is difficult to retrace, interpret and thus explain why a certain result was obtained, and that they lack robustness. Even the smallest perturbations in the input data can dramatically affect the output, leading to completely different results. This is of great importance in virtually all critical domains where we suffer from poor data quality, i.e., where we do not have available the i.i.d. data we would need for ideal learning. However, in medicine, biology, and all life-critical domains, it is about being able to trust the results and retrace them when needed [17,18].

In our next step the classification has been made explainable, i.e. those subnetworks are detected that were relevant for the classification ("disease subnetworks") - subgraphs are called "local spheres" in [20] and [40]. In order to guarantee a representative baseline comparison to the above methodology, the subnetwork detection was realised by means of a random forest [45]. Here, too, the learning process is masked by a knowledge graph. Random forests are particularly relevant in medicine due to their good interpretability. In the work [46] we enabled federated learning with the methods mentioned above. Here, the knowledge graph is divided into relevant subnetworks using explainable AI, based on which an ensemble classifier is constructed. This ensemble classifier can be efficiently learned in a federated way. In addition, a user interface was developed [2] that allows a domain expert to analyse and manipulate the detected subnetworks, delete and add nodes, and finally reintegrate them into the federated ensemble classifier. This paper is organized as follows: In Sect. 2 we provide some background and related work, in Sect. 3 we provide an overview of our implementations, and in Sect. 4 we give a frank description of what we have learned, and in Sect. 5 we conclude and provide some future outlook.

2 Background and Related Work

There is nothing more practical than a good theory (Kurt Lewin, (1890–1947)). In our work we pursued four central topics from the paper [20]: (i) Explainable AI on GNNs, (ii) Federated Learning, (iii) Knowledge Graphs, and (iv) Human-AI interaction. Consequently, we have aligned all of these topics on the application of precision medicine.

2.1 Explainable AI on Graph Neural Networks

Graph Neural Networks (GNNs) extend neural network architectures to operate on graph-based data by defining learnable functions that extract features and patterns from the graph structure to perform tasks such as node classification, graph classification, link prediction, etc. [58]. GNNs are very successful and enable efficient integration of domain-knowledge graphs to make Deep Learning interpretable and explainable [20]. Federated solutions thereof seem to occur naturally in several applications such as distributed sensors for traffic surveillance, a collaboration of hospitals for efficient solutions of complex medical tasks, distributed social media applications and so on. In the era of big data both the size of the graph datasets as well as the GNN architectures grows, making efficient and privacy-preserving information exchange and computation a challenge. What is more, since the communicating parties, whether they are servers or clients can be represented by a graph themselves, it is shown that GNN architectures can support federation in turn [33].

As is generally the case with neural networks, also GNN results are not easy to retrace and interpret. To address this shortcoming, intensive work is currently being done worldwide on GNN methods that can be explained. Examples include GNNexplainer, PGExplainer, and GNN-LRP. *GNNExplainer* [59], for example, provides *local* explanations for predictions of any graph-based model. This can be used for both node classification and graph classification. *PGExplainer* [35] is a parameterized modification of GNNexplainer. Unlike GNNexplainer, it provides model-level explanations that we find useful for graph classification tasks. *GNN-LRP* [51] is derived from higher-order Taylor expansions based on layer-wise relevance propagation (LRP) [30]. It explains the prediction by extracting paths from the input to the output of the GNN model that makes the largest contribution to the prediction. These paths correspond to *walks* on the input graph. GNN-LRP was developed for node-level explanations and has been modified to work for graph classification in a special arrangement [5]. The presented work with a method called CF-Explainer [34] is particularly interesting. Here, explanatory factors can be revealed using counterfactuals.

GCExplainer [38] stands in the forefront as the first GNN explainer that detects the *learned concepts* of a GNN. The main idea is to perform clustering after the last aggregation layer and to assume that each of the clusters corresponds to a human-recognizable concept. Users have the opportunity to parameterize the explanation process through the number of clusters and the neighbourhood size of the explained component. This approach incorporates the human-in-the-loop [16,23] and at the same time has been shown to achieve good concept purity and completeness. Furthermore, it is the basis of current work that makes GNNs explainable per design by first learning the concepts, then on that basis doing a concept-based prediction [37]. Such explainable AI methods can facilitate the discovery of disease-causing regions in networks, helping to uncover a subset of *candidate features* organized in disease-relevant network modules.

This is exactly where the human-in-the-loop concept helps, as interaction with explanations and the incorporation of conceptual knowledge can further improve the learning algorithm.

2.2 Federated Learning

Federated learning (FL) is an ML approach in which the training data is decentralized and distributed across multiple devices or locations, and the model training process is performed locally on each device or location [40]. The updates to the model are then aggregated centrally, resulting in a global model that incorporates the knowledge learned from each device or location. FL is of course useful in scenarios where the data is sensitive, private, or subject to regulatory constraints, such as medical records or financial transactions. Instead of centralizing the data and running the model training process on a single server or cloud platform, federated learning allows the data to remain on individual devices or locations, and only the model updates are transmitted for aggregation. This preserves the privacy and security of the data and reduces the risk of data breaches or leaks. FL should not be mixed up with purely decentralized learning, where local models do not automatically contribute to each other apart from manually sampling the models and updating the hyperparameters [3]; and also not with collaborative learning in various forms, where the goal is to share information about internal model building between the involved parties in a peer-to-peer manner but keep the local training data confidential. A variant could also train on decentralized features that purportedly model the same underlying instances [24]. It has been known for some time that features for one modality are learned better when multiple modalities are present at the time of feature learning. In multimodal learning, information is from multiple sources. Often, several different modalities contribute to a result. We are motivated by [1,9,19]. This brings us directly to graphs and particularly knowledge graphs.

Federation itself has evolved to be a broad topic; although the main principles are firm, different implementations realize the same goals. What is similar in all instantiations is that there is data isolation to some degree and that the information being exchanged should be minimal and privacy-preserved (i.e. encrypted). Furthermore, the i.i.d. scenario is rather the exception than the norm; several frameworks need to simulate it before the actual deployment [44]. Nonetheless, collaboration has proven to be fruitful in most cases, since no one dataset contains all representative information about a task and ML solutions lack the ability of systematic generalization and out-of-distribution (OOD) prediction even when trained with rich and diverse datasets.

In the more concrete case of Federated GNN, there are mainly three possibilities [14], as also shown in Fig. 1. In the graph-level FL, each client has its graph dataset and potentially also a GNN. In the subgraph-level FL, each of the clients has one part of the graph and in the node-level FL nodes of one graph are distributed among clients.

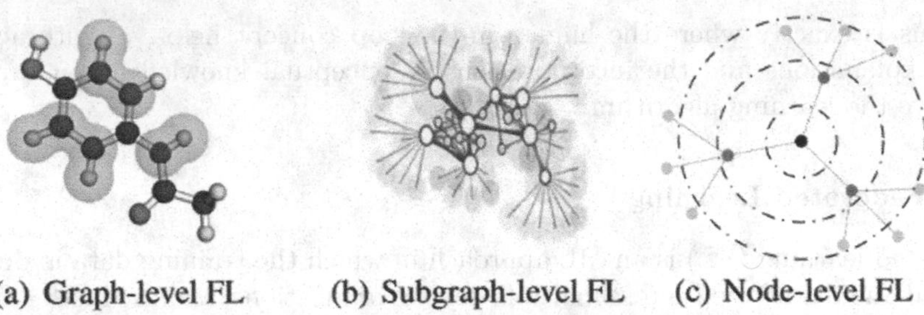

Fig. 1. Three settings of GNN federation [14].

This is following the principles of Horizontal FL (HFL) and Vertical FL (VFL). In the first case, the features of the graphs of all clients are quite similar, but their sample characteristics (data distribution) differ substantially. The opposite occurs in the second case. Both of them are viable scenarios of FL and need to be addressed either with centralized or decentralized FL. In the federated centralized strategy, it is typical that there are several synchronous or asynchronous events containing parts of the dataset, and one server is responsible for the federation (which is also called aggregation). In the federated decentralized case, many clients exchange information with each other; this is more robust as far as privacy attacks are concerned but has substantial communication and organizational overhead.

2.3 Knowledge Graphs

Knowledge graphs (KG) are a type of database that represents knowledge in a structured, interconnected format, using a graph-based data model. It typically consists of a set of nodes (also called entities) that represent concepts or things, and a set of edges (also called relationships or properties) that connect the nodes and represent the connections or interactions between them. Many phenomena from nature can be represented in graph structures, whether at the molecular level (e.g. protein-protein interaction) or at the macroscopic level (e.g. social networks) and various methods from network science [7] and computational topology [15] can be applied. Some of the most successful application areas of machine learning and knowledge extraction in recent years can be seen as learning with graph representations [57].

In a knowledge graph, each node and edge can have additional attributes or metadata associated with it, providing additional information or context about the node or edge. This metadata can include labels, descriptions, categories, or other semantic information. Knowledge graphs are often used to represent information from diverse sources and domains in a multi-modal manner. They can be used to represent both factual knowledge (such as the properties of objects or events) and conceptual knowledge (such as the relationships between abstract concepts). Knowledge graphs are also used as a foundation for various applications, such as natural language processing, semantic search, recommendation systems, and data integration. They enable efficient querying and reasoning about complex, heterogeneous data, as well as support the development of intelligent agents that can reason and learn from the knowledge represented in the graph [12]. KG's are very useful for explainability and explainable AI methods based on counterfactual queries to the trained GNN models are very promising [39,53].

2.4 Human-in-the-Loop

Human-in-the-Loop [16] refers to the process of involving a human expert interactively in the machine learning (ML) process to provide feedback, guidance, or even corrections to the model. The human is an integral part of the ML pipeline, interacting with the model/algorithm to improve its performance and ensuring that it aligns with the desired goals and values. This approach is useful in scenarios where the data is complex, ambiguous, or subject to change, and where the model's performance can benefit from human expertise or even from the experts' subjective judgment. This is because sometimes - of course not always - the human expert has domain knowledge, experience and contextual understanding, in German "Hausverstand" - what the best AI algorithms are lacking today. An additional benefit is that the human-in-the-loop approach can also improve the transparency, interpretability, and fairness of machine learning models, as it allows for human oversight and intervention in cases where the model produces biased or undesirable results. However, the human-in-the-loop approach, on the other hand, has drawbacks as it can be time-consuming, expensive, and potentially introduce bias or subjectivity into the modelling process, so it is important to carefully design and evaluate the interaction between the human and the model.

3 Methods, Solutions and Implementations

3.1 Disease Subnetwork Detection

In a publication about GNNSubNet [48], we presented a novel method for disease subnetwork detection using protein-protein interaction (PPI) networks and explainable graph neural networks (GNN). Our method leveraged the PPI knowledge to enable more reliable and biologically meaningful learning trajectories compared to classical deep learning approaches. The nodes of the induced PPI network are enriched by biological features from various modalities, such as gene expression and DNA methylation (see Fig. 2). We applied our proposed method to patients with kidney cancer and demonstrated its ability to detect disease subnetworks. The developed methodology is implemented within our GNN-SubNet Python package, freely available on GitHub (https://github.com/pievos101/GNN-SubNet). In addition, we enhance ensemble learning based on the detected networks. This makes the classifier more robust, but also more interpretable [46]. Ensemble-learning with GNNs is implemented within our Ensemble-GNN Python package (https://github.com/pievos101/Ensemble-GNN). In further updates of the package additional GNN-based explainers such as GNN-LRP and PGM-Explainer to further increase the interpretability of the detected subnetworks will be implemented.

Moreover, as a reliable baseline, in terms of classification performance and overlay interpretability, we have developed the software package DFNET (https://github.com/pievos101/DFNET) [45], which implements a network-guided random forest to derive an ensemble classifier based on any induced knowledge-graph. However, in a federated case, a local random forest would need to share the exact split values of its nodes [13]. This is of much concern and was one of the reasons why we further developed federated solutions based on deep GNNs. The shared parameter among clients in that deep learning setting is more secure with regard to privacy concerns.

3.2 Explainability

The classification of Part 1 has been made explainable, i.e. those subnetworks detected that were relevant for the classification ("disease subnetworks") - subgraphs aka "local spheres". For this purpose we have developed a modified version of the GNNexplainer [59] to compute global explanations. This is realized by sampling patient-specific input graphs while optimizing a single-node mask (see Fig. 2). From these values, edge weights are calculated and assigned to the edges of the PPI network. Finally, a weighted community detection algorithm infers the relevant subnetworks.

PPI networks generally provide crucial insights into cellular functions and processes, and alterations in these interactions often lead to diseases. Consequently, such networks are important in understanding complex diseases like cancer, which typically involve changes in the interaction patterns of proteins. Explainability can here help to understand disease mechanisms, e.g. to reveal the underlying mechanisms of diseases. By understanding which interactions contribute to the prediction and how, researchers can potentially uncover new biological insights. For example, a model might predict a certain protein as being critical to a disease because of its numerous interactions with other proteins. This could lead to further biological investigations into the role of that protein in the disease. This can help in creating personalized treatment strategies. For instance, if certain protein interactions are critical in the disease progression of a particular patient, treatments can be tailored to target these specific interactions. Identifying which features (e.g., specific proteins or interactions) are most important in the model's predictions. For example, a model might reveal that a specific protein or a set of proteins plays a significant role in a particular disease, informing further biological research.

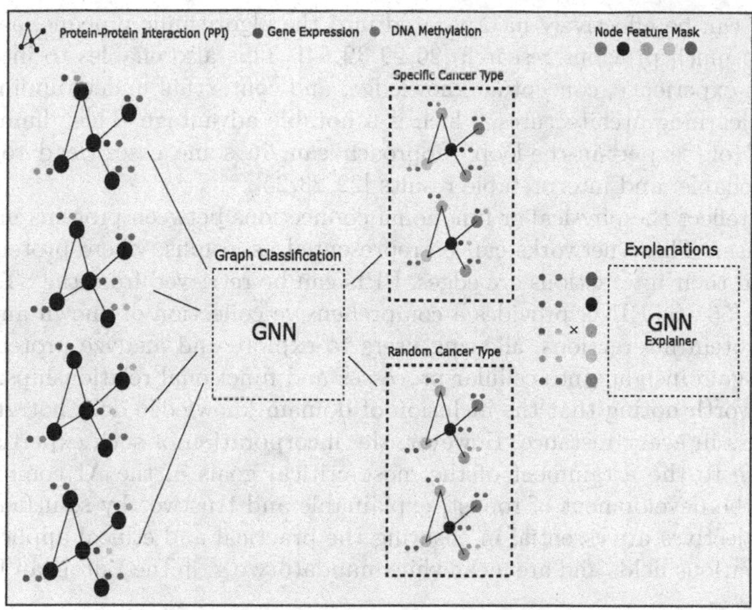

Fig. 2. Illustration of patient classification into a cancer-specific and randomized cancer group using explainable Graph Neural Networks (taken from [48]). Each patient is represented by the topology of a protein-protein interaction network (PPI). Nodes are enriched by multi-omic features from gene expression and DNA Methylation (coloured circles). The topology of each graph is the same for all patients, but the node feature values vary, reflecting the cancer-specific molecular patterns of each patient. (Color figure online)

Furthermore, *model-agnostic* counterfactual explanations and their associated counterfactual paths can be generated using our *cpath* software library (https://github.com/pievos101/cpath). The implemented methodology provides counterfactual explanations by identifying alternative paths that could have led to different predictions. The proposed method is particularly suited to generate explanations based on counterfactual paths on knowledge graphs. By exploring hypothetical changes to the input data on the knowledge graph, we can systematically validate the behaviour of the model and investigate the features, or combination of features, that are most important for the model's predictions. Our approach provides a more intuitive and interpretable explanation of the model's behaviour than traditional feature importance methods and can help to identify and mitigate biases in the model. A scientific paper about *cpath* is in progress.

3.3 Knowledge Graph

GNNs provide a crucial benefit of enabling the integration of knowledge graphs [27]. This implies that both ontologies and Protein-Protein Interaction (PPI) networks can be effectively incorporated into the algorithmic pipeline, as highlighted in much previous research [26,29,32,54]. This also enables to integrate of human experience, conceptual knowledge, and contextual understanding into machine learning architectures, which is a notable advantage. This "human-in-the-loop" or "expert-in-the-loop" approach can, in some cases, lead to more robust, reliable, and interpretable results [22,23,25].

PPIs reflect the physical or functional connections between proteins in a cell or organism. These networks can be represented as graphs, where proteins are nodes and their interactions are edges. PPIs can be retrieved from the STRING database [56]. STRING provides a comprehensive collection of known and predicted protein interactions, allowing users to explore and analyze protein networks to gain insights into cellular processes and functional relationships.

It is worth noting that the inclusion of domain knowledge does not guarantee success in every instance. However, the incorporation of such expertise can contribute to the attainment of the most critical goals of the AI community, namely, the development of robust, explainable and trustworthy solutions [18]. These objectives are essential in ensuring the practical and ethical applications of AI in various fields and are meanwhile mandatory e.g. in the European Union.

3.4 Federated Ensemble Learning with GNNs

In recent work [46] we enabled federated learning with the methods mentioned above. Here, the knowledge graph is divided into relevant subnetworks using explainable AI, based on which an ensemble classifier is constructed. This ensemble classifier can be efficiently learned in a federated way.

The main idea of the ensemble federation is depicted in Fig. 3. Each client contains several graphs and each of those graphs represents a patient. The values of the nodes and edges are different in general (as depicted by the different colours of the nodes in the upper part of Fig. 3), but the structure of the graphs is the same. Those graphs can be classified by a GNN and the GNN-SubNet method [48] can compute a set of relevant subgraphs for this classification. GNN-SubNet concentrates on providing the relevant structure or topology only; therefore the subgraphs are depicted with white in the middle of Fig. 3. The concrete values of the nodes and edges are transferred in a third step though from the original graphs (upper part of Fig. 3) to the concrete subgraphs that have the topology of the relevant subgraphs and values overtaken from the original graph (lower part of Fig. 3). By creating a new dataset for each discovered relevant subgraph where its structure is repeated and the values are taken from the original graph of all the patients in the client, a separate GNN is trained. The predictions of all those GNNs are input to a majority vote procedure that - in its non-federated version - has an acceptable local performance.

The federation is depicted in Fig. 4 and follows a decentralized strategy. The clients use local GNNs of their peers in the inter-client network, that were created with similar logic but were trained with graphs having different topologies - since the relevant subgraphs for each client are expected to vary in general. There is no exchange of the discovered relevant topologies of each client, only

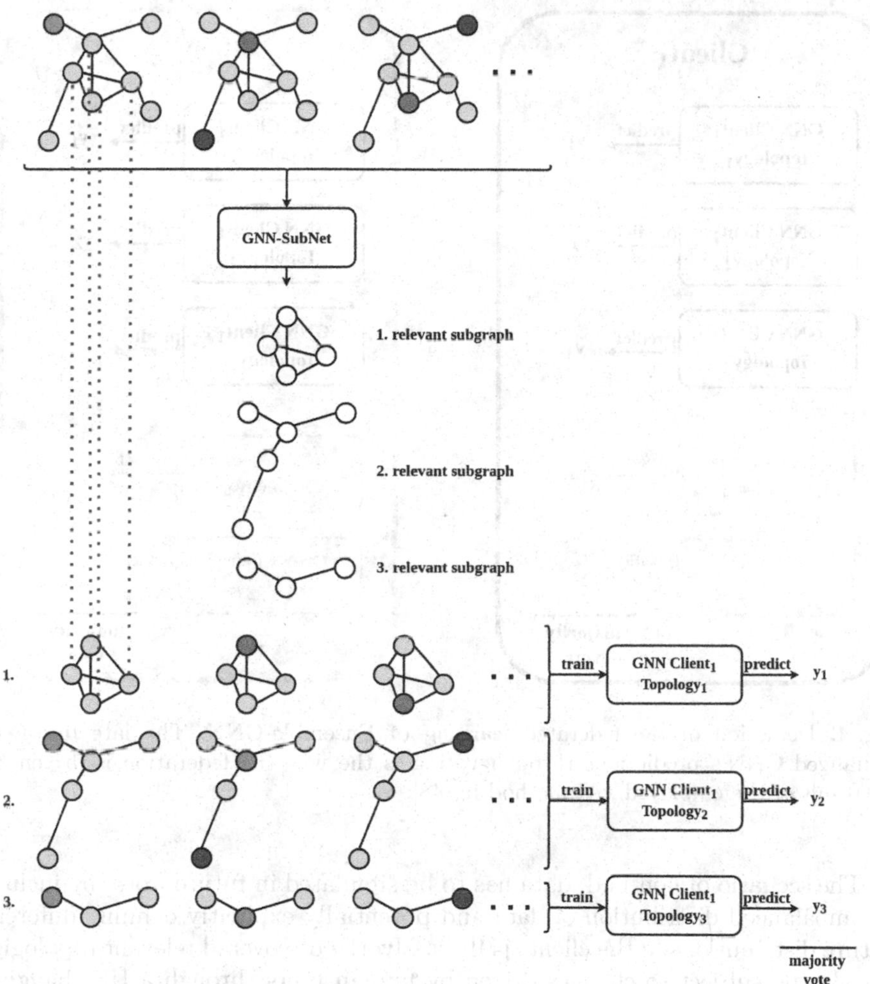

Fig. 3. The use of GNN-SubNet in one client, containing a set of graphs for classification. This method extracts a list of relevant subgraph structures (topologies) and uses them by filling the corresponding values of nodes and edges from the original graphs. The newly created datasets are used to train local GNNs and make predictions which are aggregated by majority voting.

the GNN parameters are transferred - which is as far as privacy is concerned less revealing. The majority vote over all those GNNs provided a better performance over each client's test set, but not over a test set that was isolated from all clients, as shown in [46]. The described methodology is implemented within our Python package Ensemble-GNN, freely available on GitHub (https://github.com/pievos101/Ensemble-GNN). A Feature Cloud app implementation is also available (https://github.com/pievos101/fc-ensemble-gnn).

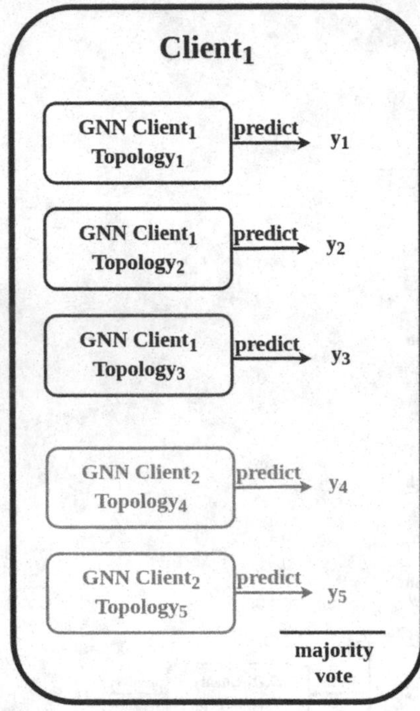

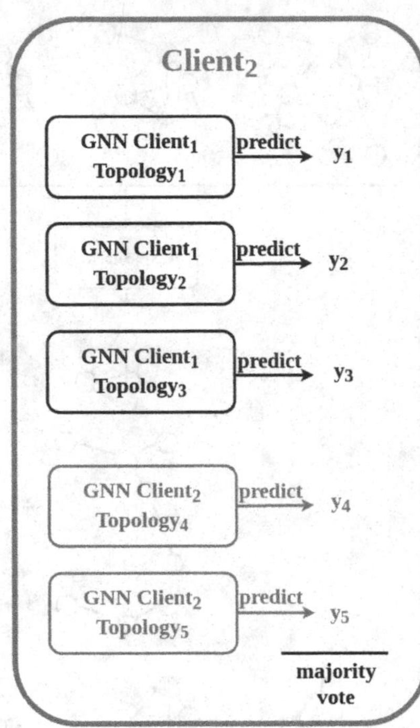

Fig. 4. Depiction of the federated learning of Ensemble-GNN. The late fusion of exchanged GNN's predictions through voting is the way the federation is driven by the result of the employed xAI method in [48].

The scenario of non-i.i.d. data has to be simulated in future work, by including imbalanced distribution of data and potentially explicitly defining different feature distributions in the clients [44]. Lastly, the discovered relevant topologies can also be subject to changes driven by human users through a UI, changing the local GNNs, and by that the whole federation process.

3.5 interaCtive expLainable plAtform for gRaph neUral networkS (CLARUS)

The CLARUS UI platform [2] is accessible under http://rshiny.gwdg.de/apps/clarus/. The goal of the UI platform is to provide any human user interactive access to prepared datasets, GNNs and several xAI methods. All necessary information about the platform usage, datasets, features and performance metrics are provided through the platform. An overview of the typical sequence of steps that a user takes is presented in Fig. 5.

For the user to be able to make informed actions [49] with the use of diverse xAI methods (GNNExplainer [59], GCExplainer [38]), all nodes and edges are presented by sorted relevance values. The colouring scheme depends on the properties of the xAI method itself; the saliency method [55], Integrated Gradients

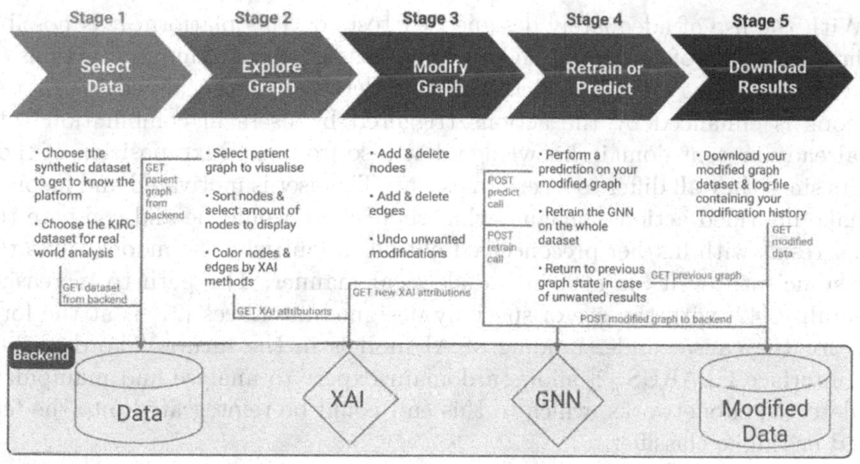

Fig. 5. The sequence of user action steps in the CLARUS platform. First, the user selects one of the prepared datasets and immediately after he/she has the opportunity to explore any graph visually by zooming and by inspecting the nodes and edges feature values. The backend has already trained a GNN with the training dataset after a stratified split of the data and presents performance results (individual and global), xAI relevance values as well as additional information that can be useful such as the degree of each node. With the help of this information, and additional acquired domain knowledge, the human user decides to take action(s) and either add or delete nodes, edges and features thereof. To see how those actions affected the task prediction of the current GNN a new prediction can be triggered. In cases where the changes are substantial a retrain from scratch can be also made, deleting by that all old information in the current GNN. This process can be repeated as many times as desired until the user conceives the decision-making process to an acceptable extent through the generated counterfactual explanations. A download of all data and model details at a particular time point, together with a unique timestamp is possible on demand.

(IG) method, and the GNNExplainer return only positive relevance values, but methods like GNN-LRP (Layer-wise Relevance Propagation) return both positive and negative values. Those two groups of relevance value ranges have discrete colourings for a better understanding of the concept of negative relevance as one denoting element in the data sample that "speak against" a class and even in a correct classification is responsible for making the confidence value smaller. Beyond that, for each sample it has to be clear if it is correctly classified or misclassified; even the exact prediction performance is present. This is because the reliability of explanations in the misclassification case is questionable and it is a subfield of xAI research itself. Therefore, several classification metrics are accessible: the confusion matrix, sensitivity, specificity and in the future Mutual Information (MI) [4,11,36]. After each retrain and prediction, those metrics are re-computed and in general they have changed values. A detailed description of the pre-selected datasets, their preprocessing, various interaction scenarios and abilities of the platform can be found in [2].

With the use of adequately designed UI tests on this platform it is possible to show the effect of counterfactual questions and corresponding user actions on user understanding of the model. The completeness of the already used xAI methods is enhanced by the actions triggered by users in combination with the already present domain knowledge, but also from the juxtaposition of their results since they all differ to a certain extent. The user is motivated and inspired to make informed actions, imagine what their effect would be and compare the actual result with his/her preconceived notions about why the model solves the task sufficiently well (or not) in a dialectical manner. The path to increasing causability [42] with the use of specially designed interfaces [21] is at the forefront for the causal understanding of AI models in the future. The described user interface CLARUS [2] allows a domain expert to analyse and manipulate the detected subnetworks, which to this end could be reintegrated into the federated ensemble classifier.

4 Lessons Learned

What was not done and why?
The implementation of other xAI methods than GNNExplainer for the detection of disease subnetworks. This is particularly relevant for ensemble-based GNN architectures. Each GNN xAI method might create different ensemble members, which to this end could be studied in terms of performance and interpretability (e.g. GO enrichment of the detected PPI subnetworks).

What problems occurred?
Other xAI methods were more difficult to integrate. Some explainers only compute relevances for edges or nodes, others like GNN-LRP [50] assign relevance values only on walks; that means that a node or edge belonging to more than one walk (which is usually the case) has not one clearly defined relevance value. Data scientists may be tempted to average all edge relevances to infer the relevance of the node or the opposite, but this is not representative of the xAI method. Furthermore, GNN-LRP provides both negative and positive relevances, which means that not only the colour map has to be distinct from the methods that provide only positive relevance, but that the relevance of the paths needs an individual visualization strategy that allows overlapping and user selection. Other methods like the GCExplainer [38] compute a representative set of subgraphs that is relevant for each relevant concept w.r.t. the accomplished task. Although this is a valuable approach which has some similarities with the detection of relevant disease subnetworks, since it does not directly return numerical relevance values for the individual components of the graphs, cannot be straightforwardly integrated into the UI framework.

What was difficult?
What was particularly difficult for both data scientists and users is the discovery of differences between the xAI methods results; this consists of the so-called "disagreement problem" [28]. Data scientists provide several xAI methods to

shed light on different aspects of the design-making process of the model, but if the results of those methods deviate from each other, this disagreement is not easy to interpret and understand. Furthermore, counter-intuitive phenomena were observed; it is assumed for example, that if a user deletes components of a graph according to decreasing (positive) relevance order, then the performance of the model will not only decrease monotonically but also that the newly computed relevance order after a new triggered prediction will remain the same. In many cases this was not experienced, making the users question the reliability of the xAI methods. Related to that, the value range of the colour map was an issue, since the minimum and maximum value of relevance change in general after a prediction is initiated.

What did we learn?
The fact that each graph has the same topology (PPI network) hinders stable and robust graph classification, especially in cases where the input graph is large. We could observe that GNNs on smaller graphs perform generally better [46]. Further, we have learned that in the herein-studied cases of the same topology graphs, using Laplacian layers might be more efficient in terms of performance. Therefore, we also included the ChebNet approach [6] as an option for GNN-SubNet and Ensemble-GNN. However, GNNs are generic models and applicable to many other related tasks. Also, we might model each patient with different graph topologies. In that case, the ChebNet approach is not applicable.

We have further learned that the quality and validity of the knowledge graph are crucial. Knowledge graphs must be further improved in order to obtain reliable and domain-specific meaningful results. Also, it has been shown that most methods for disease module discovery learn from the PPI node degrees and mostly fail to exploit the biological knowledge encoded in the edges of the PPI networks [31]. Although we believe that our proposed methodology is not biased to that described case, further investigations are needed to understand and quantify the bias induced by the network structure.

What open work remains for the future?
Heterogeneous Graphs (including text and images or different types of nodes and edges) were not included. After preliminary tests, we know that they need more resources and xAI methods need to be thoroughly tested before deployment. So far we have multi-model genomic data in tabular form, structured by a PPI network.

Until now the GNN architecture is pre-defined for every dataset and it is somehow intertwined with the characteristics of this dataset - and most of all its size. In the case where the user changes increases or decreases the size of the dataset and/or changes its characteristics substantially, the platform cannot guarantee similar performance since the GNN's architecture is not adapted. To automatically find the adequate GNN architecture is a topic of Automated Machine Learning (Auto-ML), and its incorporation in this platform will come with additional time costs which will, in turn, influence the waiting time of the users in favour of performance and better xAI results.

The existence of the aforementioned "disagreement problem" [28] drives future work in the direction of not only integrating more xAI methods but also considering the computation and presentation of several xAI quality metrics thereof to the users. Fidelity, sensitivity, clusterability, robustness and others [8,52] provide additional guidelines for the reliability of each method in cases where the top relevant features or their ordering is inconsistent. In the end, upon deployment, several UI evaluation tests have to be made to explore the extent of biased preference of xAI relevance results. In the end, the plurality of xAI methods does not necessarily consist of a problem but might be the means for a sophisticated, holistic and dialectic approach for shedding light on different aspects of the decision-making process of GNNs.

The main reason federation is used, is for the central model to learn something from the different local models, trained with their datasets. Comparing the performance of the local models with the central model: what are the differences there?

It does make a considerable difference whether we test the federated global model on an independent global test data set, or on multiple client-specific test data sets (see [46]). It still needs to be investigated which scenario is most relevant and why these differ so much in terms of the performance of the global model.

5 Conclusion and Future Outlook

In this work, we have demonstrated how to make federated deep learning more interpretable and accessible to the domain expert. First, we have incorporated domain knowledge into the deep learning process using Graph Neural Networks and Protein-Protein Interaction (PPI) networks. Second, we have decomposed the PPI knowledge graph into more interpretable smaller subnetworks using explainable AI. Based on these subnetworks an ensemble classifier is constructed which can be learned in a federated manner. The shared parameters of this deep learning ensemble are more secure compared to e.g. the shared split values of decision trees in a federated random forest. Finally, the ensemble member (subnetworks) can be analysed by a domain expert through an interactive UI.

Future work can be done from various directions. Until now, xAI methods that were used (GNNExplainer, PGExplainer, GCexplainer) return relevant values of nodes, edges and features thereof. Apart from the fact that some fundamental principles of them need to be explained to the users (f.e. that the GNN-LRP assigns relevance to walks and not directly to nodes and edges), the interpretation of those numerical values is a task that the user's mental model needs to undertake. In contrast to that, explanations in the form of rules, provide a completely different user experience and understanding. It would be interesting to research how Logical Rules (e.g. with Prolog) guide the selection of subnetworks [10], similarly or differently with the numerical relevance values.

Furthermore, a framework that asks the domain expert about their preconceived notions as far as what parts of the input data should be important, before seeing xAI results is worthwhile studying. The comparison of users' reactions after confronting relevant values vs. uninfluenced opinions derived from their knowledge before any interaction could uncover interesting effects of human-AI interaction.

Acknowledgements. The authors declare that there are no conflict of interests. This work does not raise any ethical issues. This work has received funding from the European Union's Horizon 2020 research and innovation programme under grant agreement No. 826078 (Feature Cloud). This publication reflects only the authors' view and the European Commission is not responsible for any use that may be made of the information it contains. Parts of this work have been funded by the Austrian Science Fund (FWF), Project: P-32554 (explainable Artificial Intelligence). This paper has been made open access CC-BY, freely accessible to the international research community. We are grateful for the valuable reviewer comments.

References

1. Acosta, J.N., Falcone, G.J., Rajpurkar, P., Topol, E.J.: Multimodal biomedical AI. Nat. Med. **28**(9), 1773–1784 (2022)
2. Beinecke, J., et al.: CLARUS: an interactive explainable AI platform for manual counterfactuals in graph neural networks. bioRxiv (2022). https://doi.org/10.1101/2022.11.21.517358
3. Bellavista, P., Foschini, L., Mora, A.: Decentralised learning in federated deployment environments: a system-level survey. ACM Comput. Surv. (CSUR) **54**(1), 1–38 (2021)
4. Bishop, C.M., Nasrabadi, N.M.: Pattern Recognition and Machine Learning, vol. 4. Springer, New York (2006)
5. Chereda, H., et al.: Explaining decisions of graph convolutional neural networks: patient-specific molecular subnetworks responsible for metastasis prediction in breast cancer. Genome Med. **13**(1), 1–16 (2021). https://doi.org/10.1186/s13073-021-00845-7
6. Defferrard, M., Bresson, X., Vandergheynst, P.: Convolutional neural networks on graphs with fast localized spectral filtering. arXiv:1606.09375 [cs, stat] (2016)
7. Dehmer, M., Emmert-Streib, F., Shi, Y.: Quantitative graph theory: a new branch of graph theory and network science. Inf. Sci. **418**, 575–580 (2017). https://doi.org/10.1016/j.ins.2017.08.009
8. Doumard, E., Aligon, J., Escriva, E., Excoffier, J.B., Monsarrat, P., Soulé-Dupuy, C.: A quantitative approach for the comparison of additive local explanation methods. Inf. Syst. **114**, 102162 (2023)
9. Ektefaie, Y., Dasoulas, G., Noori, A., Farhat, M., Zitnik, M.: Multimodal learning with graphs. Nat. Mach. Intell. **5**(4), 340–350 (2023)
10. Finzel, B., Saranti, A., Angerschmid, A., Tafler, D., Pfeifer, B., Holzinger, A.: Generating explanations for conceptual validation of graph neural networks. KI-Künstl. Intell. **36**, 271–285 (2022). https://doi.org/10.1007/s13218-022-00781-7
11. Géron, A.: Hands-On Machine Learning with Scikit-Learn, Keras, and TensorFlow: Concepts, Tools, and Techniques to Build Intelligent Systems. O'Reilly Media (2019)
12. Hamilton, W., Bajaj, P., Zitnik, M., Jurafsky, D., Leskovec, J.: Embedding logical queries on knowledge graphs. In: Advances in Neural Information Processing Systems, vol. 31 (2018)

13. Hauschild, A.C., et al.: Federated Random Forests can improve local performance of predictive models for various healthcare applications. Bioinformatics **38**(8), 2278–2286 (2022). https://doi.org/10.1093/bioinformatics/btac065
14. He, C., et al.: FedGraphNN: a federated learning benchmark system for graph neural networks. In: ICLR 2021 Workshop on Distributed and Private Machine Learning (DPML) (2021)
15. Holzinger, A.: On topological data mining. In: Holzinger, A., Jurisica, I. (eds.) Interactive Knowledge Discovery and Data Mining in Biomedical Informatics. LNCS, vol. 8401, pp. 331–356. Springer, Heidelberg (2014). https://doi.org/10.1007/978-3-662-43968-5_19
16. Holzinger, A.: Interactive machine learning for health informatics: when do we need the human-in-the-loop? Brain Inform. **3**(2), 119–131 (2016). https://doi.org/10.1007/s40708-016-0042-6
17. Holzinger, A.: The next frontier: AI we can really trust. In: Kamp, M., et al. (eds.) ECML PKDD 2021. CCIS, vol. 1524, pp. 427–440. Springer, Cham (2021). https://doi.org/10.1007/978-3-030-93736-2_33
18. Holzinger, A., et al.: Information fusion as an integrative cross-cutting enabler to achieve robust, explainable, and trustworthy medical artificial intelligence. Inf. Fusion **79**(3), 263–278 (2022). https://doi.org/10.1016/j.inffus.2021.10.007
19. Holzinger, A., Haibe-Kains, B., Jurisica, I.: Why imaging data alone is not enough: AI-based integration of imaging, omics, and clinical data. Eur. J. Nucl. Med. Mol. Imaging **46**(13), 2722–2730 (2019). https://doi.org/10.1007/s00259-019-04382-9
20. Holzinger, A., Malle, B., Saranti, A., Pfeifer, B.: Towards multi-modal causability with graph neural networks enabling information fusion for explainable AI. Inf. Fusion **71**(7), 28–37 (2021). https://doi.org/10.1016/j.inffus.2021.01.008
21. Holzinger, A., Müller, H.: Toward human-AI interfaces to support explainability and causability in medical AI. IEEE Comput. **54**(10), 78–86 (2021). https://doi.org/10.1109/MC.2021.3092610
22. Holzinger, A., Plass, M., Holzinger, K., Crisan, G.C., Pintea, C.M., Palade, V.: Towards interactive machine learning (iML): applying ant colony algorithms to solve the traveling salesman problem with the human-in-the-loop approach. In: Buccafurri, F., Holzinger, A., Kieseberg, P., Tjoa, A., Weippl, E. (eds.) CD-ARES 2016. LNCS, vol. 9817, pp. 81–95. Springer, Heidelberg (2016). https://doi.org/10.1007/978-3-319-45507-5_6
23. Holzinger, A., et al.: Interactive machine learning: experimental evidence for the human in the algorithmic loop. Appl. Intell. **49**(7), 2401–2414 (2019). https://doi.org/10.1007/s10489-018-1361-5
24. Hu, Y., Niu, D., Yang, J., Zhou, S.: Stochastic distributed optimization for machine learning from decentralized features, pp. 1–10. arXiv:1812.06415 (2018)
25. Hudec, M., Minarikova, E., Mesiar, R., Saranti, A., Holzinger, A.: Classification by ordinal sums of conjunctive and disjunctive functions for explainable AI and interpretable machine learning solutions. Knowl. Based Syst. **220**, 106916 (2021). https://doi.org/10.1016/j.knosys.2021.106916
26. Jeanquartier, F., Jean-Quartier, C., Holzinger, A.: Integrated web visualizations for protein-protein interaction databases. BMC Bioinform. **16**(1), 195 (2015). https://doi.org/10.1186/s12859-015-0615-z
27. Ji, S., Pan, S., Cambria, E., Marttinen, P., Philip, S.Y.: A survey on knowledge graphs: representation, acquisition, and applications. IEEE Trans. Neural Netw. Learn. Syst. **33**(2), 494–514 (2022). https://doi.org/10.1109/TNNLS.2021.3070843
28. Krishna, S., et al.: The disagreement problem in explainable machine learning: a practitioner's perspective. arXiv preprint arXiv:2202.01602 (2022)

29. Kulmanov, M., Smaili, F.Z., Gao, X., Hoehndorf, R.: Machine learning with biomedical ontologies. bioRxiv (2020). https://doi.org/10.1101/2020.05.07.082164
30. Lapuschkin, S., Binder, A., Montavon, G., Müller, K.R., Samek, W.: The LRP toolbox for artificial neural networks. J. Mach. Learn. Res. (JMLR) **17**(1), 3938–3942 (2016)
31. Lazareva, O., Baumbach, J., List, M., Blumenthal, D.B.: On the limits of active module identification. Briefings Bioinform. **22**(5), bbab066 (2021)
32. Liu, G., Wong, L., Chua, H.N.: Complex discovery from weighted PPI networks. Bioinformatics **25**(15), 1891–1897 (2009). https://doi.org/10.1093/bioinformatics/btp311
33. Liu, R., Yu, H.: Federated graph neural networks: overview, techniques and challenges. arXiv preprint arXiv:2202.07256 (2022)
34. Lucic, A., ter Hoeve, M., Tolomei, G., de Rijke, M., Silvestri, F.: CF-GNNExplainer: counterfactual explanations for graph neural networks. arXiv:2102.03322 (2021)
35. Luo, D., et al.: Parameterized explainer for graph neural network. In: Advances in Neural Information Processing Systems, vol. 33, pp. 19620–19631 (2020)
36. MacKay, D.J.: Information Theory, Inference and Learning Algorithms. Cambridge University Press, Cambridge (2003)
37. Magister, L.C., et al.: Encoding concepts in graph neural networks. arXiv e-prints arXiv:2207.13586 (2022)
38. Magister, L.C., Kazhdan, D., Singh, V., Liò, P.: GCExplainer: human-in-the-loop concept-based explanations for graph neural networks. arXiv preprint arXiv:2107.11889 (2021)
39. Mahajan, D., Tan, C., Sharma, A.: Preserving causal constraints in counterfactual explanations for machine learning classifiers. arXiv:1912.03277 (2019)
40. Malle, B., Giuliani, N., Kieseberg, P., Holzinger, A.: The more the merrier - federated learning from local sphere recommendations. In: Holzinger, A., Kieseberg, P., Tjoa, A.M., Weippl, E. (eds.) CD-MAKE 2017. LNCS, vol. 10410, pp. 367–373. Springer, Cham (2017). https://doi.org/10.1007/978-3-319-66808-6_24
41. Matschinske, J., et al.: The featurecloud AI store for federated learning in biomedicine and beyond (2021). https://doi.org/10.48550/arXiv.2105.05734. arXiv:2105.05734
42. Müller, H., Holzinger, A., Plass, M., Brcic, L., Stumptner, C., Zatloukal, K.: Explainability and causability for artificial intelligence-supported medical image analysis in the context of the European In Vitro Diagnostic Regulation. New Biotechnol. **70**, 67–72 (2022). https://doi.org/10.1016/j.nbt.2022.05.002
43. Naik, N.: Migrating from virtualization to dockerization in the cloud: simulation and evaluation of distributed systems. In: 2016 IEEE 10th International Symposium on the Maintenance and Evolution of Service-Oriented and Cloud-Based Environments (MESOCA), pp. 1–8. IEEE (2016). https://doi.org/10.1109/MESOCA.2016.9
44. Ortega, A., Frossard, P., Kovačević, J., Moura, J.M., Vandergheynst, P.: Graph signal processing: overview, challenges, and applications. Proc. IEEE **106**(5), 808–828 (2018)
45. Pfeifer, B., Baniecki, H., Saranti, A., Biecek, P., Holzinger, A.: Multi-omics disease module detection with an explainable greedy decision forest. Sci. Rep. **12**(1), 1–15 (2022). https://doi.org/10.1038/s41598-022-21417-8
46. Pfeifer, B., et al.: Ensemble-GNN: federated ensemble learning with graph neural networks for disease module discovery and classification. bioRxiv (2023). https://doi.org/10.1101/2023.03.22.533772

47. Pfeifer, B., Holzinger, A., Schimek, M.G.: Robust random forest-based all-relevant feature ranks for trustworthy AI. Stud. Health Technol. Inform. **294**, 137–138 (2022). https://doi.org/10.3233/SHTI220418
48. Pfeifer, B., Saranti, A., Holzinger, A.: GNN-SubNet: disease subnetwork detection with explainable graph neural networks. Bioinformatics **38**(S-2), ii120–ii126 (2022). https://doi.org/10.1093/bioinformatics/btac478
49. Saranti, A., et al.: Actionable explainable AI (AxAI): a practical example with aggregation functions for adaptive classification and textual explanations for interpretable machine learning. Mach. Learn. Knowl. Extract. **4**(4), 924–953 (2022). https://doi.org/10.3390/make4040047
50. Schnake, T., et al.: Higher-order explanations of graph neural networks via relevant walks. arXiv preprint arXiv:2006.03589 (2020)
51. Schnake, T., et al.: XAI for graphs: explaining graph neural network predictions by identifying relevant walks. arXiv:2006.03589 (2020)
52. Schwalbe, G., Finzel, B.: A comprehensive taxonomy for explainable artificial intelligence: a systematic survey of surveys on methods and concepts. Data Min. Knowl. Discov. 1–59 (2023). https://doi.org/10.1007/s10618-022-00867-8
53. Singh, R., et al.: Directive explanations for actionable explainability in machine learning applications. arXiv:2102.02671 (2021)
54. Staab, S., Studer, R.: Handbook on Ontologies. Springer, Heidelberg (2010). https://doi.org/10.1007/978-3-540-92673-3
55. Sundararajan, M., Taly, A., Yan, Q.: Axiomatic attribution for deep networks. In: International Conference on Machine Learning, pp. 3319–3328. PMLR (2017)
56. Szklarczyk, D., et al.: The STRING database in 2021: customizable protein-protein networks, and functional characterization of user-uploaded gene/measurement sets. Nucleic Acids Res. **49**(D1), D605–D612 (2021)
57. Veličković, P.: Everything is connected: graph neural networks. Curr. Opin. Struct. Biol. **79**, 102538 (2023). https://doi.org/10.1016/j.sbi.2023.102538
58. Wu, Z., Pan, S., Chen, F., Long, G., Zhang, C., Philip, S.Y.: A comprehensive survey on graph neural networks. IEEE Trans. Neural Netw. Learn. Syst. **32**(1), 4–24 (2021). https://doi.org/10.1109/TNNLS.2020.2978386
59. Ying, Z., Bourgeois, D., You, J., Zitnik, M., Leskovec, J.: GNNExplainer: generating explanations for graph neural networks. In: Advances in Neural Information Processing Systems, vol. 32 (2019)

The Tower of Babel in Explainable Artificial Intelligence (XAI)

David Schneeberger[1(✉)], Richard Röttger[4], Federico Cabitza[3], Andrea Campagner[3], Markus Plass[1], Heimo Müller[1], and Andreas Holzinger[1,2]

[1] Medical University of Graz, Graz, Austria
david.schneeberger@medunigraz.at
[2] University of Natural Resources and Life Sciences, Vienna, Austria
[3] University of Milano-Bicocca, Milan, Italy
[4] South Denmark University (SDU), Odense, Denmark

Abstract. As machine learning (ML) has emerged as the predominant technological paradigm for artificial intelligence (AI), complex black box models such as GPT-4 have gained widespread adoption. Concurrently, explainable AI (XAI) has risen in significance as a counterbalancing force. But the rapid expansion of this research domain has led to a proliferation of terminology and an array of diverse definitions, making it increasingly challenging to maintain coherence. This confusion of languages also stems from the plethora of different perspectives on XAI, e.g. ethics, law, standardization and computer science. This situation threatens to create a "tower of Babel" effect, whereby a multitude of languages impedes the establishment of a common (scientific) ground. In response, this paper first maps different vocabularies, used in ethics, law and standardization. It shows that despite a quest for standardized, uniform XAI definitions, there is still a confusion of languages. Drawing lessons from these viewpoints, it subsequently proposes a methodology for identifying a unified lexicon from a scientific standpoint. This could aid the scientific community in presenting a more unified front to better influence ongoing definition efforts in law and standardization, often without enough scientific representation, which will shape the nature of AI and XAI in the future.

Keywords: Artificial Intelligence · AI · Machine Learning · ML · Explainable AI · XAI · explainability · interpretability · transparency · ethics · law · GDPR · DSA · Artificial Intelligence Act · standardization · ISO · IEC · IEEE

1 Introduction and Motivation

With the (nearly) ubiquitous spread of complicated black box models like GPT-4, explainable AI (XAI) has gained importance in both science and industry as a counterbalancing force. XAI refers to the development of artificial intelligence

A. Holzinger et al. (Eds.): CD-MAKE 2023, LNCS 14065, pp. 65–81, 2023.
https://doi.org/10.1007/978-3-031-40837-3_5

(AI) systems that can provide clear, understandable, and interpretable explanations for their advice and decisions. The very definition of explanation, and of its mentioned desirable properties, is, however, often not straightforward from a scientific point of view, leaving intuitive understanding aside.

Indeed, with the expansion of this research area the definition of terms and the variety of definitions is growing so fast that it is becoming extremely difficult to follow. This confusion of languages also stems from the plethora of different perspectives on XAI, e.g. ethics, law, standardization and computer science. There is no community-based agreement about central terms like explanation, explainability or interpretability and, in the scientific domain, the context of these definitions is often not clear. We are therefore facing, as mentioned in the Introduction, the threat of a "tower of Babel" effect, i.e. a confusion of languages and terminologies which makes it hard to find common (scientific) ground.

To counter this linguistic ambiguity, this paper maps the perspectives of ethics guidelines, law and standardization and in these fields. In comparison to the scientific perspective, these fields are often driven by the quest for standardized, uniform definitions. It shows that despite this goal, there is still no common vocabulary in these fields. Subsequently, it proposes a method to focus the diverging perspectives in the XAI field in the search for a common "vocabulary", i.e. a unified lexicon from a scientific standpoint. Such a unified lexicon could aid the scientific community in presenting a more unified front to better influence ongoing definition efforts in law and standardization, which will shape the nature of AI and XAI in the future but are often marred by a lack of scientific participation and democratic legitimacy.

2 Ethics Guidelines and XAI

Law (e.g. the Artificial Intelligence Act, see Sect. 3.3) and standards are often informed by relevant documents and reports, i.e. soft law or ethics guidelines. For example, the OECD (Organisation for Economic Co-operation and Development) [46] defines the principle of transparency and explainability in the following way: AI actors "should provide meaningful information, appropriate to the context [...] to foster a general understanding of AI systems, to make stakeholders aware of their interactions with AI systems [...] to enable those affected by an AI system to understand the outcome, and, to enable those adversely affected by an AI system to challenge its outcome based on plain and easy-to-understand information on the factors, and the logic that served as the basis [...]".

This illustrates that terms like transparency and explainability are often used without drawing clear boundaries. Documents often refer to them as umbrella terms comprising several distinct elements, i.e. more general information (e.g. information on the interaction with an AI system), but also elements, which could necessitate the implementation of XAI approaches (e.g. "information on the factors and the logic that served as basis"). This muddled language makes it harder to derive clear implementation measures for XAI.

In contrast, the ethics guidelines of the high-level expert group on AI [28], set up by the European Commission, differentiate between several elements of transparency, which itself is linked with the principle of explicability, i.e. traceability (concerning the documentation of data sets, algorithms, and the processes that yield the decision), explainability (mainly concerning the ability to explain both the technical processes of an AI system and the related human decisions; information of the degree to which an AI system influences and shapes the organisational decision-making process, design choices of the system, and the rationale for deploying it) and communication (i.e. humans have the right to be informed that they are interacting with an AI system; capabilities and limitations should be communicated). Mainly the second element, explainability concerning the technical process, is linked with the implementation of XAI but again does not state concrete measures.

This problem of the use of vague umbrella terms, illustrated by the OECD example above, also exists on a macro level. As meta studies on ethics guidelines [39] show, "transparency" is the most often mentioned principle, but the interpretation, what transparency entails, varies widely in these guidelines, concerning what should be transparent (e.g. data use, human-AI-interaction, automated decisions, purpose of data use/application of the AI system) or the goal of transparency (e.g. minimize harm, improve AI, legal reasons, foster trust, principle of democracy). To achieve transparency, disclosure of information is often suggested but there is no agreement what should be disclosed (e.g. use of AI, source code, data use, evidence base, limitations, laws, responsibility for AI, investments, impact).

Ienca and Vayena [31] differentiate between two main thematic families of transparency mentioned in guidelines: Firstly, transparency of algorithms and data processing methods (which refers to the implementation of XAI approaches) and secondly transparency of human practices related to the design, development and deployment of AI systems (e.g. disclosing relevant information to data subjects, avoiding secrecy, forbidding conflicts of interest between AI actors and oversight bodies).

These divergent interpretations of transparency lead to divergences in the implementation strategies proposed to achieve transparency. Generally, a major problem lies in deducing concrete technological implementations from the very abstract ethical values and principles described in ethics guidelines [25].

As a brief mapping of these guidelines has illustrated, they seem to contribute to the "tower of Babel" effect concerning XAI terms as they often - which partly lies in the nature of ethics guidelines - only set out abstract principles without describing concrete implementation strategies.

3 Law and XAI

3.1 GDPR

Switching to the perspective of law and XAI, as AI specific regulation has only recently come into the focus of national and international legislators, the legal framework currently does not contain explicit legal definitions of "explainability" or "transparency". This could change when the proposed Artificial Intelligence Act (AIA) comes into force (see Sect. 3.3).

Of course, at the EU and the national level there are (older) laws, which were not written with AI in mind, but which are also applicable to AI systems and contain transparency obligations (with further references [3,48]).

For example, the General Data Protection Regulation (GDPR) [21] has wide implications for the use of AI and it has become a model law for AI regulation. The processing of data in the context of (fully) automated individual decision-making, i.e. without (substantial) human involvement, is principally forbidden by Art. 22 GDPR - which has been in the center of the "right to an explanation" debate (with further references [6,40,45,49]) – but fully automated decision-making is allowed if one of three exceptions (necessary for entering into/performance of a contract, authorisation by EU/member state law, explicit consent) applies.

In such a case, specific information has to be proactively provided (Art. 13, 14) and the data subject also has a right to access this information on request (Art. 15). This includes information about the "existence of automated decision-making", about "the logic involved" and "the significance and the envisaged consequences".

The passage "the logic involved" has been interpreted in different ways, e.g. as a subject-specific local explanation of a specific decision [24,44,50] or as variant of a general (global) explanation (mainly concerning the features employed) [59]. Explaining the logic involved could therefore necessitate the implementation of a feature-importance based XAI approach.

A recent opinion (16 March 2023, C-634/21, ECLI:EU:C:2023:220) (with further references [47,54]) of the attorney general Pikamäe could clarify the interpretation. These opinions are often but not always adopted by the European Court of Justice. The opinion states that the "logic involved" does not necessitate the disclosure of the algorithm used. According to the opinion only "general information, in particular on the factors taken into account in the decision-making process and their weighting at an aggregated level", i.e. a form of a global feature-importance explanation, has to be provided. But as the opinion also states that "sufficiently detailed explanations on the method used to calculate the score and on the reasons that led to a certain result" have to be provided, this seems contradictory as the wording "a certain result" seems to imply a local explanation. This contradiction will have to be clarified by the court of Justice but it seems more likely that "logic involved" will be interpreted as a more general (global) explanation, mainly based on aggregated features.

Recital 71 also mentions a right "to obtain an explanation of the decision reached after such assessment" as part of suitable measures to safeguard the data subject (Art. 22 para. 3) but this right is only mentioned in the recital. Recitals mainly function as guidelines on how to interpret law but can not create law themselves. Therefore, the existence and the content of a "right to (an) explanation" is still disputed in scholarship (e.g. [49,59]).

3.2 Digital Services Act (DSA)

The new Digital Services Act (DSA) [22], which for example comes into play if an information society service provider (e.g. a social network) uses AI to moderate content (for an overview, see [18,42]), also contains transparency provisions. Providers of intermediary services have to include information "on any policies, procedures, measures and tools used for the purpose of content moderation, including algorithmic decision-making" in their terms and conditions (Art. 14 para. 1 DSA). They are also subject to yearly transparency public reporting obligations on content moderation. This includes information on "any use made of automated means for the purpose of content moderation" (Art. 15 para. 1(e) DSA). These obligations do not seem to directly relate to the implementation of XAI methods, but they require transparency on an abstract, global level, i.e. a qualitative description and information about the purpose and performance metrics (i.e. accuracy and error rates) of these systems.

Online platforms displaying advertising must also ensure that the recipients of the service can identify meaningful information "about the main parameters used to determine the recipient to whom the advertisement is presented and, where applicable, about how to change those parameters" (Art. 26 para. 1(d) DSA). This requires a form of explanation on the main features used in displaying advertisements, i.e. a feature-importance explanation, which seems to have a local ("used to determine the recipient") and a counterfactual element ("how to change those parameters"). This obligation could therefore necessitate the implementation of a XAI approach, which provides this local feature-importance and counterfactual information.

3.3 The (Proposed) Artificial Intelligence Act (AIA)

In April 2021, the European Commission proposed the so-called Artificial Intelligence Act (AIA) [19]. Since then several amendments have been suggested by the EU co-legislators, the Council [11] and the European Parliament [20]. Even though there were some remaining issues (e.g. AI definition, regulation of general-purpose AI/foundational models like GPT-4) the European Parliament held a positive plenary vote on 14 June 2023 [60]. Therefore, the final phase of the law-making process, the so-called trilogue, has started.

The AIA (for a general introduction see [57]) follows a risk-based approach. AI systems with an "unacceptable risk" (Art. 5 AIA e.g. social scoring modelled on China) will be banned, while high-risk AI systems will be subjected to strict regulation and must undergo an ex-ante conformity assessment. Concerning systems which pose a limited risk, these are subject to specific transparency obligations (Art. 52 AIA, e.g. chatbots must identify themselves).

The AIA addresses two different forms of high-risk AI systems (Art. 6 AIA): First, AI systems that are products or a safety component of a product already covered by EU harmonisation legislation requiring a third-party conformity assessment (e.g. medical devices). Second, in Annex III AIA eight categories of stand-alone AI systems are listed which are also considered high-risk (e.g. migration, asylum and border control management).

The AIA contains a specific transparency obligation for high-risk AI systems. According to Art. 13 para. 1 AIA high-risk AI systems must be "be designed and developed in such a way to ensure that their operation is sufficiently transparent to enable users to interpret the system's output and use it appropriately". The appropriate type and degree of transparency seems to be relative, its goal is achieving compliance with (other) relevant obligations of the AIA (recital 47: "a certain degree of transparency").

Crucially, the AIA does not offer (legal) definitions (Art. 3 AIA) for the central terms "sufficiently transparent" or "to interpret". The AIA does not mention the concept of "explainability" and therefore does not differentiate between interpretability and explainability [17]. This lack of definitions could lead to legal uncertainty and the mentioned "tower of Babel" effect.

As has been stated in legal scholarship, this leaves the interpretation of Art. 13 para. 1 AIA and the level of transparency/interpretability required unclear [4,15]. Therefore, it has been argued that the question of how to make AI systems interpretable is left to the discretion of the AI system provider, i.e. the AI developer [17].

In conclusion, this leaves the interpretation, whether Art. 13 AIA necessitates the implementation of XAI techniques and which approach has to be chosen, e.g. if a local or global explanation is required, open. It can also be argued that only a general form of transparency, mainly through the provision of "instructions", which have to be proved according to Art. 13 para. 2 seq., will suffice to satisfy this requirement. These instructions must for example contain the purpose, the level of accuracy, robustness, circumstances, which may lead to risks, performance metrics regarding the use groups, specification for the input data or on training/validation/testing data.

For example according to [5] Art. 13 para. 1 AIA does not imply the necessity of explainability in the sense that the way in which data have been processed must be entirely traceable, but a more general form of transparency of the system's functioning and output generation. Furthermore, a study [52] on request of the European Commission stated that XAI techniques are not the "only means available to understand and interpret AI systems outputs" and therefore not required for all high-risk AI systems. Instead "documentation approaches, scenarios, principles of operations, as well as interactive training materials" will fulfill the requirements of Art. 13 AIA. This indicates that the implementation of XAI approaches is not a core component of this transparency obligation.

Several attempts to define the terminology used in Art. 13 AIA illustrate the struggle to find uniform definitions, which shape how XAI will be used in the future. For example, the Council [11] proposed to simplify this obligation, i.e. to use the term "understand" instead of "interpret", which in our opinion is equally vague and has no real benefits.

The second co-legislator, the European Parliament [20], also tries to fill this vague terminology with life. In the version of the Parliament, AI systems must be "sufficiently transparent to enable providers and users to reasonably understand the system's functioning." In our opinion the addition of "functioning" suggests a more general level of transparency, which also "shall be ensured in accordance with the intended purpose of the AI system", again indicating that the level of transparency is context sensitive. As a very important step in the direction of a precise terminology, the Parliament suggested to define "transparency", which shall "mean that, at the time the high-risk AI system is placed on the market, all technical means available in accordance with the generally acknowledged state of art are used to ensure that the AI system's output is interpretable by the provider and the user." As this refers to the state of the art, which is always in flux, this could mean that XAI approaches will become mandatory as they become state of the art and if they provide a clear benefit in helping the user interpret the output. On the other hand, the Parliament in our opinion seems to suggest a high-level, global form of transparency, based on a simplified understanding of the system and the features used ("The user shall be enabled to understand and use the AI system appropriately by generally knowing how the AI system works and what data it processes [...]"). This reduced obligation of "generally knowing" does not seem to necessitate the implementation of XAI techniques. This should in turn allow "the user to explain the decisions taken by the AI system to the affected person [...]". In our opinion this clarification is an important step in the right direction as it minimizes legal uncertainty regarding how "transparency" must be interpreted.

As a point of criticism, in the original AIA proposal the output must be interpretable only for the (professional) user (i.e. a doctor) and not the person who is affected by an AI system (i.e. a patient). But professional users are seldom the only ones put at risk by AI systems [7]. Therefore, Art. 13. AIA is sometimes referred to as a form of "user-empowering explainability" [53]. Critically, people who are affected by high-risk AI systems, are left without a new right to information [17]. This lack of a "human-centred approach" has been a major point of criticism [55].

To solve this oversight, the European Parliament [20] proposed the introduction of "A right to explanation of individual decision-making" (Art. 68c AIA). This would give "[a]ny affected person subject to a decision which is taken [...] on the basis of the output from an high-risk AI system" (e.g. a diagnosis by a doctor) "which produces legal effects or similarly significantly affects him or her" (e.g. it affects the health of a patient) a "right to request [...] clear and meaningful explanation [...] on the role of the AI system in the decision-making procedure, the main parameters of the decision taken and the related input data." In our opinion, this suggests a form of a local feature-importance explanation (main parameters of the decision, related input data), which could necessitate the implementation of XAI approaches, and additionally an explanation of the role of the AI system (e.g. diagnostic aid). This explanation must also be target appropriate (recital 84b "[...] they should take into account the level of expertise and knowledge of the average consumer or individual"). If this focus on the explanation of an individual decision is held up in the trilogue, this could necessitate the implementation of a XAI approach, which can produce a local feature-importance explanation.

Thematically linked, Art. 14 AIA on human oversight also requires the implementation of measures that enable the individuals, to whom human oversight is assigned, to "be able to correctly interpret the high-risk AI system's output". In this regard, "the characteristics of the system and the interpretation tools and methods available", i.e. the implementation of XAI techniques, have to be taken into account.

Even though the amendments by the European Parliament described above are a step in the right direction and could lead to a more precise terminology, there is still a high level of legal uncertainty in interpreting these transparency obligations. This leads to economic risk for AI providers, who have to interpret the provision themselves when assessing the conformity with the AIA. Of course, the jurisprudence of the European Court of Justice could lead to clarification, but this will only be on a case-to-case basis and will take years. Therefore, the third layer, standardization, could play an important role in defining these abstract concepts set out by law.

4 Standardization and XAI

As law, even AI-specific regulation, must be applicable to many different categories of automated/autonomous software systems, these instruments must be in a sense "technology-agnostic" as law can not be easily amended in lockstep with every novel technological development. Therefore, legal rules are by-design often written from an abstract perspective, i.e. they only set out high-level principles and goals like "security" or "transparency". The concrete technical implementation is often defined by standards, which are (often) developed by (private) organizations, so-called SDOs (Standards Development Organizations). To ensure a uniform level of AI safety, several SDOs are drafting AI standards to fill existing regulatory gaps.

At the international and EU level, the most important SDOs are the International Organization for Standardization (ISO), the International Electrotechnical Commission (IEC), the Institute of Electrical and Electronics Engineers (IEEE), the International Telecommunication Union (ITU), the Internet Engineering Task Force (IETF), the European Committee for Standardization (CEN), the European Committee for Electrotechnical Standardization (CENELEC) and the European Telecommunications Standards Institute (ETSI) [16].

In the upcoming part of the paper, we aim to give a brief overview of the standards concerning explainability/interpretability. As a caveat, most of these standards are still in development and as (most) of the drafts can not be publicly accessed, we do not aim to give an in-depth analysis.

ISO and IEC created the joint technical committee JTC 1/SC 42 which serves as "the focus and proponent [...] (for the) standardization program on Artificial Intelligence". Several working groups exist which are focused on different aspects (e.g. WG 1 foundational standards; WG 2 data; WG 3 trustworthiness) [37].

On the one hand, ISO/IEC AWI 12792, which is still in development, aims to create a transparency taxonomy describing "the semantics of the information elements and their relevance to the various objectives of different AI stakeholders" [34].

On the other hand, the technical specification ISO/IEC AWI TS 6254 "Objectives and approaches for explainability of ML models and AI systems", which is also still in a drafting state, "describes approaches and methods that can be used to achieve explainability objectives of stakeholders with regards to ML models and AI systems' behaviours, outputs, and results" [36].

It identifies characteristics of explainability (explanation needs, form, approaches, and technical constraints) and uses them to categorise existing approaches. As a limitation, according to a report [52], it does not discuss or compare the technological maturity and known limitations of the methodologies (i.e. if methods are trustworthy and reflect the actual decision-making process).

The ongoing discussions about these two standards illustrate the central aim and struggle of defining "transparency" and "explainability", which are the cornerstones of these standards [1]. Transparency was broadly defined as the "availability in relation to stakeholders of meaningful, faithful, comprehensive, accessible and understandable information about a relevant aspect of an AI system". XAI approaches could help in generating this necessary information. Interpretability concerning algorithms was defined as the "ease with which a stakeholder can comprehend in a timely manner the objective of an AI system, the reasons for the system's behavior, and whether it is working given its purpose and in line with stakeholder expectations, and how different inputs could lead to different outcomes". Interpretability can be reached through technical approaches like explainability methods or other analysis or visualization methods. Similar to the ethics guidelines of the high-level expert group on AI (see Sect. 2) two levels of explainability were differentiated. Explainability concerning policy as the "ability to provide stakeholders of an AI system with concise, accessible, sufficient and useful explanatory information beyond the AI system's results", which refers to the wider socio-economic context of an AI system, and explainability concerning algorithms as the "capability of an AI system to correctly produce the reasons for its own behavior in a timely manner, allowing scrutiny of whether it is working given its purpose and in line with stakeholder expectations, and how different inputs could lead to different outcomes", which refers to the implementation of XAI techniques.

Additionally, the terms explainability and/or interpretability are also mentioned in ISO/IEC 22989:2022 [33] on "Artificial intelligence concepts and terminology" and in ISO/IEC AWI TS 29119-11 [35] concerning testing of AI systems and in the ISTQB (International Software Testing Qualifications Board) syllabus [38] for "Certified Tester AI Testing" [13].

At the level of the IEEE, the P7000 series of standards is being developed as part of the Global Initiative on Ethics of Autonomous and Intelligent Systems. In contrast to more traditional standards, these standards aim to address "specific issues at the intersection of technological and ethical considerations" [30].

Regarding transparency, the already published standard IEEE P7001 [29] sets out transparency requirements without defining how to achieve them, i.e. which XAI techniques or solution to use. It (only) describes different levels of transparency with an increasing range of sophistication and complexity [52].

At the national level, the German SDOs DIN (Deutsches Institut für Normung) and DKE (Deutsche Kommission für Elektrotechnik Elektronik Informationstechnik) have released the second version of an extensive "Standardization Roadmap AI", which maps the existing standards and analyses the need for new AI standards [13]. As the roadmap states, there is a need to specify formal requirements for XAI methods (i.e. formulation of concrete operationalizable/testable requirements). It also states that additional basic research in XAI is required because available methods have not yet been fully and widely researched and applied. To fill these gaps, DIN is also working on a standard concerning explainability [12].

Besides these standards for explainability/interpretability, a whole range of standards for AI systems and related technologies is being developed at the national and international level (see [13,16]).

In comparison to the perspectives of ethics and law, the field of standardization illustrates even better the quest for a standardized, uniform terminology, which is still ongoing. But as the mapping above indicates, the contours of central terms are becoming sharper and sharper.

5 The Link Between Law and Standardization

Law and standardization are thematically interlinked. As a study regarding the AIA states: "Standards are set to bring the necessary level of technical detail into the essential requirements prescribed in the legal text, defining concrete processes, methods and techniques that AI providers can implement in order to comply with their legal obligations" [52]. Co-Regulation through standardization based on the new Legislative Framework (NLF) is a cornerstone of the AIA. The essential requirements contained in law are given concrete form by standards [16].

Instead of interpreting obligations like the transparency obligation Art. 13 AIA discussed in Sect. 3.3, which could take time and expertise and also lead to legal risk, AI providers can mitigate uncertainty and follow (harmonized) standards. This leads to the presumption, that an AI system conforms with the requirements of the AIA. Therefore, in practice (harmonised) standards will play an important role in shaping the technical requirements and therefore the XAI landscape.

These harmonised standards are developed on demand of the European Commission and are published in the official journal of the EU. At the EU level CEN, CENELEC and ETSI (see Sect. 4) function as the SDOs which can either transpose existing standards into European standards if they comply with European values, standards and legislation, or they can develop own standards. At the moment of writing the European Commission has already started the process to adopt a standardization request providing a formal mandate to European SDOs to develop the necessary standards [52].

Even though these standards could bring the necessary clarity to high-level obligations contained in ethics guidelines or the AIA by defining essential XAI terms, this heightened role of standardization has some disadvantages. Besides other general problems of standardizing AI (e.g. rapid change of the underlying technology, ongoing debate on ethical and legal questions [2]) there are numerous points for criticism: SDOs like the IEC and ISO typically work on a subscription model and retain copyright [56], creating a monetary barrier especially for small AI developers to access these standards. The standardization process is susceptible to lobbying [56] and large, global players could therefore try to influence the definition of central terms to shape the XAI landscape. Regulation by standards shifts the law-making power to private bodies, which, compared to national or EU legislation, lack in options for democratic control and participation [16,17,23,43,57]. This also reduces the possibility for the scientific community to influence the ongoing AI governance discussion.

The European Parliament has seemingly recognized his problem in their AIA amendments stating that it is necessary "to ensure a balanced representation of interests by involving all relevant stakeholders in the development of standards." (Recital 61) Therefore, the Commission must consult with the AI Office and the Advisory Forum (Art. 40 AIA, Recital 61a), which "should ensure varied and balanced stakeholder representation and should advise the AI Office" (Recital 76).

6 A Proposed Solution

As our analysis of ethics guidelines, law and standardization has shown, the quest for a precise terminology is still ongoing. In turn, XAI scientists cannot rely on the vague, partially contradictory, and overly numerous definitions. Furthermore, especially in standardization there is often very low participation of representatives of academy and scientific researchers. Methods of democratic representation are often lacking.

A first step to counter this development is to be aware of the definition problem and to create sensitivity about the opacity of the standards drafting mechanism. This position papers aims to contribute in building such an awareness in the scientific community.

As a second step, we then pose the opposite problem: how can scientists and XAI scholars inform the process of law-making and standardization so as to provide guidance for the conformity assessment that will be so crucial in evaluating the legality of the next AI systems disseminated to the general public or adopted in sensitive areas such as health care or public safety?

We therefore created a simple and feasible method so that, at least the community of scholars who are most interested in these issues, can converge in a lexicographic and definitional effort that brings order and gains the necessary visibility and credibility to inform standard and policy making.

In recent years, scientists active in the field of XAI have produced several reviews (e.g., [8,10,14,26,27,32,41,58]), both systematic and more narrative and exploratory ones, to understand the lexical and definition variety in the field and, in some ways, help reduce the linguistic babel, since this is seen as an obstacle for the diffusion and wide adoption of successful design patterns, and sound evaluation methods. Nonetheless, while all of these contributions primarily consist of taxonomies or similar hierarchical categorizations that attempt to represent, and somehow systematize, the above mentioned variety, we note that their aims (and, thus, the set of concepts and definitions they document and attempt to map out) differ. Indeed, while some of the referenced surveys [10,26,32] largely aimed at categorizing existing XAI techniques from the methodological point of view, with a consequent focus on notions related to presentation modality or explanation type; others have also considered a more user-oriented perspective, and thus focused on definitions and notions related to the evaluation, validation and effects of explanations [14,27]; or also to a more general investigation of the understanding of the notion of explanation itself [8,58]. Thus, it is easy to see that the above mentioned contributions can only be understood as a starting point for our proposed initiative, which is still far from being an exhausted topic.

What we are proposing, indeed, is to activate a truly communal initiative that can lead a set of representative scholars to 1) collect all the major definitions proposed in the highest impact articles or most comprehensive reviews 2) invite all the authors of these articles and registered participants at major conferences in the field (e.g. the International Conference on eXplainable Artificial Intelligence, the IJCAI Workshop on Explainable Artificial Intelligence, the Actionable Explainable AI Session at the Cross Domain Conference for Machine Learning and Knowledge Extraction, CD-MAKE) to vote about the precision, clarity and comprehensiveness of definitions of concepts such as explanation, explainability, transparency, causability, understandability on opportune ordinal scales, 3) to aggregate the results with state-of-the art methods, such as the one used in [9]; and 4) to return the results to the community, possibly iterating a few times so as to reduce variability and facilitate consensus building, in a manner not unlike a Delphi method involving the most motivated people in the field and mediated by asynchronous collaboration tools such as online questionnaires [51] and shared papers.

7 Conclusion

This paper mapped the ongoing efforts to define central XAI terms in ethics, law and standardization. It illustrates that the quest for a common vocabulary is still ongoing but there is the danger that the essential vocabulary and therefore the XAI landscape could be defined by efforts marred by a lack of scientific participation. After describing these challenges, the authors propose to start a consolidation process at the Cross Domain Conference for Machine Learning and Knowledge Extraction, CD-MAKE conference and systematically close the gap between scientific publications on one side and ethics guidelines, law and standards on the other side. A unified lexicon could aid the scientific community in presenting a more unified front to better influence ongoing definition efforts which will shape the nature of AI and XAI in the future. Instead, all areas should strengthen each other and learn from each other.

Acknowledgements. The authors declare that there are no conflict of interests. This work does not raise any ethical issues. This work has received funding from the European Union's Horizon 2020 research and innovation programme under grant agreement No. 826078 (Feature Cloud). This publication reflects only the authors' view and the European Commission is not responsible for any use that may be made of the information it contains. Parts of this work have been funded by the Austrian Science Fund (FWF), Project: P-32554 (explainable Artificial Intelligence). This paper has been made open access CC-BY, freely accessible to the international research community. We are grateful for the valuable reviewer comments.

References

1. AI Standards Hub: Output from workshop on ISO/IEC standards for AI transparency and explainability. https://aistandardshub.org/forums/topic/output-from-workshop-on-iso-iec-standards-for-ai-transparency/-and-explainability/
2. Beining, L.: Vertrauenswürdige KI durch Standards? (2020). https://www.stiftung-nv.de/sites/default/files/herausforderungen-standardisierung-ki.pdf
3. Bibal, A., Lognoul, M., de Streel, A., Frénay, B.: Legal requirements on explainability in machine learning. Artif. Intell. Law **29**, 149–169 (2021). https://doi.org/10.1007/s10506-020-09270-4
4. Bomhard, D., Merkle, M.: Europäische KI-Verordnung. Recht Digit. **1**(6), 276–283 (2021)
5. Bordt, S., Finck, M., Raidl, E., von Luxburg, U.: Post-hoc explanations fail to achieve their purpose in adversarial contexts. In: FAccT 2022: 2022 ACM Conference on Fairness, Accountability, and Transparency, pp. 891–905. ACM, New York (2022). https://doi.org/10.1145/3531146.3533153
6. Brkan, M.: Do algorithms rule the world? Algorithmic decision-making and data protection in the framework of the GDPR and beyond. Int. J. Law Inf. Technol. **27**(2), 91–121 (2019). https://doi.org/10.1093/ijlit/eay017
7. Busuioc, M., Curtin, D., Almada, M.: Reclaiming transparency: contesting the logics of secrecy within the AI act. Eur. Law Open **2**, 1–27 (2022). https://doi.org/10.1017/elo.2022.47
8. Cabitza, F., et al.: Quod erat demonstrandum?-Towards a typology of the concept of explanation for the design of explainable AI. Expert Syst. Appl. **213**, 118888 (2023)
9. Cabitza, F., Ciucci, D., Locoro, A.: Exploiting collective knowledge with three-way decision theory: cases from the questionnaire-based research. Int. J. Approximate Reasoning **83**, 356–370 (2017)
10. Cambria, E., Malandri, L., Mercorio, F., Mezzanzanica, M., Nobani, N.: A survey on XAI and natural language explanations. Inf. Process. Manage. **60**(1), 103111 (2023)
11. Council: Proposal for a regulation of the European Parliament and of the Council laying down harmonised rules on Artificial Intelligence (Artificial Intelligence Act) and amending certain union legislative acts - general approach, 14954/22 (2022). https://www.kaizenner.eu/post/aiact-part3

12. DIN: SPEC 92001-3 Künstliche Intelligenz - Life Cycle Prozesse und Qualitätsanforderungen - Teil 3: Erklärbarkeit. https://www.din.de/de/forschung-und-innovation/din-spec/alle-geschaeftsplaene/wdc-beuth:din21:354291453
13. DIN, DKE: Normungsroadmap Künstliche Intelligenz: Version 2 (2022). https://www.dke.de/de/arbeitsfelder/core-safety/normungsroadmap-ki
14. Ding, W., Abdel-Basset, M., Hawash, H., Ali, A.M.: Explainability of artificial intelligence methods, applications and challenges: a comprehensive survey. Inf. Sci. (2022)
15. Ebers, M.: Standardisierung Künstlicher Intelligenz und KI-Verordnungsvorschlag. Recht Digit. **2**, 588–597 (2021)
16. Ebers, M.: Standardizing AI: the case of the european commission's proposal for an 'artificial intelligence act'. In: The Cambridge Handbook of Artificial Intelligence: Global Perspectives on Law and Ethics, pp. 321–344. Cambridge University Press, Cambridge (2022). https://doi.org/10.1017/9781009072168.030
17. Ebers, M., Hoch, V.R.S., Rosenkranz, F., Ruschemeier, H., Steinrötter, B.: The European Commission's proposal for an Artificial Intelligence Act- a critical assessment by members of the Robotics and AI Law Society (RAILS). J **4**(4), 589–603 (2021). https://doi.org/10.3390/j4040043
18. Eifert, M., Metzger, A., Schweitzer, H., Wagner, G.: Taming the giants: the DMA/DSA package. Common Mark. Law Rev. **58**(4), 987–1028 (2021). https://doi.org/10.54648/cola2021065
19. European Commission: Proposal for a Regulation of the European Parliament and of the Council laying down harmonised rules on Artificial Intelligence (Artificial Intelligence Act) and amending certain union legislative acts, COM(2021) 206 final. https://eur-lex.europa.eu/legal-content/EN/TXT/?uri=celex:52021PC0206
20. European Parliament: Amendments adopted by the european parliament on 14 June 2023 on the proposal for a regulation of the European Parliament and of the Council on laying down harmonised rules on Artificial Intelligence (Artificial Intelligence Act) and amending certain union legislative acts (COM(2021) 0206 - C9–0146/2021 - 2021/0106(COD))1. https://www.kaizenner.eu/post/aiact-part3
21. European Parliament, Council: Regulation (EU) 2016/679 of the European Parliament and of the Council of 27 April 2016 on the protection of natural persons with regard to the processing of personal data and on the free movement of such data, and repealing directive 95/46/EC (General Data Protection Regulation), OJ L 2016/119, 1. https://eur-lex.europa.eu/eli/reg/2016/679/oj
22. European Parliament, Council: Regulation (EU) 2022/2065 of the European Parliament and of the Council of 19 October 2022 on a Single Market For Digital Services and amending Directive 2000/31/EC (Digital Services Act), OJ L 2022/277, 1. https://data.europa.eu/eli/reg/2022/2065/oj
23. Guijarro Santos, V.: Nicht besser als nichts: Ein Kommentar zum KI-Verordnungsentwurf. Zeitschrift Digitalisierung Recht **3**(1), 23–42 (2023)
24. Hacker, P., Passoth, J.H.: Varieties of AI explanations under the law. From the GDPR to the AIA, and beyond. In: Holzinger, A., Goebel, R., Fong, R., Moon, T., Müller, K.R., Samek, W. (eds.) xxAI 2020. LNCS, vol. 13200, pp. 343–373. Springer, Cham (2022). https://doi.org/10.1007/978-3-031-04083-2_17
25. Hagendorff, T.: The ethics of AI ethics: an evaluation of guidelines. Mind. Mach. **30**(1), 99–120 (2020). https://doi.org/10.1007/s11023-020-09517-8
26. Hanif, A., Zhang, X., Wood, S.: A survey on explainable artificial intelligence techniques and challenges. In: 2021 IEEE 25th International Enterprise Distributed Object Computing Workshop (EDOCW), pp. 81–89. IEEE (2021)

27. Haque, A.B., Islam, A.N., Mikalef, P.: Explainable artificial intelligence (XAI) from a user perspective: a synthesis of prior literature and problematizing avenues for future research. Technol. Forecast. Soc. Chang. **186**, 122120 (2023)
28. High-level Expert Group on AI: Ethics guidelines for trustworthy AI. https://op.europa.eu/en/publication-detail/-/publication/d3988569-0434-11ea-8c1f-01aa75ed71a1
29. IEEE: IEEE 7001–2021: IEEE standard for transparency of autonomous systems. https://standards.ieee.org/ieee/7001/6929/
30. IEEE: The IEEE global initiative on ethics of autonomous and intelligent systems. https://standards.ieee.org/industry-connections/ec/autonomous-systems/
31. Ienca, M., Vayena, E.: AI ethics guidelines: European and global perspectives. In: Towards Regulation of AI Systems, pp. 38–60. Council of Europe (2020). https://rm.coe.int/prems-107320-gbr-2018-compli-cahai-couv-texte-a4-bat-web/1680a0c17a
32. Islam, M.R., Ahmed, M.U., Barua, S., Begum, S.: A systematic review of explainable artificial intelligence in terms of different application domains and tasks. Appl. Sci. **12**(3), 1353 (2022)
33. ISO, IEC: 22989:2022: Information technology - artificial intelligence - artificial intelligence concepts and terminology. https://www.iso.org/standard/74296.html
34. ISO, IEC: AWI 12792: Information technology - artificial intelligence - transparency taxonomy of AI systems. https://www.iso.org/standard/84111.html
35. ISO, IEC: AWI TS 29119-11: Software and systems engineering - software testing - part 11: Testing of AI systems. https://www.iso.org/standard/84127.html
36. ISO, IEC: AWI TS 6254: Information technology - artificial intelligence - objectives and approaches for explainability of ML models and AI systems. https://www.iso.org/standard/82148.html
37. ISO, IEC: JTC 1/SC 42: Artificial intelligence. https://www.iso.org/committee/6794475.html
38. ISTQB: Certified tester AI testing (CT-AI). https://www.istqb.org/certifications/artificial-inteligence-tester
39. Jobin, A., Ienca, M., Vayena, E.: The global landscape of AI ethics guidelines. Nat. Mach. Intell. **1**(9), 389–399 (2019). https://doi.org/10.1038/s42256-019-0088-2
40. Kaminski, M.E.: The right to explanation, explained. Berkeley Technol. Law J. **34**, 189–218 (2019). https://doi.org/10.15779/Z38TD9N83H
41. Kargl, M., Plass, M., Müller, H.: A literature review on ethics for AI in biomedical research and biobanking. Yearb. Med. Inform. **31**(01), 152–160 (2022)
42. Knyrim, R., Urban, L.: DGA, DMA, DSA, DA, AI-Act, EHDS - ein Überblick über die europäische Datenstrategie (Teil I). Dako **3**, 55–58 (2023)
43. Laux, J., Wachter, S., Mittelstadt, B.: Three pathways for standardisation and ethical disclosure by default under the European Union Artificial Intelligence Act. Elsevier Preprint SSRN (2023). https://doi.org/10.2139/ssrn.4365079
44. Malgieri, G., Comandé, G.: Why a right to legibility of automated decision-making exists in the general data protection regulation. Int. Data Priv. Law **7**(4), 243–265 (2017). https://doi.org/10.1093/idpl/ipx019
45. Malgieri, G.: Automated decision-making and data protection in Europe. In: Research Handbook on Privacy and Data Protection Law, pp. 433–448. Edward Elgar, Cheltenham/Northampton (2022). https://doi.org/10.4337/9781786438515
46. OECD: Transparency and explainability. https://oecd.ai/en/dashboards/ai-principles/P7

47. Palmiotto Ettorre, F.: Is credit scoring an automated decision? The opinion of the AG Pikamäe in the case C-634/21 (2023). https://digi-con.org/is-credit-scoring-an-automated-decision-the-opinion-of-the-ag-//pikamae-in-the-case-c-634-21/
48. Schneeberger, D.: Der Einsatz von Machine Learning in der Verwaltung und die Rolle der Begründungspflicht. Ph.D. thesis, Graz (2023)
49. Schneeberger, D., Stöger, K., Holzinger, A.: The European legal framework for medical AI. In: Holzinger, A., Kieseberg, P., Tjoa, A.M., Weippl, E. (eds.) CD-MAKE 2020. LNCS, vol. 12279, pp. 209–226. Springer, Cham (2020). https://doi.org/10.1007/978-3-030-57321-8_12
50. Selbst, A.D., Powles, J.: Meaningful information and the right to explanation. Int. Data Priv. Law **7**(4), 233–242 (2017). https://doi.org/10.1093/idpl/ipx022
51. Shinners, L., Aggar, C., Grace, S., Smith, S.: Exploring healthcare professionals' perceptions of artificial intelligence: validating a questionnaire using the e-Delphi method. Digit. Health **7**, 20552076211003430 (2021)
52. Soler Garrido, J., et al.: AI watch: artificial intelligence standardisation landscape update (2023). https://publications.jrc.ec.europa.eu/repository/handle/JRC131155
53. Sovrano, F., Sapienza, S., Palmirani, M., Vitali, F.: Metrics, explainability and the European AI Act proposal. J **5**(1), 126–138 (2022). https://doi.org/10.1093/idpl/ipx022
54. Strassemeyer, L.: Externes Scoring kann, muss aber nicht unter Art. 22 Abs. 1 DSGVO fallen. Datenschutz-Berater (4), 102–106 (2023)
55. Van Kolfschooten, H.: EU regulation of artificial intelligence: challenges for patients' rights. Common Mark. Law Rev. **59**(1), 81–112 (2022). https://doi.org/10.54648/cola2022005
56. Veale, M., Matus, K., Robert, G.: AI and global governance: modalities, rationales, tensions. Annu. Rev. Law Soc. Sci. (2023). https://doi.org/10.31235/osf.io/ubxgk
57. Veale, M., Zuiderveen Borgesius, F.: Demystifying the draft EU artificial intelligence act. Comput. Law Rev. Int. **22**, 97–112 (2021). https://doi.org/10.9785/cri-2021-220402
58. Vilone, G., Longo, L.: Notions of explainability and evaluation approaches for explainable artificial intelligence. Inf. Fusion **76**, 89–106 (2021)
59. Wachter, S., Mittelstadt, B., Floridi, L.: Why a right to explanation of automated decision-making does not exist in the general data protection regulation. Int. Data Priv. Law **7**(2), 76–99 (2017). https://doi.org/10.1093/idpl/ipx005
60. Żenner, K.: Documents and timelines: the artificial intelligence act (part 3) (2023). https://www.kaizenner.eu/post/aiact-part3

Hyper-Stacked: Scalable and Distributed Approach to AutoML for Big Data

Ryan Dave[1], Juan S. Angarita-Zapata[2,3], and Isaac Triguero[1,4](✉)

[1] School of Computer Science, University of Nottingham, Nottingham, UK
ryand28612@gmail.com
[2] Aimsun SLU, Barcelona, Spain
juan.angarita@aimsun.com
[3] DeustoTech, Faculty of Engineering, University of Deusto, Bilbao, Spain
[4] Department of Computer Science and Artificial Intelligence, Andalusian Research Institute in Data Science and Computational Intelligence (DaSCI), University of Granada, Granada, Spain
triguero@decsai.ugr.es

Abstract. The emergence of Machine Learning (ML) has altered how researchers and business professionals value data. Applicable to almost every industry, considerable amounts of time are wasted creating bespoke applications and repetitively hand-tuning models to reach optimal performance. For some, the outcome may be desired; however, the complexity and lack of knowledge in the field of ML become a hindrance. This, in turn, has seen an increasing demand for the automation of the complete ML workflow (from data preprocessing to model selection), known as Automated Machine Learning (AutoML). Although AutoML solutions have been developed, Big Data is now seen as an impediment for large organisations with massive data outputs. Current methods cannot extract value from large volumes of data due to tight coupling with centralised ML libraries, leading to limited scaling potential. This paper introduces Hyper-Stacked, a novel AutoML component built natively on Apache Spark. Hyper-Stacked combines multi-fidelity hyperparameter optimisation with the Super Learner stacking technique to produce a strong and diverse ensemble. Integration with Spark allows for a parallelised and distributed approach, capable of handling the volume and complexity associated with Big Data. Scalability is demonstrated through an in-depth analysis of speedup, sizeup and scaleup.

Keywords: AutoML · Big Data · Apache Spark · Supervised learning

1 Introduction

Automated Machine Learning (AutoML) is an emerging area that seeks to automate the Machine Learning (ML) workflow from data preprocessing to model

Published by Springer Nature Switzerland AG 2023
A. Holzinger et al. (Eds.): CD-MAKE 2023, LNCS 14065, pp. 82–102, 2023.
https://doi.org/10.1007/978-3-031-40837-3_6

validation [7]. Such automation provides robust AutoML methods that enable people, with either little or no specialised knowledge, to integrate ML solutions into the daily activities of business organisations. The latter is known as the democratisation of ML [7] and it is aligned with the actual purpose of Artificial Intelligence: to learn and act automatically without human intervention [23].

With AutoML, ML solutions are now easily accessible by expert and non-expert ML users. Those methods usually search for the most suitable ML methods and their best hyperparameters (known as the Combined Algorithm Selection and Hyperparameter problem, CASH problem [26]) using an online search strategy; that is, a process takes place after the input dataset has been provided. This online search can be purely based on optimisation approaches that test different promising combinations of algorithms from a predefined base of ML classifiers to minimise or maximise a performance measure [17].

Alternatively, there are AutoML methods whose online search is complemented with learning strategies like meta-learning [27]. These techniques first extract meta-features of the input dataset at hand (e.g., number of instances, features, classes). From these meta-features, meta-learning identifies good candidates of pipeline structures from a predefined knowledge base that stores meta-features for different datasets and ML models that are likely to perform well on them. Then, the candidate models are typically used for a warm-start optimisation approach. In addition, other AutoML methods use ensemble learning to build diverse sets of classifiers from predefined portfolios of ML algorithms [5,11]. These ensemble approaches have proven to be more robust than other AutoML methods, such as the case of Auto-Gluon, which is the state-of-the-art in AutoML thanks to its ensemble learning strategy based on multi-layer stacking [4].

In recent years, both the number of data sources and the scale of such data have increased exponentially [29], wherein such data is referred to by the term'Big Data' [3]. The challenge of computing large amounts of data resides in the simple principle that volume increases complexity [18]. Furthermore, as the training time of an ML algorithm is heavily dependent on the number of data points, efficient ML algorithms must exploit parallelism to achieve sufficient scalability. Without this, operations on large datasets become infeasible.

Open source AutoML solutions fail to handle the size and variety of Big Data [28]. Popular tools are often coupled with ML libraries that rely on centralised data and processing and will only work on a single machine [1]. Consequently, these cannot scale up as a single machine is limited in terms of parallelism due to restrictions in hardware. Some commercial products claim to scale AutoML workloads over multiple nodes; however, many fail to take advantage of superior Big Data frameworks, such as Apache Spark or Dask. Those that are built to run on Spark implement outdated solutions, e.g. TransmogrifAI (grid search) [15], or attempt to integrate it into such frameworks as a second thought (e.g., H2O's Sparkling Water that is an interface between Spark and H2O), which does not fully leverage the framework's abilities. Therefore, a gap exists for a novel solution that implements an efficient, scalable solution that can run natively on Spark. In that context, we take the AutoML state-of-the-art and build further upon the stacking ensemble concept by integrating k-fold cross validation to conceive a Super Learner [10], which is implemented natively on Apache Spark. Thus, we can propose an AutoML method to handle Big Data based on the simple concept that more diversity is introduced with more models, leading to increases in stacking performance.

This article introduces a novel approach to scalable AutoML in Big Data, named Hyper-Stacked. It combines the strength of the Super Learner stacking ensemble, the efficiency of Greedy K-Fold hyperparameter optimisation, and Apache Spark's scalability. The main contributions of this work are:

- A novel AutoML component design, Hyper-Stacked, is presented. This design efficiently integrates greedy k-fold and the Super Learner stacking approach to produce a high-performant ensemble. The approach automates the search for a diverse set of models and combines them to bolster the overall performance. Results showed that the ensemble consistently outperforms the best-performing individual model.
- This approach was implemented natively on Apache Spark to produce a distributed and scalable model capable of dealing with the volume, variety and complexity associated with Big Data. To validate these claims, parallelism and scalability were critically evaluated in speedup, sizeup and scaleup experiments. Results showed that Hyper-Stacked can handle data growth significantly better than sequential processing and single node parallelisms.

The rest of this paper is structured as follows. Section 2 presents background and related work about AutoML with emphasis on the CASH problem, Ensemble learning, Meta-learning, and Spark. Then, Sect. 3 introduces Hyper-Stacked. Section 4 exposes the experimental framework, and Sect. 5 presents and analyses the results obtained. Finally, conclusions are discussed in Sect. 6.

2 Background and Related Work

2.1 Problem Definition

In AutoML, when algorithm selection is combined with hyperparameter optimisation, it is often referred to as the CASH problem. It can be defined as follows [5]. Let γ denote the loss that an algorithm $A^{(j)}$ (where j is just an identifier for the algorithm) returns on $D^{(i)}_{valid}$ when trained on $D^{(i)}_{train}$, with hyperparameters λ. Given the set of algorithms A, their respective hyperparameters Λ, and sets of cross validation folds D_{train} and D_{test}, CASH focuses on determining the joint algorithm $A^{(j)}$ and hyperparameter $\Lambda^{(j)}$ that minimises the loss γ.

$$A^{*}, \lambda = \underset{A^{(j)} \in A, \lambda^{(j)} \in \Lambda}{argmin} \frac{1}{K} \sum_{i=1}^{k} \gamma\left(A^{(j)}_{\lambda}, D^{(i)}_{train}, D^{(i)}_{test}\right) \quad (1)$$

As an alternative to using the *argmin* operator with respect to a single algorithm $A^{(j)}$, we can instead construct a set E, where E represents an ensemble.

In this instance, more than a single algorithm can be chosen and individual predictions are combined to produce a final output. The new representation is presented below.

$$A^*, \lambda = \underset{E \subseteq A, H \subseteq \Lambda}{argmin} \frac{1}{K} \sum_{i=1}^{k} \gamma \left(E_H, D_{train}^{(i)}, D_{test}^{(i)} \right) \tag{2}$$

2.2 CASH Methods

This section introduces the most representative approaches to solve the CASH problem presented in Eq. 1.

Black Box Optimisation Approaches

The most basic optimisation technique for hyperparameter tuning is grid search. It involves performing an exhaustive search given a subset of the hyperparameter space [13]. For example, an algorithm may require the tuning of 3 distinct parameters. Parameter values are selected in uniform or exponential intervals to form sets of candidate parameters. The algorithm then iterates over each possible combination of the three parameter subsets to return the best. In this context, a clear limitation exists, as it remains essentially a brute force approach. As the number of distinct parameters increase, the number of possible combinations will increase exponentially and therefore is not viable for larger datasets.

An improvement on grid search emerged, known as random search. Random search aims to trial a number of random hyperparameter configurations and has been proven to return models that are equivalent or better, within a fraction of the computation time [2]. The method of randomly sampling the space, rather than brute force allows the exploration of a lager search space given the same computational budget.

Alternatively, Bayesian Optimization (BO) provides an adaptive approach to black box optimisation. It works by first building a surrogate model (a cheaper approximation function). Then, the uncertainty in that surrogate is then evaluated. Finally, desirable sample spaces are proposed by an acquisition function defined from this surrogate. As samples are selected, the surrogate is updated iteratively and the uncertainty is re-quantified. Auto-Sklearn, a python based AutoML library, adopts this as the primary optimisation method. However, in the context of Big Data, the standard approach to BO fails to succeed in high dimensional environments and new approaches are required [16]. AutoML tools, such as Auto-SKlearn, are yet to incorporate these new approaches.

Multi-fidelity Approaches

Multi-fidelity optimisation seeks to speed up the optimisation process by using performance estimates from lower-fidelity models [14]. In general, these techniques rely on first training lower-fidelity models (e.g., models trained with a low computational budget) to reveal promising configurations. These models can then be allocated additional computational budget to continue training and give a "higher" fidelity model. It is relevant to mention that in this approach, the computational budget must be easily measurable (e.g., training time).

State-of-the-art multi-fidelity methods are built on the successive halving algorithm that was first proposed by Karnin et al. [8]. Successive halving initially randomly samples a set of hyperparameter configurations and models are trained with a specified budget and evaluated to return a metric. The configurations are ranked based on the metric, and the worst performing half is discarded. Then, rounds of successful halving are performed until one configuration remains. In this sense, successive halving can give the effect of early stopping which can heavily reduce computation.

Other relevant methods under the multi-fidelity approach are Hyperband and Greedy k-fold. Hyperband [12] works on the main principle of multi-fidelity by randomly distributing budget values and performing rounds of successive halving. This allows the exploration of different convergence behaviours and ensures that configurations are not discarded too early. On the other hand, Greedy k-fold [24] applies a similar mechanism to k-fold cross validation. It works by first evaluating a single fold for all configurations, then proceeds with a greedy approach. Again, low fidelity models are initially trained, but rather than performing rounds of evaluations, it pursues only the most promising candidate model. Evaluations in the original paper show results to perform significantly better than the successive halving approach, on average 70% faster.

The main limitation of multi-fidelity evaluation is that using low fidelity approximations to perform early stopping may remove an optimal configuration. This is not usually seen as a concern for most, as the performance speedup often heavily outweighs the approximation error [7]. Efficiency is important to combat volume when limited to a computational budget.

2.3 Ensemble Learning

This section provides an overview of the existing literature surrounding ensemble learning to solve Eq. 2.

Majority Voting and Stacking

Majority voting remains one of the simplest methods in ensemble learning. Within this ensemble approach, a set of base (heterogenous or homogenous) classifiers are all trained on the same data set. When making a prediction, every data point results in a prediction from every classifier. Then, a final prediction is made by selecting the class that had the most "votes" from the set of classifiers.

Stacking learning builds upon the weighted ensemble by training a meta-learner on model predictions. Specifically, a set of base learners are first trained on a training set and each output a prediction. Afterward, predictions are aggregated to construct a new dataset where each data point holds the predictions from each base model. Thus, the meta-learner can be trained to learn complex behaviours of the base models. Rather than a user-given weighting, the meta-learner will determine the importance of base models empirically.

A significant limitation exists with the stacking method. It is easily susceptible to overfitting. If a single base model is seen to overfit, the meta-learner may result in a heavy reliance on that same model and will harm the overall model performance. This limitation has been reduced by techniques that centre around the use out of fold predictions, i.e. base models will predict on unseen data. These methods commonly produce the highest accuracy out of any individual model or ensemble methods and hence remain popular [19], but are often overlooked due to their added complexity. Stacking methods are especially applicable here, as research has shown that stacking can handle high-dimensional datasets [22], a common attribute of Big Data.

Super Learner

Super Learner builds further upon the stacking ensemble by integrating k-fold cross validation, such as is done in H2O's AutoML framework [11]. In the H2O implementation, heterogenous ensembles (different types of learners) are used. Then, a mix of random and fixed grids are used to diversify, and two super learners are trained. One of the super learner is optimised for model performance by including all model configurations. The latter is based on the simple concept that more diversity is introduced with more models, leading to an increase in stacking performance.

Meanwhile, the second super learner is optimised for production uses. In this case, this super learning considers only the best model from each algorithm to output faster predictions [11]. The benefit of this twofold approach is that the two super Learners perform "asymptotically as well as the best possible weighted combination" [21], and therefore both will perform at least as effectively as the best performing base model.

K-Fold Repeated Bagging

In the context of AutoML, there is another competitive approach coupled with ensemble learning, which is K-fold repeated bagging. This approach can be observed in the AutoML framework named Auto-Gluon [4]. Auto-gluon implements an improved method to prevent overfitting in their approach to multi-layer stacking. Multi-layer stacking passes predictions through multiple sets of models, rather than a single set of base models. These ensembles have the potential to perform better than single layer models, however tend to suffer more from overfitting as the effect is amplified through layers.

To combat this, k-fold repeated bagging was introduced. K-fold repeated bagging makes additional o-of-fold predictions on n different random partitions of the training data and takes an average. The value of n is determined by dividing the total allotted time between an estimate of the time taken for a given partition. The overall approach is therefore heavily dependent on the given budget. It is important to say that Auto-gluon was shown to outperform H2O's framework and 99% of data scientists in a Kaggle benchmark, however it remains a centralised approach. In a distributed context, hurdles such as the non-trivial task of time estimation need to be considered.

2.4 Meta-learning

This section focuses on aspects of meta-learning that rely on learning from prior tasks rather than outputs coming within the same task. The interest reader can be referred to [27] to consult other approaches of meta-learning within AutoML. To learn from prior tasks, a learning algorithm may be run a number of times and the related data from the training (model evaluations, hyperparameter configurations, training time etc.) can be stored as features in a new dataset [7]. It raises the layer of abstraction above traditional ML in two main approaches.

The first approach is learning from model evaluations, which has demonstrated effective results in warm-start optimisation [25]. Warm-start optimisation removes the exploration of search spaces that have been explored in similar tasks and provides a starting point for hyperparameter optimisation. Conversely, the second approach is learning from task properties, which looks at the CASH problem from a different perspective. Instead of tuning every algorithm, the search space can be reduced by selecting a few of the most promising. Learning tasks can be characterized by meta-features and a meta-model can provide a way of associating these meta-features to a subset of algorithms based on prior experience. This has produced promising results in the context of Big Data, when combined with multi-fidelity optimisation [1]. The primary limitation is that only a finite amount of information can be captured in the meta features [27].

2.5 Spark

Apache Spark is an open-source, distributed processing framework used for large scale workloads. This section provides relevant concepts related to Apache Spark.

Data Locality

Typically, to deal with larger datasets, the user may be required to scale their resources, that is, additional memory or cores may be added. Nevertheless, this approach of scaling up on a single node, known as vertical scaling, fails to continue to scale as it is limited to the hardware capacity and eventually will reach a hard upper limit.

An alternative paradigm to vertical scaling is horizontal scaling, which allows nodes to be added to an existing pool of resources. As more machines can be introduced, the user is no longer bound by the hardware limits. The traditional approach to high performance computing (HPC) relies on communication between storage nodes and compute nodes. In this sense, a bottleneck exists between storage and compute in data intensive jobs, as network I/O becomes the limiting factor and node computation remains low and unused [6].

The bottleneck mentioned above can be resolved by distributing data across the compute nodes and storing it on local disks. This allows each node to perform operations on their subset of the data, reducing cross-switch network traffic and leading to a performance gain [6]. This is the concept of data locality and is essential to the scalability of data processing. In summary, regardless of how the code is written, scalability is also heavily dependent on the architecture and where your data is situated. In this context, Spark implements data locality to facilitate the efficient compute of operations.

Parallelism

When performing parallel operations on shared data, the data itself must also support parallelism. Concurrent, or parallel data structures allow data to be accessed by multiple threads. Spark implements resilient distributed datasets (RDDs) to accomplish data parallelism by organising data as a collection of partitions that can be held over one or more machines [30]. This can then be operated in parallel via a low-level API, through actions and transformations. For example, data may be partitioned into 20 distinct partitions, across 2 nodes in a cluster. Spark can then run a single task per partition in that RDD concurrently, up to the number of cores in that cluster. If each machine has 4 cores, it is possible to run 8 concurrent tasks on 8 partitions. This allows scalability as tasks can be run independently across hundreds of nodes in a cluster.

Spark ML

Spark ML is Apache Spark's ML library that implements ML algorithms and utility functions. This scalable library is in some sense a basic AutoML library. It allows a pipeline to string together pre-processing operations and a Cross-Validator class to perform grid search and return the best model. There currently exists no available ensemble learning methods, aside from common ML algorithms such as Random Forest and Gradient Boosted Trees that are ensembles as themselves.

Spark ML is important to scalability as implementations overcome the curse of modularity. The curse of modularity states that there is an assumption behind ML algorithms that the data can fit, in its entirety, in memory on a single machine [9]. In other words, some algorithms have been developed using modular strategies that, when used outside of the scope of in-memory data, will break. This explains why many popular libraries are inherently unable to scale. Opposite to such situation, Spark ML implements these algorithms in a way that can be broken down and distributed across multiple machines.

3 Hyper-Stacked: A Scalable and Distributed Approach to AutoML for Big Data

In this section, we introduce Hyper-Stacked[1]. First, Sect. 3.1 motivates the need for the proposed method. Then, Sect. 3.2 presents the general architecture of Hyper-Stacked and details of its inner workflow.

3.1 Motivation

As we stated before, the core aims of existing AutoML methods are (1) High Computational Efficiency, (2) Good Performance, and (3) Reduced Human Interaction. Nevertheless, the current solutions suffer from the following issues, which have motivated the design of Hyper-Stacked.

- **Centralised data approach:** Open source AutoML solutions fail to handle the size and variety of Big Data [28]. Popular tools are often coupled with ML libraries that rely on centralised data and processing and will only work on a single machine [1]. Consequently, these are unable to scale as a single machine is limited in terms of parallelism due to restrictions in hardware. Some commercial products claim to scale AutoML workloads over multiple nodes; however, many of them fail to take advantage of superior Big Data frameworks, such as Apache Spark.
- **Optimisation and reduced scalability:** When optimisation deals with small- or medium-size datasets, many algorithms can be generated, tuned and tested because the complexity of the learning task at hand is influenced by its data size. On the other hand, the latter may stop happening as the data size grows. In this scenario, it is harder to tune and test multiple algorithms, as the optimisation becomes expensive and the set of candidate ML methods could decrease, affecting the final solutions' performance.

Considering the motivations presented above, we want to conceive a new AutoML method that relies on a distributed approach. In this sense, we introduce Hyper-Stacked, a new AutoML method based on greedy k-fold and Super Learner stacking to produce a high-performant ensemble. The approach automates the search of a diverse set of models and combines them to bolster the overall performance, which allows to path the way towards three main goals of AutoML: (1) High Computational Efficiency, (2) Good Performance, and (3) Reduced Human Interaction. Firstly, the Super Learner was chosen as the base mechanism as it hosts the high performance of stacking (goal 2) and reduces overfitting. Secondly, The overarching concept of Hyper-Stacked focuses on finding strong heterogeneous learners amongst the search space to achieve a high stacking performance, while keeping the number of base models low to remain efficient (goal 1). Thirdly, Hyper-Stacked aims to succeed in both through effective hyperparameter optimisation. In doing so, we are guaranteed to automatically (goal 3) return the best individual model (the aim of the CASH problem) and an ensemble that returns an equal or higher predictive performance.

[1] https://github.com/jsebanaz90/Hyper-Stacked.

In summary, Hyper-Stacked combines the strength of the Super Learner stacking ensemble, the efficiency of Greedy K-Fold hyperparameter optimisation, and the scalability of Spark. To the best of our knowledge, at this time, the Super Learner has not yet been implemented on Apache Spark and no solution yet exists that combines the Super Learner and Greedy k-fold optimisation.

3.2 Hyper-Stacked's Design and Workflow

Hyper-Stacked was implemented as a Scala package that can run on top of a distributed Spark cluster. We also used the MLlib, to make practical machine learning scalable and manageable. From this library, we selected the following classifiers to be part of the algorithms that can be part of the ensemble built in the inner structure of Hyper-Stacked: Random Forest (RF), Gradient Boosted Trees (GBT), LinearSVC (LSVC), Logistic Regression (LR), and Naïve Bayes (NB). Besides, the meta-learner is chosen from this portfolio of methods.

The architecture of Hyper-Stacked is shown in Fig. 1. Besides, pseudocode 1 presents the step-by-step followed by Hyper-Stacked's Super learner and Greedy k-fold components; wherein lines 1–19 illustrate the generation of the base models and the meta-learner selection, lines 20–25 represent the training process of a Hyper-Stacked model, and lines 26–28 summarize how the final predictions are made. More details of this process are presented as follows.

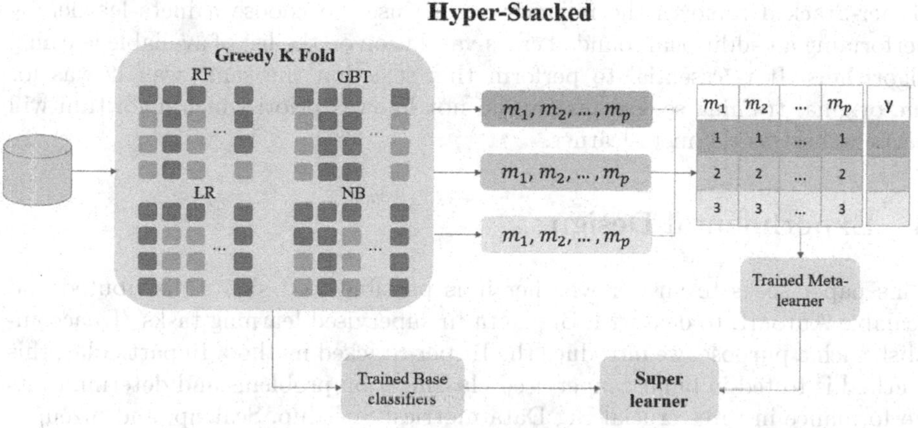

Fig. 1. General architecture of Hyper-Stacked based on greedy k-fold and Super Learner stacking

Hyper-Stacked uses a list of descriptor tuples instead of a set of candidate models. Each tuple contains a learning algorithm and the number of random hyperparameter configurations to generate. Random parameters are generated and fed into the greedy k-fold method for each algorithm type. The latter allows us to run greedy k-fold multiple times, returning different model types. The specified number is important to vary the number of configurations for the different hyperparameter search spaces an algorithm may have.

The result of the original greedy k-fold would contain the hyperparameter configuration of the best model, and a model could then be trained on the full dataset with those optimal hyperparameters. In Hyper-Stacked, we return models that were trained during hyperparameter optimisation, reuse them and the k-fold data splits later. Instead of a configuration, a list of trained models is returned, one for each fold. The final result of greedy k-fold will be a list of lists of fold models, where each element represents a different hyperparameter configuration that was fully evaluated; this is flattened for ease of iteration. In addition, to minimise the movement of data, each fold is considered one at a time, and we iterate through each hyperparameter configuration before continuing to the next fold. Each fold model will output a prediction, and we aggregate them in the same way as the original to train the meta learner.

To output a prediction from the Super Learner, features are first passed into each base model to construct a set of features for the meta-model. The meta-model can then produce a final output based on the results of the base models. The meta-classifier can therefore learn complex behaviours of the base classifiers. In this case, rather than a user-given weighting, as defined in the weighted ensemble, the meta-classifier will empirically determine the importance of base classifiers.

Finally, Pseudocode 1 shows the automatic selection of the meta-learner. Hyper-Stacked removes the reliance on the user to choose a meta-learner by performing an additional round of cross-validation on the list of available learning algorithms. It is essential to perform this selection the same way it was for the original training set because we do not know a priori which algorithm will perform best as the meta-learner.

4 Experimental Design

This paper seeks to answer whether it is possible to design a distributed and scalable AutoML to deal with Big Data in supervised learning tasks. To accomplish such a purpose, we introduce the Hyper-Stacked method. In particular, this method is tested in binary supervised classification problems and determines its performance in three crucial Big Data metrics: Speedup, Scaleup, and Sizeup.

This section shows the factors and issues related to the experimental study. First, we provide details of the problems chosen for the experimentation (Sect. 4.1). Then, we introduce details about the big data architectures considered to test Hyper-Stacked in Sect. 4.2. Finally, we present the three experiments on Speedup, Scaleup, and Sizeup metrics.

4.1 Binary Supervised Learning Problems

For this experimentation we chose four representative datasets, which are shown in Table 1. These datasets represent binary classification problems and their composition vary on size and dimensions.

Algorithm 1: Super Learner with Greedy k-Fold

Data:
$D_{train} \longleftarrow$ training dataset
$D_{test} \longleftarrow$ test dataset
$n \longleftarrow$ number of k folds
$T \longleftarrow$ list of tuples *(a,r)* where a is a learning algorithm and r is the number of random parameters to generate for that algorithm
$M \longleftarrow$ a list of learning algorithms included in meta-learner selection
Result: Set of predictions corresponding to input D_{test}

k_folds $\longleftarrow$ divide D_{train} into n number of approximately equal partitions;
$L_{best} = []$;
for *t in T* **do**
  $C_a \longleftarrow$ generate random candidate models *(a,r)*;
  $\lambda_{best} \longleftarrow$ **do** Greedy K-Fold(C_a) where $|\lambda_{best}| \geq 1$;
  append λ_{best} to L_{best};
end
flatten L_{best};
for k_i *in k_folds* **do**
  $k_{valid} \longleftarrow k_i$;
  $k_{train} \longleftarrow$ remaining k folds;
  for *l in* L_{best} **do**
    $OOF \longleftarrow$ predict on k_{valid} with l and store result and label;
  end
  $i_{OOF} \longleftarrow$ concatenate OOF (D_{train} *. length x l . length*);
end
meta-features $\longleftarrow$ union all i_{OOF};
meta_k_folds $\longleftarrow$ divide *meta-features* into n number of approximately equal partitions;
$m_{best} \longleftarrow$ **do** K-fold Cross-Validation(*meta_k_folds*);
base_models = [];
for *l in L* **do**
  base_model $\longleftarrow$ train l on D_{train};
  append *base_model* to *base_models*;
end
meta_model $\longleftarrow$ train m_{best} on *meta-features*;
base_layer_output $\longleftarrow$ **for** *base_model* in *base_models* **do** transform D_{test} with *base_model*;
predictions $\longleftarrow$ transform *base_layer_output* with *meta_model*;
return *predictions on* D_{test}

Table 1. Binary classification datasets

Dataset	Number of Training instances	Number of features
FLIGHT	516513	29
SUSY	3500066	18
HEPMASS	4899792	28
HIGGS	7701355	28

It is important to mention that the number of instances in these datasets may be lower than traditionally seen in current real-world Big Data. The intention is to keep the number of instances in a permissible range to run on a single node and return in a 'reasonable' time. In reality, a single node would take a significant amount of compute time and could take days, weeks or months. With mid-large size datasets, we can run trials comfortably on different-sized clusters to demonstrate the scalability and strength of the approach. These datasets are deemed suitable for ML problems by the research community and are commonly used in benchmarks and research papers.

4.2 Experimental Setups

All experiments done with Hyper-Stacked were set up on the Databricks platform, which allows clusters of variable size and configurations to be easily instantiated. To measure the time of experiments, we implemented a logger object. Specifically, the time before is recorded, then the function is executed, and the time is once again recorded to calculate the difference between them finally. The log function is only used around larger functions, as small operations are likely to return inaccurate results due to lazy evaluation. A run-time limit of ten hours was applied to all trials, and the leader board of model performances was recorded for every trial.

The specifications for the three chosen cluster sizes to be used in these experiments are summarised in Table 2. All cluster sizes contain a single driver node with 4 cores with 14GB memory. The memory of each worker remains the same (28 GB).

Table 2. Specifications for the Spark clusters to be used in the experiments

Cluster	Number of workers	Number of cores per worker	Total number of cores
1	1	1	1
2	1	8	8
3	3	8	24

4.3 Experiment Speedup, Sizeup, Scaleup

In this set of experiments, the primary aim is to demonstrate the scalability, efficiency and effective parallelism of Hyper-Stacked. Therefore, we step up experiments around speedup, sizeup and scaleup metrics to accomplish such an aim. These experiments focused on measuring the relative change as an element of the system changes. The three experiments are described as follows.

- In speedup, the size of the data remains constant and the number of cores are increased. Speedup shows how much faster the same data can be processed with n cores instead of 1. This metric can be estimated by calculating the ratio of the time taken for a sequential execution versus a parallel execution. For the experiments carried out in this work, the data is kept at 33% as this value was achievable by all cluster sizes.
- In sizeup, the number of cores remains constant and the size of the data is increased. Sizeup allows us to see how time scales with increasing intervals of data size. This metric can be found by calculating the ratio of execution time between an initial dataset versus a dataset n-times larger. For this study, the intervals were chosen to be 10%, 20%, 40% and 80%.
- In scaleup, both the number of cores and size of the data are increased (by the same factor). Scaleup combines the two to measure how a program performs as a system gets larger and has to process larger datasets. The following configurations were chosen: 100/24 ($\approx$4.16%) with 1 core, 100/3 ($\approx$33.33%) with 8 cores and 100% with 24 cores.

5 Analysis of Results

This section analyses the experimentation results from the following angles.

- Speedup: to evaluate the ability of Hyper-Stacked of improving its execution time through parallelism (using additional cores).
- Sizeup: to assess the ability of Hyper-Stacked to handle increasing amounts of data within a parallel environment.
- Scaleup: to evaluate the ability of Hyper-Stacked to handle both increase the amount of data and the size of the system. It can be found by calculating the ratio of execution time between an initial dataset and system, versus a dataset m-times larger with an m-times larger system.

5.1 Speedup

Figures 2a and 2b show the speedup for each dataset with a single core, eight cores, and twenty four cores. As it can be seen, the speedup achieved for the FLIGHT dataset was only 1.899 despite parallelising over 24 cores. Furthermore, there was almost no speedup between eight cores and twenty-four (speedup increase of 0.084). In contrast, the speedup achieved for the HIGGS dataset was 4.413, but again, despite the addition of sixteen cores, the speedup increase was

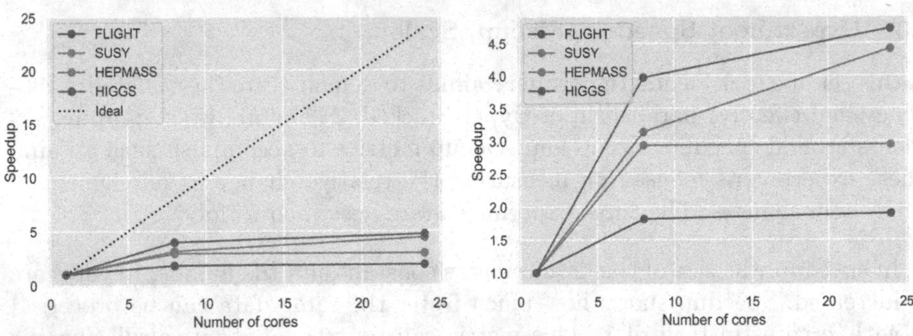

(a) Hyper-Stacked vs Linear speedup (b) Hyper-Stacked's speedup by dataset

Fig. 2. Speedup analysis in supervised binary classification problems

1.125. This can be clearly seen in Fig. 2a where comparisons were done againts linear speedup.

Through these results, it is hard to pinpoint the main cause of the lack of speedup. It should be noted that between the datapoints at eight and twenty-four cores is the introduction of a distributed cluster (increase from 1 to 3 worker nodes). Despite this, the speedup is not expected to be as low as observed in a Spark application. Through additional investigation, the training of base models (layer one models) and meta models (layer two models) were compared in terms of speedup. Figure 3 shows that the speedup of these are almost identical, failing to identify any existing limiting component.

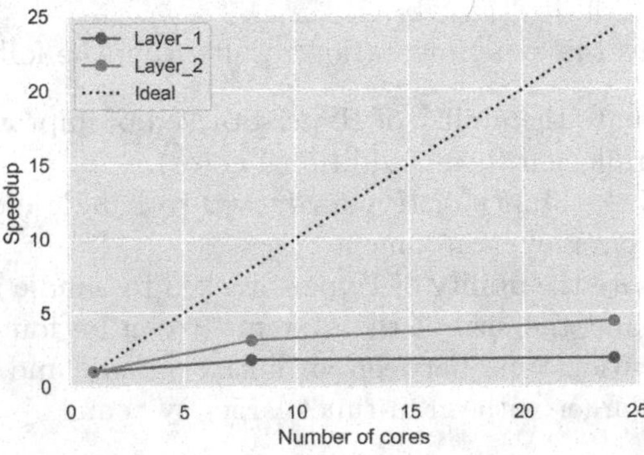

Fig. 3. Speedup comparison of Layer 1 and Layer 2 models for HIGGS dataset

Furthermore, Fig. 2b shows a clearer comparison between the datasets. A trend can be clearly seen as the larger datasets (HIGGS and HEPMASS) achieve a significantly better speedup than the smaller datasets (FLIGHT and SUSY). In fact, they are sorted in size order. As a result of these findings, an additional run was performed, which is shown in Fig. 4.

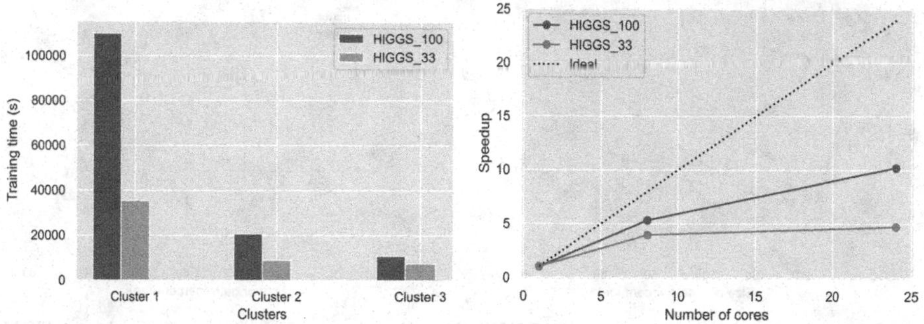

(a) Training time comparison between clusters in HIGGS dataset

(b) Speedup comparison in HIGGS dataset

Fig. 4. Speedup experiment with HIGGS dataset using different data sizes

In this experiment, the size of the data was increased from 33% to 100% of the dataset. As seen in Figs. 4a and 4b, a greater difference is seen when using a larger amount of data. The speedup analysis in Fig. 4a shows a speedup increase from 4.413 (seen previously) to 10.210. This can be considered a reasonable speedup and significantly better than seen when using 33% of the data. Therefore, we can say the speedup can be heavily dependent on the size of the data, and clearly, the amount of the data used in the initial experiments was insufficient. It is possible that some rate-limiting factors exist within the component; however, without additional experiments using larger datasets, it is difficult to determine at this time.

5.2 Sizeup

The results in Figs. 5 and 6 depict the sizeup for each dataset. Specifically, Fig. 5a shows the sizeup results for the smallest of the four datasets (FLIGHT dataset). It can be observed that Cluster 1 can execute eight times the size of the original data with a size up of 3.753. In comparison, Cluster 3 executes eight times the size of the original data with a size up of 1.964. The latter means that although the data increased eightfold, the execution time of Cluster 1 increased by around a factor of 4, whereas Cluster 3 only increased by a factor of 2.

In the other datasets, the overall data size increases and Cluster 1 is unable to execute within the given budget. Using SUSY, Cluster 1 is able to execute 40% of the data with a sublinear size up of 3.753. Unfortunately, problems arise when executing HEPMASS and HIGGS (Figs. 6a and 6b). Due to the 10-hour constraint, we are only able to gather two datapoints, 10% and 20%. For the 20% size up, cluster 1 begins to exceed a linear size up for HEPMASS and HIGGS achieving a size up 2.617 and 2.310 respectively. The gradients of the lines, shown in Figs. 6a and 6b, show that as the datasets become larger, systems with sequential processing are unable to deal with the growth of data.

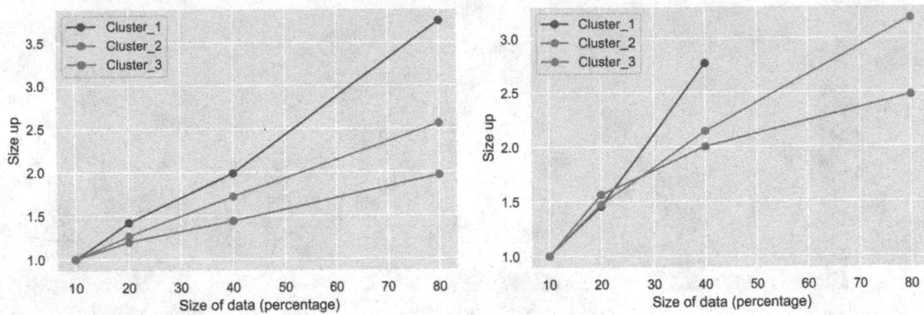

(a) Sizeup comparison for the FLIGHT dataset (b) Sizeup comparison for the SUSY dataset

Fig. 5. Sizeup comparison for FLIGHT and SUSY datasets

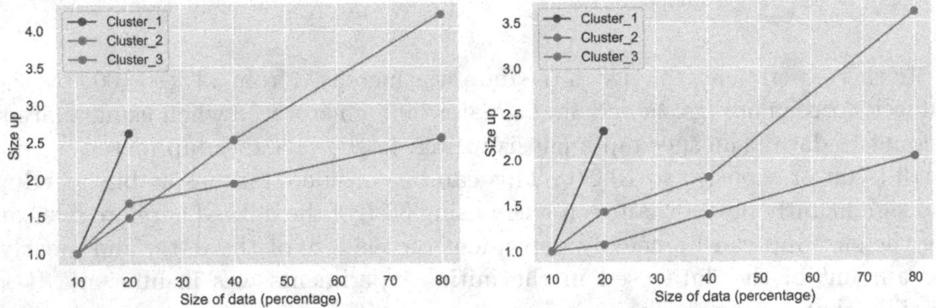

(a) Sizeup comparison for the HEPMASS dataset (b) Sizeup comparison for the HIGGS dataset

Fig. 6. Sizeup comparison for HEPMASS and HIGGS datasets

Additionally, both clusters 2 and 3 are able to return the results for all datasets in a reasonable time. Size up values between clusters 2 and 3 are similar with smaller datasets, as seen in Figs. 5a and 5b. These two clusters are also consistently similar at low data percentages with all of the datasets. However, the difference is seen when the growth of the larger datasets reaches 80%. In the largest dataset (HIGGS), shown in Fig. 6b, an eightfold data size increase caused the execution time of Cluster 2 to increase by a factor of 3.602, whereas Cluster 3 only increased by a factor of 2.024. This shows that the parallelism within Hyper-Stacked allows greater sized clusters to effectively handle the growth of data.

5.3 Scaleup

Figure 7 shows the scaleup analysis for Hyper-Stacked. It displays a comparison to the ideal scaleup value. The ideal value is where the execution time is kept equal as the system and data size grows by the same factor, and in reality is unattainable. Although Hyper-Stacked does not achieve an 'ideal' result, all results appear to taper off in an acceptable range between 0.684 and 0.792. The decrease between 8 and 24 cores is assumed to be partially down to the fact the processing is now distributed and communication (shuffle read and write) between workers is necessary.

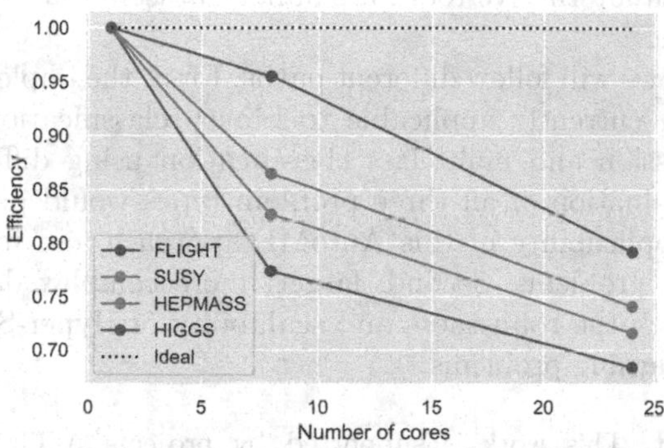

Fig. 7. Hyper-Stacked vs ideal scaleup by dataset

From the graphs presented, we can see that the scaleup exhibits a similar trend to speedup. The lines appear sorted in size order, HIGGS achieving a significantly better efficiency value than FLIGHT. FLIGHT achieves a scale up of 0.774 then 0.684, whereas HIGGS achieves a better result of 0.995 then 0.792. Similar to speedup, it is hard to determine the true metric through these experiments as the sizes of the datasets chosen seem to be insufficient. Nevertheless, the findings show that Hyper-Stacked is still shown to be scalable when increasing the size of the data, number of cores and number of nodes.

6 Conclusions

In this work, we introduced Hyper-Stacked. This new AutoML method combines the strength of the Super Learner stacking ensemble, the efficiency of Greedy K-Fold hyperparameter optimisation, and the scalability of Apache Spark. Hyper-Stacked was implemented natively on Apache Spark to produce a distributed and scalable model capable of dealing with the volume, variety and complexity of Big Data. Parallelism and scalability were critically evaluated in speedup, sizeup and scaleup through different experiments to validate the general architecture of Hyper-Stacked. The experiments focused on binary classification problems, using datasets varying in size and dimensions.

From the results obtained, we extracted interesting conclusions. A limitation was uncovered during the speedup experiments in Sect. 5.1, where the addition of workers resulted in a minimal speedup. This can be explained by Amdahl's law, which states that the performance increase from parallelisation cannot exceed the inverse of the non-parallelisable element of work [20]. The mechanism that Spark uses to distribute work (setting up a job on the driver, scheduling and data shuffles) results in overhead, i.e. a nonzero amount of non-parallelisable work. The relative amount of parallelisable work is then bounded by the size of the dataset, as larger datasets reduce the significance of the overhead. As Hyper-Stacked was developed to perform efficiently with Big Data, it may be preferable to use an alternative tool for smaller datasets that fit in the memory of a single machine.

In future work, we will follow different paths. First, the implementation itself, despite only being currently applicable to binary classification, can be easily extended to regression and multiclass classification using different Spark ML algorithms. An evaluation of all three problem types would provide additional insight into the applicability of this AutoML approach and its ability to generalise to different problems. Second, larger, more complex datasets could be approached to check the robustness and scalability of Hyper-Stacked in multiclass supervised learning problems.

Acknowledgement. This work is supported by projects A-TIC-434-UGR20 and PID2020-119478GB-I00).

References

1. Abd Elrahman, A., El Helw, M., Elshawi, R., Sakr, S.: D-SmartML: a distributed automated machine learning framework. In: 2020 IEEE 40th International Conference on Distributed Computing Systems (ICDCS), pp. 1215–1218 (2020). https://doi.org/10.1109/ICDCS47774.2020.00115
2. Bergstra, J., Bengio, Y.: Random search for hyper-parameter optimization. J. Mach. Learn. Res. **13**(10), 281–305 (2012). https://jmlr.org/papers/v13/bergstra12a.html
3. Christiansen, B.: Ensemble averaging and the curse of dimensionality. J. Clim. **31**(4), 1587–1596 (2018)
4. Erickson, N., et al.: Autogluon-tabular: robust and accurate automl for structured data. arXiv preprint arXiv:2003.06505 (2020)
5. Feurer, M., Klein, A., Eggensperger, K., Springenberg, J., Blum, M., Hutter, F.: Efficient and robust automated machine learning. In: Cortes, C., Lawrence, N.D., Lee, D.D., Sugiyama, M., Garnett, R. (eds.) Advances in Neural Information Processing Systems, pp. 2962–2970. Curran Associates, Inc. (2015)
6. Guo, Z., Fox, G., Zhou, M.: Investigation of data locality in MapReduce. In: 2012 12th IEEE/ACM International Symposium on Cluster, Cloud and Grid Computing (CCGRID 2012), pp. 419–426 (2012). https://doi.org/10.1109/CCGrid.2012.42
7. Hutter, F., Kotthoff, L., Vanschoren, J. (eds.): Automated Machine Learning: Methods, Systems, Challenges. Springer, Heidelberg (2018). https://doi.org/10.1007/978-3-030-05318-5
8. Karnin, Z., Koren, T., Somekh, O.: Almost optimal exploration in multi-armed bandits. In: Proceedings of the 30th International Conference on International Conference on Machine Learning, pp. III-1238–III-1246. JMLR.org (2013)

9. Kumar, K.A., Gluck, J., Deshpande, A., Lin, J.: Hone: "scaling down" hadoop on shared-memory systems. Proc. VLDB Endow. **6**(12), 1354–1357 (2013). https://doi.org/10.14778/2536274.2536314
10. van der Laan Mark, J., Polley, E.C., Hubbard, A.E.: Super learner. Stat. Appl. Genet. Mol. Biol. **6**(1), 1–23 (2007)
11. LeDell, E., Poirier, S.: H2O AutoML: scalable automatic machine learning. In: 7th ICML Workshop on Automated Machine Learning (AutoML) (2020)
12. Li, L., Jamieson, K., DeSalvo, G., Rostamizadeh, A., Talwalkar, A.: Hyperband: a novel bandit-based approach to hyperparameter optimization. J. Mach. Learn. Res. **18**, 1–52 (2018)
13. Liashchynskyi, P.B., Liashchynskyi, P.B.: Grid search, random search, genetic algorithm: a big comparison for NAS. ArXiv abs/1912.06059 (2019)
14. March, A., Willcox, K.: Constrained multifidelity optimization using model calibration. Struct. Multidisc. Optim. **46**, 93–109 (2012). https://doi.org/10.1007/s00158-011-0749-1
15. Moore, K., et al.: TransmogrifAI (2017). https://github.com/salesforce/TransmogrifAI
16. Moriconi, R., Deisenroth, M.P., Sesh Kumar, K.S.: High-dimensional Bayesian optimization using low-dimensional feature spaces. Mach. Learn. **109**(9–10), 1925–1943 (2020). https://doi.org/10.1007/s10994-020-05899-z
17. Olson, R.S., Bartley, N., Urbanowicz, R.J., Moore, J.H.: Evaluation of a tree-based pipeline optimization tool for automating data science. In: Proceedings of the Genetic and Evolutionary Computation Conference 2016, pp. 485–492 (2016)
18. Parker, C.: Unexpected challenges in large scale machine learning. In: Proceedings of the 1st International Workshop on Big Data, Streams and Heterogeneous Source Mining: Algorithms, Systems, Programming Models and Applications, BigMine 2012, pp. 1–6. Association for Computing Machinery, New York (2012). https://doi.org/10.1145/2351316.2351317
19. Pavlyshenko, B.: Using stacking approaches for machine learning models. In: 2018 IEEE Second International Conference on Data Stream Mining and Processing (DSMP), pp. 255–258 (2018). https://doi.org/10.1109/DSMP.2018.8478522
20. Pei, S., Kim, M.S., Gaudiot, J.L.: Extending Amdahl's law for heterogeneous multicore processor with consideration of the overhead of data preparation. IEEE Embed. Syst. Lett. **8**(1), 26–29 (2016). https://doi.org/10.1109/LES.2016.2519521
21. Polley, E.C., van der Laan, M.J.: Super learner in prediction. U.C. Berkeley Division of Biostatistics Working Paper Series. Working Paper 266 (2010). https://biostats.bepress.com/ucbbiostat/paper266
22. Sharma, S.R., Singh, B., Kaur, M.: A novel approach of ensemble methods using the stacked generalization for high-dimensional datasets. IETE J. Res. 1–16 (2022). https://doi.org/10.1080/03772063.2022.2028582
23. Song, H., Triguero, I., Özcan, E.: A review on the self and dual interactions between machine learning and optimisation. Progr. Artif. Intell. **8**, 1–23 (2019)
24. Soper, D.S.: Greed is good: rapid hyperparameter optimization and model selection using greedy k-fold cross validation. Electronics **10**(16), 1973 (2021). https://doi.org/10.3390/electronics10161973
25. Swersky, K., Snoek, J., Adams, R.P.: Multi-task Bayesian optimization. In: Burges, C., Bottou, L., Welling, M., Ghahramani, Z., Weinberger, K. (eds.) Advances in Neural Information Processing Systems, vol. 26. Curran Associates, Inc. (2013)
26. Thornton, C., Hutter, F., Hoos, H.H., Leyton-Brown, K.: Auto-WEKA. In: Proceedings of the 19th International Conference on Knowledge Discovery and Data Mining, pp. 847–855 (2013)

27. Vanschoren, J.: Meta-learning. In: Hutter et al. [7], pp. 39–68 (2018)
28. Waring, J., Lindvall, C., Umeton, R.: Automated machine learning: review of the state-of-the-art and opportunities for healthcare. Artif. Intell. Med. **104**, 101822 (2020). https://doi.org/10.1016/j.artmed.2020.101822
29. Yao, Q., et al.: Taking human out of learning applications: a survey on automated machine learning. CoRR (2019)
30. Zaharia, M., Chowdhury, M., Franklin, M.J., Shenker, S., Stoica, I.: Spark: cluster computing with working sets. In: Proceedings of the 2nd USENIX Conference on Hot Topics in Cloud Computing, HotCloud 2010, p. 10. USENIX Association, USA (2010)

Transformers are Short-Text Classifiers

Fabian Karl and Ansgar Scherp[(✉)]

Universität Ulm, Ulm, Germany
{fabian.karl,ansgar.scherp}@uni-ulm.de

Abstract. Short text classification is a crucial and challenging aspect of Natural Language Processing. For this reason, there are numerous highly specialized short text classifiers. A variety of approaches have been employed in short text classifiers such as convolutional and recurrent networks. Also many short text classifier based on graph neural networks have emerged in the last years. However, in recent short text research, State of the Art (SOTA) methods for traditional text classification, particularly the pure use of Transformers, have been unexploited. In this work, we examine the performance of a variety of short text classifiers as well as the top performing traditional text classifier on benchmark datasets. We further investigate the effects on two new real-world short text datasets in an effort to address the issue of becoming overly dependent on benchmark datasets with a limited number of characteristics. The datasets are motivated from a real-world use case on classifying goods and services for tax auditing. NICE is a classification system for goods and services that divides them into 45 classes and is based on the Nice Classification of the World Intellectual Property Organization. The Short Texts Of Products and Services (STOPS) dataset is based on Amazon product descriptions and Yelp business entries. Our experiments unambiguously demonstrate that Transformers achieve SOTA accuracy on short text classification tasks, raising the question of whether specialized short text techniques are necessary. The NICE dataset showed to be particularly challenging and makes a good benchmark for future advancements.

A preprint can be also found on arXiv [14]. Source code is available here: https://github.com/FKarl/short-text-classification.

Keywords: Text Classification · Transformer · BERT · GNN

1 Introduction

Text classification is a crucial aspect of Natural Language Processing (NLP), and extensive research in this field is being conducted. Many researchers are working to improve the speed, accuracy, or robustness of their algorithms. Traditional text classification, however, does not take some traits into account that appear in numerous real-world applications, such as short text. Therefore, studies have been conducted specifically on short texts [38,47]. From user-generated content like social media to business data like accounting records, short text covers a

A. Holzinger et al. (Eds.): CD-MAKE 2023, LNCS 14065, pp. 103–122, 2023.
https://doi.org/10.1007/978-3-031-40837-3_7

wide range of topics. For example, the division into goods and services (see Sect. 4.1) is an important part of the tax audit. Currently, an auditor checks whether the element descriptions match the appropriate class of good or service. Since this can be very time-consuming, it is desirable to bring it into a semi-automatic context with the help of classifiers. Also, the subdivision into more specific classes can be useful for determining whether a given amount for an entry in the accounting records is reasonable.

Since short texts are typically only one to two sentences long, they lack context and therefore pose a challenge for text classification. In order to get better results, many short text classifiers also operate in a transductive setup [38,41,43], which includes the test set during training. However, as they need to be retrained each time new data needs to be classified, those transductive models are not very suitable for real-world applications. The results of both transductive and the generally more useful inductive short text classifier are typically unsatisfactory due to the challenge that short text presents. Recent studies on short texts have emphasized specialized models [33,36,38,41,43,47] to address the issues associated with the short text length. However, State of the Art (SOTA) text classification methods, particularly the pure use of Transformers, have been unexploited. In this work, the effectiveness on short texts is examined and tested by means of benchmark datasets. We also introduce two new, realistic datasets in the domain of goods and services descriptions. Our contributions are in summary:

- We provide a comparison of various modern text classification techniques. In particular, specialized short text methods are compared with the top performing traditional text classification models.
- We introduce two new real-world datasets in the goods and services domain to cover additional dataset characteristics in a realistic use-case.
- Transformers achieve SOTA accuracy on short text classification tasks. This questions the need of specialized short text classifier.

Below, we summarize the related work. Section 3 provides a description of the models that were selected for our experiments. The experimental apparatus is described in Sect. 4. An overview of the achieved results is reported in Sect. 5. Section 6 discusses the results, before we conclude.

2 Related Work

Despite the fact that Bag of Words (BoW)-based models have long represented the cutting edge in text classification, attention has recently shifted to sequence-based and, more recently, graph-based concepts. However, BoW-based models continue to offer a solid baseline [7]. For example in fastText [12] the average of the trained word representations are used as text representation and then fed into a linear classifier. This results in an efficient model for text classification. To give an overview of the various concepts, Sect. 2.1 provides various works in the field of sequence-based models, Sect. 2.2 discusses graph-based models, and Sect. 2.3 examines how these concepts are applied to short text. Finally, a summary of the findings from the related work is presented in Sect. 2.4.

2.1 Sequence-Based Models

For any NLP task, Recurrent Neural Networks (RNN) and Long short-term memory (LSTM) are frequently used and a logical choice because both models learn historical information while taking location information for all words into account [17,23]. Since RNNs must be computed sequentially and cannot be computed in parallel, the use of Convolutional Neural Networks (CNNs) is also common [17,34]. The text must be represented as a set of vectors that are concatenated into a matrix in order to be used by CNNs. The standard CNN convolution and pooling operations can then be applied to this matrix. TextCNN [15] uses this in combination with pretrained word embeddings for sentence-level classification tasks. While CNN-based models extract the characteristics from the convolution kernels, the relationship between the input words is captured by RNN-based models [17]. An important turning point in the advancement of NLP technologies was the introduction of Bidirectional Encoder Representations from Transformers (BERT) [35]. By performing extensive pre-training in an unsupervised manner and automatically mining semantic knowledge, BERT learns to produce contextualized word vectors that have a global semantic representation.

The effectiveness of BERT-like models for text classification is demonstrated by Galke and Scherp [7].

2.2 Graph-Based Models

Recently, text classification has paid a lot of attention to graph-based models, particularly Graph Neural Networks (GNNs) [3,28,37]. This is due to the fact that tasks with rich relational structures benefit from the powerful representation capabilities of GNNs, which preserve global structure information [37]. The task of text classification offers this rich relational structure because text can be modeled as edges and nodes in a graph structure. There are different ways to represent the documents in a graph structure, but two main approaches have emerged [37,38]. The first approach builds a graph for each document using words as nodes and structural data, such as word co-occurence data, as edges. However, only local structural data is used. The task is constructed as a whole graph classification problem in order to classify the text. A popular *document-level* approach is HyperGAT [5] which uses a dual attention mechanism and hypergraphs applied to documents to learn text embeddings. The second approach creates a graph for the entire corpus using words and documents as nodes. The text classification task is now a node classification task for the unlabeled document nodes. The drawback of this method is that models using it are inherently transductive. For example, TextGCN [42] uses this concept by employing a standard Graph Convolutional Networks (GCN) on this heterogeneous graph. Following TextGCN, Lin et al. [19] propose BertGCN, a model that makes use of BERT to initialize representations for the document nodes in order to combine the benefits of both the large-scale pretraining of BERT and the transductive TextGCN. However, the increase provided by this method is limited to datasets with long average text lengths. Zeng et al. [44]

also experiment with combining TextGCN and BERT in the form of TextGCN-Bert-serial-SB, a Simplified-Boosting Ensemble, where BERT is only trained on the TextGCN's misclassification. Which model is applied to which document is determined by a heuristic based on the node degree of the test document. However, TextGCN-CNN-serial-SB, which substitutes TextCNN for BERT, yields better results. By using a joint training mechanism, TextING [46] and BERT are trained on sub-word tokens and base their predictions on the results of the two models. In contrast to applying each model separately, this produces better results. Another approach combining graph classifiers with BERT is ContTextING [11]. ContTextING utilizes a joint training mechanism to create a unified model that incorporates both document-wise contextual information from a BERT-style model and node interactions within a document through the use of a GNN module. The predictions for the text classification task are determined by combining the output from both of these modules.

2.3 Short Text Models

Of course, short texts can also be classified using the methods discussed above. However, this is challenging because short texts tend to lack context and adhere to less strict syntactic structure [38]. This has led to the emergence of specialized techniques that focus on improving the results for short text. Early works focused on sentence classification using methods like Support Vector Machines (SVM) [29]. A survey by Galke et al. [6] compared SVM and other classical methods like Naive Bayes and kNN with multi-layer perceptron models (MLP) on short text classification. Other works on sentence classification used Convolutional Neural Networks (CNN) [13,36,45], which showed strong performance on benchmark datasets. Recently, also methods exploiting graph neural networks were adopted to the needs of short text. For instance, Heterogeneous Graph Attention networks (HGAT) [41] is a powerful semi-supervised short text classifier. This was the first attempt to model short texts as well as additional information like topics gathered from a Latent Dirichlet Allocation (LDA) [1] and entities retrieved from Wikipedia with a Heterogeneous Information Network (HIN). To achieve this, a HIN embedding with a dual-level attention mechanism for nodes and their relations was used. For the semantic sparsity of short text, both the additional information and the captured relations are beneficial. A transductive and an inductive HGAT model were released, with the transductive model being better on every dataset. NC-HGAT [33] expands the HGAT model to produce a more robust variant. Neighbor contrastive learning is based on the premise that documents that are connected have a higher likelihood of sharing a class label and, as a result, should therefore be closer in feature space. In order to represent the additional information, SHINE [38] also makes use of a heterogenous graph. In contrast, SHINE generates component graphs in the form of word, entity, and Part Of Speech (POS) graphs and creates a dynamically learned short document graph by employing hierarchical pooling over all component graphs. In the semi-supervised setting, SHINE outperforms HGAT as a strong transductive model. SimpleSTC (Simple Short Text Classification) [48] is a graph-based method for short-text classification similar to SHINE. But instead of constructing the word-graph only over the data corpus itself, SimpleSTC employs a global corpus to create a reference graph that shall enrich and help to understand the short text in the smaller corpus. As global corpus, articles from Wikipedia are used. The authors sample 20 labeled documents per class as training set and validation set. Short-Text Graph Convolutional Networks (STGCN) [43] is an additional short text classifier. A graph of topics, documents, and unique words is the foundation of STGCN. Although the STGCN results by themselves are not particularly strong, the impact of pre-trained word vectors obtained by BERT was also examined. The classification of the STGCN is significantly enhanced by the combination of STGCN with BERT and a Bi-LSTM.

2.4 Summary

Graph neural network-based methods are widely used in short text classification. However, in recent short text research, SOTA text classification methods, particularly the pure use of Transformers, have been unexploited. The majority of short text models are transductive. The crucial drawback of being transductive is that every time new data needs to be classified, the model must be retrained.

3 Selected Models for Our Comparison

We begin with models for short text classification in Sect. 3.1 and then Sect. 3.2 introduces a selection of top-performing models for text classification. Following Galke and Scherp [7], we have excluded works that employ non-standard datasets only, use different measures, or are otherwise not comparable.For example, regarding short text classification there are works that are applied on non-standard datasets only [10,49].

3.1 Models for Short Text Classification

The models listed below either make claims about their ability to categorize short texts or were designed with that specific goal. The **SECNN** [36] is a text classification model built on CNNs that was created specifically for short texts with few and insufficient semantic features. Wang et al. [36] suggested four components to address this issue. In order to achieve better coverage on the word vector table, they used an improved Jaro-Winkler similarity during preprocessing to identify any potential spelling mistakes. Second, they use a CNN model built on the attention mechanism to look for words that are related. Third, in order to accomplish the goal of short text semantic expansion, the external knowledgebase *Probase* [39] is used to enhance the semantic features of short text. Finally, the classification process is performed using a straightforward CNN with a Softmax output layer.

The Sequential Graph Neural Network (**SGNN**) [47] is a GNN-based model that emphasizes the propagation of features based on sequences. By training each document as a separate graph, it is possible to learn the words' local and sequential features. GloVe's [24] pre-trained word embedding is utilized as a semantic feature of words. In order to update the feature matrix for each document graph, a Bi-LSTM is used to extract the contextual feature of each word. After that, a simplified GCN aggregates the features of neighboring word nodes. Additionally, Zhao et al. [47] introduce two variants: Extended-SGNN (**ESGNN**), in which the initial contextual feature of words is preserved, and **C-BERT**, in which the Bi-LSTM is swapped for BERT.

The Deep Attention Diffusion Graph Neural Network (**DADGNN**) [22] is a graph-based method that combats the oversmoothing problem of GNNs and allows stacking more layers by utilizing attention diffusion and decoupling techniques. This decoupling technique is also very advantageous for short texts because it obtains distinct hidden features in deep graph networks.

The Long short-term memory (LSTM) [9], which is frequently used in text classification, has a bidirectional variant called **Bi-LSTM** [20]. Due to its strong results for short texts [23,47] and years of use as the SOTA method for many tasks, this model is a good baseline for our purpose.

3.2 Top-Performing Models for Text Classification

An overview of the top text classification models that excel on texts of all lengths and were not specifically created with short texts in mind is provided in this section. We employ the base models for the Transformers.

The Bidirectional Encoder Representations from Transformers (**BERT**) [4] is a language representation model that is based on the Transformer architecture [35]. Encoder-only models, such as BERT, rely solely on the encoder component of the Transformer architecture, whereby the text sequences are converted into rich numerical representations [34]. These models are well suited for text classification due to this representation. BERT is designed to incorporate a token's left and right contexts into its computed representation. This is commonly referred to as bidirectional attention.

The Robustly optimized BERT approach (**RoBERTa**) [21] is a systematically improved BERT adaptation. In the RoBERTa model, the pre-training strategy was changed and training was done on larger batches with more data, to increase BERT's performance.

To improve BERT and RoBERTa models, Decoding-enhanced BERT with disentangled attention (**DeBERTa**) [8] makes two architectural adjustments. The first is the disentangled attention mechanism, which encodes the content and location of each word using two vectors. The content of the token at position i is represented by H_i and the relative position $i|j$ between the token at position i and j are represented by $P_{i|j}$. The equation for determining the cross attention score is as follows: $A_{i,j} = H_i H_j^T + H_i P_{j|i}^T + P_{i|j} H_j^T + P_{i|j} P_{j|i}^T$. The second adjustment is an enhanced mask decoder that uses absolute positions in the decoding layer to predict masked tokens during pre-training. For masked token prediction, DeBERTa includes the absolute position after the transform layers but before the softmax layer. In contrast, BERT incorporates the position embedding into the input layer. As a result, DeBERTa is able to capture the relative position in all Transformer layers.

Sun et al. [32] proposed **ERNIE 2.0**, a continuous pre-training framework that builds and learns pre-training tasks through continuous multi-task learning. This allows the extraction of additional valuable lexical, syntactic, and semantic information in addition to co-occurring information, which is typically the focus.

The concept behind **DistilBERT** [26] is to leverage knowledge distillation to produce a more compact and faster version of BERT while retaining most of its language understanding capacities. DistilBERT reduces the size of BERT by 40%, is 60% faster, and still retains 97% of its language understanding capabilities. In order to accomplish this, DistilBERT optimizes the following three objectives while using the BERT model as a teacher: (1) Distillation loss: The model was trained to output probabilities equivalent to those of the BERT base model. (2) Masked Language Modeling (MLM): As described by Devlin et al. [4] for the BERT model, the common pre-training using masked language modeling is being used.(3) Cosine embedding loss: The model was trained to align the DistilBERT and BERT hidden state vectors.

A Lite BERT (**ALBERT**) [16] is a Transformer that uses two parameter-reduction strategies to save memory and speed up training by sharing the weights of all layers across its Transformer. This model is therefore particularly effective for longer texts. During pretraining, ALBERTv2 employs MLM and Sentence-Order Prediction (SOP), which predicts the sequence of two subsequent text segments.

WideMLP [7] is a BoW-Based Multilayer Perceptron (MLP) with a single wide hidden layer of 1,024 Rectified Linear Units (ReLUs). This model serves as a useful benchmark against which we can measure actual scientific progress.

InducTive Graph Convolutional Networks for Text classification (**InducT-GCN**) [37] is a GCN-based method that categorically rejects any information or statistics from the test set. To achieve the inductive setup, InducT-GCN represents document vectors with a weighted sum of word vectors and applies TF-IDF weights instead of representing document nodes with one-hot vectors. A two-layer GCN is employed for training, with the first layer learning the word embeddings and the second layer in the dimension of the dataset's classes outputs into a softmax activation function.

4 Experimental Apparatus

4.1 Datasets

First, we describe the benchmark datasets. Second, we introduce our new datasets in the domain of goods and services. The characteristics are denoted in Table 1.

Table 1. Characteristics of short text datasets. #C refers to the number of classes. Avg. L is the average document length.

Benchmarks	#Doc	#Train	#Test	#C	Avg. L
R8	7,674	5,485	2,189	8	65.72
MR	10,662	7,108	3,554	2	20.39
SearchSnippets	12,340	10,060	2,280	8	18.10
Twitter	10,000	7,000	3,000	2	11.64
TREC	5,952	5,452	500	6	10.06
SST-2	9,613	7,792	1,821	2	20.32
Goods & Services	#Doc	#Train	#Test	#C	Avg. L
NICE-45	9,593	6,715	2,878	45	3.75
NICE-2	9,593	6,715	2,878	2	3.75
STOPS-41	200,341	140,238	60,103	41	5.64
STOPS-2	200,341	140,238	60,103	2	5.64

Benchmark Datasets. Six short text benchmark datasets, namely R8, MR, SearchSnippets, Twitter, TREC, and SST-2, are used in our experiments. The following gives a detailed description of them. **R8** is an 8-class subset of the Reuters 21578 news dataset[1]. It is not a classical short text scenario with an average length of 65.72 tokens but offers the ability to set the methods in comparison to traditional text classification. **MR**[2] is a widely used dataset for text classification. It contains movie-review documents with an average length of 20.39 tokens and is therefore suitable for short text classification. The dataset **SearchSnippets**[3], which is made up of snippets returned by a search engine and has an average length of 18.10 tokens, was released by Phan et al. [25]. **Twitter**[4] is a collection of 10,000 tweets that are split into the categories negative and positive based on sentiment. The length of those tweets is on average 11.64 tokens. **TREC**[5], which was introduced by Li and Roth [18], is a question type classification dataset with six classifications for questions. It provides the shortest texts in our collection of benchmark datasets, with an average text length of 10.06 tokens. **SST-2**[6] [30] or SST-binary is a subset of the Stanford Sentiment Treebank, a fine-grained sentiment analysis dataset, in which neutral reviews have been removed and the data has either a positive or negative label. The average number of tokens in the texts is 20.32.

1 http://www.daviddlewis.com/resources/testcollections/reuters21578/.
2 https://www.cs.cornell.edu/people/pabo/movie-review-data/.
3 http://jwebpro.sourceforge.net/data-web-snippets.tar.gz.
4 https://www.nltk.org/howto/twitter.html#Using-a-Tweet-Corpus.
5 https://cogcomp.seas.upenn.edu/Data/QA/QC/.
6 https://nlp.stanford.edu/sentiment/.

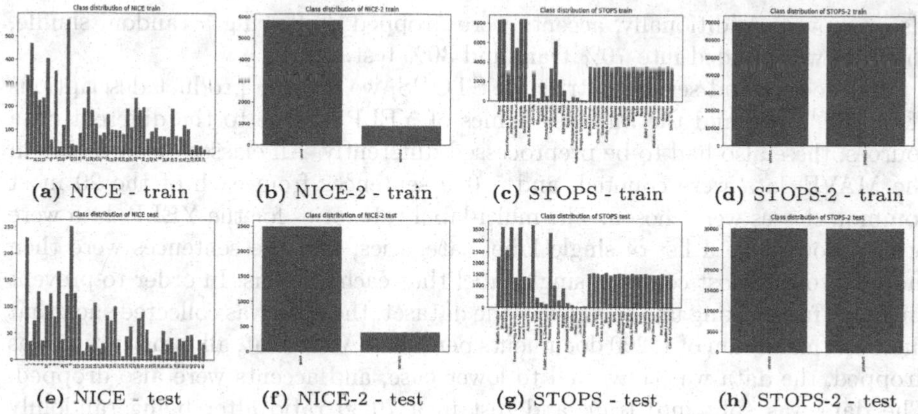

(a) NICE - train (b) NICE-2 - train (c) STOPS - train (d) STOPS-2 - train

(e) NICE - test (f) NICE-2 - test (g) STOPS - test (h) STOPS-2 - test

Fig. 1. Class distribution of our new datasets (separated by train and test split)

Goods and Services Datasets. In order to evaluate the performance on data with real world applications, we introduce two new datasets that are focused on the distinction between goods and services. Although there are already datasets for product classification, such as the WDC-LSPM[7], to the best of our knowledge, our datasets are the first to combine goods and services. **NICE** is a classification system for goods and services that divides them into 45 classes and is based on the Nice Classification[8] of the World Intellectual Property Organization (WIPO). There are 11 classes for various service types and 34 categories for goods. With 9,593 documents, NICE-45 is comparable in size to the benchmark datasets. This dataset, which has texts with an average length of 3.75 tokens, is an excellent example of extremely short text. For the division into goods and services, there is also the binary version NICE-2. **Short Texts Of Products and Services (STOPS)** is the second dataset we offer. With 200,341 documents and an average length of 5.64 tokens, STOPS-41 is a reasonably large dataset. The data set was derived from a potential use case in the form of Amazon descriptions and Yelp business entries, making it the most realistic. Like NICE, STOPS has a binary version STOPS-2. Both datasets provide novel characteristic properties that the benchmark datasets did not cover. In particular, the number of fine-granular classes presents a challenge that is not addressed by common benchmarks. For details on the class distribution of these datasets, please refer to Fig. 1.

4.2 Preprocessing

To create NICE, the WIPO[9] classification data was converted to lower case, all punctuation was removed, and side information that was enclosed in brackets was

[7] http://webdatacommons.org/largescaleproductcorpus/.

[8] https://www.wipo.int/classifications/nice/en/.

[9] https://www.wipo.int/nice/its4nice/ITSupport_and_download_area/20220101/MasterFiles/index.html.

also removed. Additionally, accents were dropped. Following a random shuffle, the data was divided into 70% train and 30% test.

As product and service entries for STOPS, we use the product descriptions of MAVE[10] [40] and the business names of YELP[11]. Due to the different data sources, these also had to be preprocessed differently. All classes' occurrences in the MAVE data were counted, and 5,000 sentences from each of the 20 most common classes were chosen. The multi-label categories for the YELP data were broken down into a list of single label categories, and the sentences were then mapped to the most common single label that each one has. In order to prevent any label from taking up too much of the dataset, the data was collected such that there is a maximum of 1,200 documents per label. After that, all punctuation was dropped, the data was converted to lower case, and accents were also dropped. The data was split into train and test in a 70:30 ratio after being randomly shuffled.

4.3 Procedure

The best short text classifier and text classification models were retrieved from the literature (see description of the models in Sect. 3). The accuracy scores were extracted in order to establish a comparison. Own experiments, particularly using various Transformers, were conducted in order to compare them. Investigations into the impacts of hyperparameters on short texts were performed. More details about these are provided in Sect. 4.4. In order to test the methods in novel contexts, we also created two new datasets, whereby STOPS stands out due to its much higher quantity of documents.

4.4 Hyperparameter Optimization

Our experiments for BERT, DistilBERT, and WideMLP used the hyperparameter from Galke and Scherp [7]. The parameters for BERT and DistilBERT are a learning rate of $5 \cdot 10^{-5}$, a batch size of 128, and fine-tuning for 10 epochs. WideMLP was trained for 100 epochs with a learning rate of 10^{-3}, a batch size of 16, and a dropout of 0.5. For ERNIE 2.0 and ALBERTv2, we make use of the SST-2 values that Sun et al. [32] and Lan et al. [16], respectively, published. For our hyperparameter selection for DeBERTa and RoBERTa, we used the BERT values from Galke and Scherp [7] as a starting point and investigated the effect of smaller learning rates. This resulted in learning rates of $2 \cdot 10^{-5}$ for DeBERTa and $4 \cdot 10^{-5}$ for RoBERTa while maintaining the other parameters. For comparison, we followed the same procedure to create ERNIE 2.0 (optimized), which yields a learning rate of $25 \cdot 10^{-6}$. The Bi-LSTM values from Zhao et al. [47] were used for both the LSTM and the Bi-LSTM model. We used DADGNN with the default parameters of 0.5 dropout, 10^{-6} weight decay, and two attention heads for all datasets.

[10] https://github.com/google-research-datasets/MAVE.

[11] https://www.yelp.com/dataset/download.

4.5 Metrics

Accuracy is used to measure the classification of short text. For multi-class cases, the subset accuracy is calculated.

5 Results

The accuracy scores for the text classification models on the six benchmark datasets are shown in Table 2. The findings demonstrate that the relatively straightforward models LSTM, Bi-LSTM, and WideMLP provide a strong baseline across all datasets. This comparison clearly demonstrates the limitations of some models, with InducT-GCN falling short in all datasets except SearchSnippets, SECNN underperforming on TREC, and DADGNN producing weak MR results in our own experiment. The Transformer models, on the other hand, are the best performing across all datasets with the exception of SearchSnippets. With consistently strong performance across all datasets, DeBERTa stands out in particular. The graph-based models from Zhao et al. [47], SGNN, ESGNN, and C-BERT, all perform well for the datasets for which results are available and ESGNN even outperforms all other models for SearchSnippets. It is important to note that Zhao et al. [47] used a modified training split and additional preprocessing. While an increase of about 5 percentage points for MR could be obtained by extending ESGNN with BERT in C-BERT, the increase is not noticeable for other datasets. When applied to short texts, the inductive models even outperform transductive models. On Twitter, ERNIE 2.0 and ALBERTv2 reach a performance of 99.97%, and when using BERT on the TREC dataset, a performance of 99.4% is obtained. Non-Transformer models also perform well on TREC, although Transformers outperform them. For the graph-based models SHINE and InducT-GCN, we also calculated the mean and standard deviation of the accuracy scores across 5 runs. This is motivated from the observation that models based on graph-neural networks are susceptible to the initialization of the embeddings [27]. SHINE had a generally high standard deviation of up to nearly 5 points, indicating greater variance in its performance. In comparison, InducT-GCN has a rather small variance of always below 1 point.

The accuracy results for our newly introduced datasets, NICE and STOPS, are shown in Table 3. New characteristics covered by NICE and STOPS include shorter average lengths and the ability to distinguish between classes at a fine-granular level in NICE-45 and STOPS-41. The investigation of more documents is also conducted in the case of STOPS. As a result, NICE-45 and STOPS-41 reveal that DADGNN encounters issues when dealing with more classes, even falling around 20 and 60 percent points behind the baseline models. While still performing worse than the baseline models, InducT-GCN outperforms DADGNN on all four datasets. Transformers once again demonstrate their strength and rank as the top performing models across all datasets on this dataset. There are also significant drops. ERNIE 2.0 performs worse than

Table 2. Accuracy on short text classification datasets. The "Short?" column indicates whether the model makes claims about its ability to categorize short texts. Provenance refers to the source of the accuracy scores.

Inductive Models	Short?	R8	MR	Snippets	Twitter	TREC	SST-2	Provenance
Transformer Models								
BERT	N	98.17[a]	86.94	88.20	99.96	**99.4**	91.37	Own experiment
RoBERTa	N	98.17[a]	89.42	85.22	99.9	98.6	94.01	Own experiment
DeBERTa	N	98.45[a]	**90.21**	86.14	99.93	98.8	**94.78**	Own experiment
ERNIE 2.0	N	98.04[a]	88.97	89.12	**99.97**	98.8	93.36	Own experiment
ERNIE 2.0 (optimized)	N	98.17[a]	89.53	89.17	**99.97**	99	94.07	Own experiment
DistilBERT	N	97.98[a]	85.31	89.69	99.96	99	90.49	Own experiment
ALBERTv2	N	97.62	86.02	87.68	**99.97**	98.6	91.54	Own experiment
BoW Models								
SVM	Y	—	—	—	—	95[f]	—	Silva et al. [29]
WideMLP	N	96.98	76.48	67.28	99.86	97	82.26	Own experiment
fastText	N	96.13	75.14	88.56[d]	—	—	—	Zhao et al. [47]
Graph-based Models								
HGAT[b]	Y	—	62.75	82.36	63.21	—	—	Yang et al. [41]
NC-HGAT[b]	Y	—	62.46	—	63.76	—	—	Sun et al. [33]
SGNN[c]	Y	98.09	80.58	90.68[d]	—	—	—	Zhao et al. [47]
ESGNN[c]	Y	98.23	80.93	**90.80**[d]	—	—	—	Zhao et al. [47]
C-BERT (ESGNN+BERT)[c]	Y	98.28	86.06	90.43[d]	—	—	—	Zhao et al. [47]
DADGNN	Y	98.15	78.64	—	—	97.99	84.32	Liu et al. [22]
DADGNN	Y	97.28	74.54	84.91	98.16	97.54	82.81	Own experiment
HyperGAT	N	97.97	78.32	—	—	—	—	Ding et al. [5]
InducT-GCN	N	96.68	75.34	76.67	88.56	92.50	79.97	Own experiment
ConTextING-BERT	N	97.91	86.01	—	—	—	—	Huang et al. [11]
ConTextING-RoBERTa	N	98.13	89.43	—	—	—	—	Huang et al. [11]
CNN and LSTMs								
SECNN[c]	Y	—	83.89	—	—	91.34	87.37	Wang et al. [36]
MGNC-CNN	Y	—	—	—	—	95.52	88.30[g]	Zhang et al. [45]
DCNN	Y	—	86.80[h]	—	—	93	—	Kalchbr. et al. [13]
LSTM (BERT)	Y	94.28	75.10	65.13	99.83	97	81.38	Own experiment
Bi-LSTM (BERT)	Y	95.52	75.30	66.79	99.76	97.2	80.83	Own experiment
LSTM (GloVe)	Y	96.34	74.99	67.67	95.23	97.4	79.95	Own experiment
Bi-LSTM (GloVe)	Y	96.84	75.32	68.15	95.53	97.2	80.17	Own experiment
Bi-LSTM (GloVe)	Y	96.31	77.68	84.81[d]	—	—	—	Zhao et al. [47]
Transductive Models	Short?	R8	MR	Snippets	Twitter	TREC	SST-2	Provenance
Graph-based Models								
SHINE[e]	Y	—	64.58	82.39	72.54	—	—	Wang et al. [38]
SHINE	Y	79.80	62.05	82.14	70.64	79.90	61.71	Own experiment
STGCN	Y	97.2	78.2	—	—	—	—	Ye et al. [43]
STGCN+BiLSTM	Y	—	78.5	—	—	—	—	Ye et al. [43]
STGCN+BERT+BiLSTM	Y	98.5	82.5	—	—	—	—	Ye et al. [43]
SimpleSTC[i]	Y	—	62.27	80.96	62.19	—	—	Zheng et al. [48]
TextGCN	N	97.07	76.74	83.49	—	—	—	Zhao et al. [47]
TextGCN	N	97.07	76.74	—	—	91.40	81.02	Liu et al. [22]
BertGCN	N	98.1	86.0	—	—	—	—	Lin et al. [19]
RoBERTaGCN	N	98.2	89.7	—	—	—	—	Lin et al. [19]
TextGCN-BERT-serial-SB	N	97.78	86.69	—	—	—	—	Zeng et al. [44]
TextGCN-CNN-serial-SB	N	**98.53**	87.59	—	—	—	—	Zeng et al. [44]

[a] With a batch size of 32 and for DeBERTa of 16.

[b] With only 40 randomly selected documents per class.

[c] Not reproducible. Authors have been contacted twice without a response.

[d] Using a modified training split of 8,636 training and 3,704 test documents and further preprocessing.

[e] Employing very low train ratios (0.38% to 6.22%).

[f] Uni-gram model with extensive pre-processing, use of WordNet, etc. and 60 hand-coded rules

[g] Removed phrases of length less than 4 from the training set

[h] Using a slightly different split of 6,920 sentences for training, 872 for development, and 1,821 for test

[i] Samples 20 labeled documents per class as training set and validation set

Table 3. Accuracy on our own short text classification datasets. The "Short?" column indicates whether the model makes claims about its ability to categorize short texts. Provenance refers to the source of the accuracy scores.

Inductive Models	Short Text	NICE-45	NICE-2	STOPS-41	STOPS-2
Transformer Models					
BERT	N	**72.79**	99.72	89.4	99.87
RoBERTa	N	66.09	**99.76**	89.56	99.86
DeBERTa	N	59.42	99.72	**89.73**	99.85
ERNIE 2.0	N	45.55	99.69	89.39	99.85
ERNIE 2.0 (optimized)	N	67.65	99.72	89.65	**99.88**
DistilBERT	N	69.28	99.75	89.32	99.85
ALBERTv2	N	59.24	99.51	88.58	99.83
BoW Models					
WideMLP	N	58.99	96.76	88.2	97.05
Graph-based Models					
DADGNN	Y	28.51	91.15	26.75	97.48
InducT-GCN	N	47.06	94.98	86.08	97.74
CNN and LSTMs					
LSTM (BERT)	Y	47.81	96.63	86.27	96.05
Bi-LSTM (BERT)	Y	52.39	96.63	85.93	98.54
LSTM (GloVe)	Y	52.64	96.17	87.4	99.46
Bi-LSTM (GloVe)	Y	55.35	95.93	87.38	99.43

the baseline models with 45.55% on NICE-45. However, ERNIE 2.0 (optimized), which uses different hyperparameter values (see Sect. 4.4), comes in third with 67.65%.

6 Discussion

Graph-based models are computationally expensive because they require not only the creation of the graph but also its training, which can be resource- and time-intensive, especially for word-document graphs with $\mathcal{O}(N^2)$ space [7]. On STOPS, this drawback becomes very apparent. We could observe that DADGNN required roughly 30 hours of training time, while BERT only took 30 minutes to fine-tune with the same resources. Although in the case of BERT, the pre-training was already very expensive, transfer learning allows this effort to be used for a variety of tasks. Nevertheless, the Transformers outperform the inductive graph-based models as well as the short text models, with just one exception. The best model for SearchSnippets is ESGNN, but additional preprocessing and a modified training split were employed. Our Bi-LSTM results, obtained without additional preprocessing, differ by 16.66 percentage points from the Bi-LSTM results from Zhao et el. [47]. This indicates that preprocessing, and not a better model, is primarily responsible for the strong outcomes of the SearchSnippets experiments. Another interesting discovery can be made using the sentiment datasets. In comparison to other datasets, the Transformers outperform graph-based models that do not utilize a Transformer themselves by a large margin. This demonstrates that graph-based models may not be as effective at sentiment prediction tasks. In contrast, the CNN-based models show strong performance on the sentiment analysis task SST-2. Still, the best CNN model is more than 6 points below the best transformer. However, it should be noted that not all Transformers are consistently excellent. For instance, for NICE-45, one can observe a lower performance with ERNIE 2.0. But the absence of this performance decrease in our optimized version of ERNIE 2.0 (optimized) suggests that choosing suitable hyperparameters is crucial in this case.

6.1 Key Results

Our experiments unambiguously demonstrate that Transformers achieve SOTA accuracy on short text classification tasks. This raises the question of whether specialized short text techniques are necessary given that the performance of the existing models is insufficient. This observation is especially interesting because many of the short text models used are from 2021 [22,36,41,47] or 2022 [33]. Most short text models attempt to enrich the documents with some kind of external context, such as a knowledge base or POS tags. However, one could argue that Transformers implicitly contain context in their weights through their pre-training.

Those short text models that compare themselves to Transformers assert that they outperform them. For instance, Ye et al. [43] claim to outperform BERT by 2.2 percentage points on MR, but their fine-tuned BERT only achieves 80.3%. In contrast, our own experiments show that BERT achieves 86.94%. With 85.86% on MR, Zhao et al. [47] achieve better BERT results, but only to beat it by a meager 0.2% with C-BERT. Given the low surplus, they would no longer outperform it with a marginally better selection of hyperparameters for BERT. Therefore, it is reasonable to assume that the importance of good hyperparameters for Transformers is underestimated and that they are often not properly optimized. ERNIE 2.0 (optimized), which outperforms ERNIE 2.0 on every dataset, also demonstrates the effect of better hyperparameters. Finally, Zhao et al. [47] is already outperformed by other transformers like RoBERTa and DeBERTa by 3 and 4 points, respectively.

Additionally, there is a need for new short text datasets because the widely used benchmark datasets share many characteristics and fall short in many use cases. The common benchmark datasets all contain around 10,000 documents, distinguish only a few classes, and frequently have a similar average length. Furthermore, many of them cover the same tasks. For instance, MR, Twitter, and SST-2 all perform sentiment prediction, which makes sense given how much short text is produced by social media. In this paper, we introduce two new datasets with distinctive attributes to cover more cases in NICE and STOPS.

New and intriguing findings are produced by the new characteristics that are investigated using these datasets. Particularly, the ability to distinguish between classes at a fine-granular level reveals the shortcomings of various models, like DADGNN or ERNIE 2.0. NICE-45 in particular proved to be challenging for all models, making it a good benchmark for future advancements.

6.2 Threats to Validity

In our study, each experiment was generally conducted once. The rationale is the extremely low standard deviation for text classification tasks observed in previous studies [7,22,47]. However, it has been reported in the literature on models using graph neural networks (GNN) that they generally have high standard deviation in their performance, which has been attributed among others to the influence of the random initialization in the evaluation [27]. Thus, we have run our experiments for SHINE and InducT-GCN five times and report averages and standard deviation. The high standard deviation observed in SHINE's performance adds to the evidence of the need for caution when interpreting the results of GNNs [27].

We acknowledge that STOPS contains user-generated labels, some of which may not be entirely accurate. However, given that this occurs frequently in numerous use cases, it is also crucial to test the models in these scenarios.

6.3 Parameter Count of Models

Table 4 lists the parameter counts of selected Transformer models, the BoW-based baseline methods WideMLP, and graph-based methods used in our experiments. Generally, the top performing Transformer models have a similar size between 110M to 130M parameters. Although DistilBERT is only have of that size and ALBERTv2 only about a tens, our experiments show still comparable accuracy scores on R8, Snippets, Twitter, and TREC. ALBERTv2 with its 12M parameters outperforms the WideMLP baseline with 31.3M parameters on all datasets, some with a large margin. The graph-based model ConTextING-RoBERTa has a similar parameter count compared to the pure Transformer models, since the RoBERTa transformer is used internally. It is the top-performer among the graph-based models on R8 and MR but cannot outperform the pure Transformer models.

6.4 Generalization

As we cover in our experiments a range of diverse domains, with sentiment analysis on various themes (MR, SST-2, Twitter), question type classification (TREC), news (R8), and even search queries (SearchSnippets), we expect to find equivalent results on other short text classification datasets. Additionally, the categorization of goods and services is covered by our new datasets NICE and STOPS. They include additional features not covered by the benchmark datasets, including a significantly larger amount of training data in STOPS, a

Table 4. Parameter counts for selected methods used in our experiments

Model	#parameters
Transformer models	
BERT	110M
RoBERTA	123M
DeBERTA	134M
ERNIE 2.0	110M
DistilBERT	66M
ALBERTv2	12M
BoW-based methods	
WideMLP	31.3M
Graph-based methods	
HyperGAT	LDA parameters + 3.1M
ConTextING-RoBERTa	129M

shorter average length, and the capacity to differentiate between a wider range of classes. By using an example from a business problem, STOPS specifically demonstrates how the knowledge gained here can be applied in corporate use.

In this work, we cover a variety of models for each architecture, particularly the most popular and best-performing ones. Our findings are consistent with the studies by Galke and Scherp [7], which demonstrate the tremendous power of Transformers for traditional text classification.

7 Conclusion and Future Work

Our experiments unequivocally demonstrate the outstanding capability of Transformers for short text classification tasks. Additional research on our newly released datasets, NICE and STOPS, supports these findings and highlights the issue of becoming overly dependent on benchmark datasets with a limited number of characteristics. In conclusion, our study raises the question of whether specialized short text techniques are required given the lower performance of current models.

Future research on improving the performance of Transformers on short text could be to do pre-training on short texts or on in-domain texts (i.e., pre-training in the same domain as the target task) [2,31,34], multi-task fine-tuning [31,34], or an ensemble of multiple Transformer models [50].

Acknowledgement. This work is co-funded under the 2LIKE project by the German Federal Ministry of Education and Research (BMBF) and the Ministry of Science, Research and the Arts Baden-Württemberg within the funding line Artificial Intelligence in Higher Education. We thank Till Blume and Felix Krieger from Ernst & Young (EY) for the discussion of the problem statement that motivated this work. We are grateful to Liu et al. [22] for providing the unreleased DADGNN source code.

References

1. Blei, D.M., Ng, A.Y., Jordan, M.I.: Latent dirichlet allocation. J. Mach. Learn. Res. **3**, 993–1022 (2003). https://jmlr.org/papers/v3/blei03a.html
2. Brinkmann, A., Bizer, C.: Improving hierarchical product classification using domain-specific language modelling. IEEE Data Eng. Bull. **44**(2), 14–25 (2021)
3. Deng, Z., Sun, C., Zhong, G., Mao, Y.: Text classification with attention gated graph neural network. Cogn. Comput. **14**, 1–10 (2022). https://doi.org/10.1007/s12559-022-10017-3
4. Devlin, J., Chang, M., Lee, K., Toutanova, K.: BERT: pre-training of deep bidirectional transformers for language understanding. In: Burstein, J., Doran, C., Solorio, T. (eds.) Proceedings of the 2019 Conference of the North American Chapter of the Association for Computational Linguistics: Human Language Technologies, NAACL-HLT 2019, Minneapolis, MN, USA, 2–7 June 2019, Volume 1 (Long and Short Papers), pp. 4171–4186. Association for Computational Linguistics (2019). https://doi.org/10.18653/v1/n19-1423
5. Ding, K., Wang, J., Li, J., Li, D., Liu, H.: Be more with less: hypergraph attention networks for inductive text classification. In: Proceedings of the 2020 Conference on Empirical Methods in Natural Language Processing (EMNLP), pp. 4927–4936. Association for Computational Linguistics (2020). https://doi.org/10.18653/v1/2020.emnlp-main.399, https://aclanthology.org/2020.emnlp-main.399
6. Galke, L., Mai, F., Schelten, A., Brunsch, D., Scherp, A.: Using titles vs. full-text as source for automated semantic document annotation. In: Corcho, Ó., Janowicz, K., Rizzo, G., Tiddi, I., Garijo, D. (eds.) Proceedings of the Knowledge Capture Conference, K-CAP 2017, Austin, TX, USA, 4–6 December 2017, pp. 20:1–20:4. ACM (2017). https://doi.org/10.1145/3148011.3148039
7. Galke, L., Scherp, A.: Bag-of-words vs. graph vs. sequence in text classification: questioning the necessity of text-graphs and the surprising strength of a wide MLP. CoRR abs/2109.03777 (2021). https://arxiv.org/abs/2109.03777
8. He, P., Liu, X., Gao, J., Chen, W.: DeBERTa: decoding-enhanced BERT with disentangled attention. In: 9th International Conference on Learning Representations, ICLR 2021, Virtual Event, Austria, 3–7 May 2021. OpenReview.net (2021). https://openreview.net/forum?id=XPZIaotutsD
9. Hochreiter, S., Schmidhuber, J.: Long short-term memory. Neural Comput. **9**(8), 1735–1780 (1997)
10. Hu, Y., Ding, J., Dou, Z., Chang, H.: Short-text classification detector: a BERT-based mental approach. Comput. Intell. Neurosci. **2022** (2022). https://doi.org/10.1155/2022/8660828
11. Huang, Y.H., Chen, Y.H., Chen, Y.S.: ConTextING: granting document-wise contextual embeddings to graph neural networks for inductive text classification. In: Proceedings of the 29th International Conference on Computational Linguistics, pp. 1163–1168. International Committee on Computational Linguistics, Gyeongju, Republic of Korea (2022). https://aclanthology.org/2022.coling-1.100
12. Joulin, A., Grave, E., Bojanowski, P., Mikolov, T.: Bag of tricks for efficient text classification. In: Proceedings of the 15th Conference of the European Chapter of the Association for Computational Linguistics: Volume 2, Short Papers, pp. 427–431. Association for Computational Linguistics, Valencia, Spain (2017). https://aclanthology.org/E17-2068
13. Kalchbrenner, N., Grefenstette, E., Blunsom, P.: A convolutional neural network for modelling sentences. In: Proceedings of the 52nd Annual Meeting of the Association for Computational Linguistics, ACL 2014, 22–27 June 2014, Baltimore,

MD, USA, Volume 1: Long Papers, pp. 655–665. The Association for Computer Linguistics (2014). https://doi.org/10.3115/v1/p14-1062
14. Karl, F., Scherp, A.: Transformers are short text classifiers: a study of inductive short text classifiers on benchmarks and real-world datasets. CoRR abs/2211.16878 (2022). https://doi.org/10.48550/arXiv.2211.16878
15. Kim, Y.: Convolutional neural networks for sentence classification. In: Proceedings of the 2014 Conference on Empirical Methods in Natural Language Processing (EMNLP), pp. 1746–1751. Association for Computational Linguistics, Doha, Qatar (2014). https://doi.org/10.3115/v1/D14-1181
16. Lan, Z., Chen, M., Goodman, S., Gimpel, K., Sharma, P., Soricut, R.: ALBERT: a lite BERT for self-supervised learning of language representations. CoRR abs/1909.11942 (2019). https://arxiv.org/abs/1909.11942
17. Li, Q., et al.: A survey on text classification: from traditional to deep learning. ACM Trans. Intell. Syst. Technol. **13**(2), 1–41 (2022). https://doi.org/10.1145/3495162
18. Li, X., Roth, D.: Learning question classifiers. In: COLING 2002: The 19th International Conference on Computational Linguistics (2002). https://aclanthology.org/C02-1150
19. Lin, Y., Meng, Y., Sun, X., Han, Q., Kuang, K., Li, J., Wu, F.: BertGCN: transductive text classification by combining GNN and BERT. In: Findings of the Association for Computational Linguistics: ACL-IJCNLP 2021, pp. 1456–1462. Association for Computational Linguistics (2021). https://doi.org/10.18653/v1/2021.findings-acl.126, https://aclanthology.org/2021.findings-acl.126
20. Liu, P., Qiu, X., Huang, X.: Recurrent neural network for text classification with multi-task learning. CoRR abs/1605.05101 (2016). https://arxiv.org/abs/1605.05101
21. Liu, Y., et al.: RoBERTa: a robustly optimized BERT pretraining approach. CoRR abs/1907.11692 (2019). https://arxiv.org/abs/1907.11692
22. Liu, Y., Guan, R., Giunchiglia, F., Liang, Y., Feng, X.: Deep attention diffusion graph neural networks for text classification. In: Proceedings of the 2021 Conference on Empirical Methods in Natural Language Processing, pp. 8142–8152 (2021)
23. Mai, F., Galke, L., Scherp, A.: Using deep learning for title-based semantic subject indexing to reach competitive performance to full-text. CoRR abs/1801.06717 (2018). https://arxiv.org/abs/1801.06717
24. Pennington, J., Socher, R., Manning, C.D.: GloVe: global vectors for word representation. In: Proceedings of the 2014 Conference on Empirical Methods in Natural Language Processing (EMNLP), pp. 1532–1543 (2014)
25. Phan, X.H., Nguyen, L.M., Horiguchi, S.: Learning to classify short and sparse text & web with hidden topics from large-scale data collections. In: Proceedings of the 17th International Conference on World Wide Web, pp. 91–100 (2008)
26. Sanh, V., Debut, L., Chaumond, J., Wolf, T.: DistilBERT, a distilled version of BERT: smaller, faster, cheaper and lighter. CoRR abs/1910.01108 (2019). https://arxiv.org/abs/1910.01108
27. Shchur, O., Mumme, M., Bojchevski, A., Günnemann, S.: Pitfalls of graph neural network evaluation. CoRR abs/1811.05868 (2018). https://arxiv.org/abs/1811.05868
28. Shi, J., Wu, X., Liu, X., Lu, W., Li, S.: Inductive light graph convolution network for text classification based on word-label graph. In: Shi, Z., Zucker, J.D., An, B. (eds.) Intelligent Information Processing XI, IIP 2022. IFIP Advances in Information and Communication Technology, vol. 643, pp. 42–55. Springer, Cham (2022). https://doi.org/10.1007/978-3-031-03948-5_4

29. da Silva, J.P.C.G., Coheur, L., Mendes, A.C., Wichert, A.: From symbolic to subsymbolic information in question classification. Artif. Intell. Rev. **35**(2), 137–154 (2011)
30. Socher, R., et al.: Parsing with compositional vector grammars. In: EMNLP (2013)
31. Sun, C., Qiu, X., Xu, Y., Huang, X.: How to fine-tune BERT for text classification? In: Sun, M., Huang, X., Ji, H., Liu, Z., Liu, Y. (eds.) Chinese Computational Linguistics - 18th China National Conference, CCL 2019, Kunming, China, 18–20 October 2019, Proceedings. Lecture Notes in Computer Science, vol. 11856, pp. 194–206. Springer, Cham (2019). https://doi.org/10.1007/978-3-030-32381-3_16
32. Sun, Y., et al.: ERNIE 2.0: a continual pre-training framework for language understanding. CoRR abs/1907.12412 (2019). https://arxiv.org/abs/1907.12412
33. Sun, Z., Harit, A., Cristea, A.I., Yu, J., Shi, L., Al Moubayed, N.: Contrastive learning with heterogeneous graph attention networks on short text classification. In: 2022 International Joint Conference on Neural Networks (IJCNN), pp. 1–6 (2022). https://doi.org/10.1109/IJCNN55064.2022.9892257
34. Tunstall, L., von Werra, L., Wolf, T.: Natural language processing with Transformers. O'Reilly Media, Inc. (2022)
35. Vaswani, A., et al.: Attention is all you need. In: Advances in Neural Information Processing Systems, vol. 30 (2017)
36. Wang, H., Tian, K., Wu, Z., Wang, L.: A short text classification method based on convolutional neural network and semantic extension. Int. J. Comput. Intell. Syst. **14**(1), 367–375 (2021)
37. Wang, K., Han, S.C., Poon, J.: Induct-GCN: inductive graph convolutional networks for text classification. arXiv preprint arXiv:2206.00265 (2022)
38. Wang, Y., Wang, S., Yao, Q., Dou, D.: Hierarchical heterogeneous graph representation learning for short text classification. CoRR abs/2111.00180 (2021). https://arxiv.org/abs/2111.00180
39. Wu, W., Li, H., Wang, H., Zhu, K.Q.: Probase: a probabilistic taxonomy for text understanding. In: Candan, K.S., Chen, Y., Snodgrass, R.T., Gravano, L., Fuxman, A. (eds.) Proceedings of the ACM SIGMOD International Conference on Management of Data, SIGMOD 2012, Scottsdale, AZ, USA, 20–24 May 2012, pp. 481–492. ACM (2012). https://doi.org/10.1145/2213836.2213891
40. Yang, L., et al.: MAVE: a product dataset for multi-source attribute value extraction. In: Proceedings of the Fifteenth ACM International Conference on Web Search and Data Mining, pp. 1256–1265 (2022)
41. Yang, T., Hu, L., Shi, C., Ji, H., Li, X., Nie, L.: HGAT: heterogeneous graph attention networks for semi-supervised short text classification. ACM Trans. Inf. Syst. **39**(3) (2021). https://doi.org/10.1145/3450352
42. Yao, L., Mao, C., Luo, Y.: Graph convolutional networks for text classification. In: Proceedings of the AAAI Conference on Artificial Intelligence, vol. 33(01), pp. 7370–7377 (2019)
43. Ye, Z., Jiang, G., Liu, Y., Li, Z., Yuan, J.: Document and word representations generated by graph convolutional network and BERT for short text classification. In: ECAI 2020, pp. 2275–2281. IOS Press (2020)
44. Zeng, F., Chen, N., Yang, D., Meng, Z.: Simplified-boosting ensemble convolutional network for text classification. Neural Process. Lett. **54**, 1–16 (2022)
45. Zhang, Y., Roller, S., Wallace, B.C.: MGNC-CNN: a simple approach to exploiting multiple word embeddings for sentence classification. In: Knight, K., Nenkova, A., Rambow, O. (eds.) NAACL HLT 2016, The 2016 Conference of the North American Chapter of the Association for Computational Linguistics: Human Language

Technologies, San Diego California, USA, 12–17 June 2016, pp. 1522–1527. The Association for Computational Linguistics (2016). https://doi.org/10.18653/v1/n16-1178
46. Zhang, Y., Yu, X., Cui, Z., Wu, S., Wen, Z., Wang, L.: Every document owns its structure: inductive text classification via graph neural networks. arXiv preprint arXiv:2004.13826 (2020)
47. Zhao, K., Huang, L., Song, R., Shen, Q., Xu, H.: A sequential graph neural network for short text classification. Algorithms **14**(12), 352 (2021)
48. Zheng, K., Wang, Y., Yao, Q., Dou, D.: Simplified graph learning for inductive short text classification. In: Goldberg, Y., Kozareva, Z., Zhang, Y. (eds.) Proceedings of the 2022 Conference on Empirical Methods in Natural Language Processing, EMNLP 2022, Abu Dhabi, United Arab Emirates, 7–11 December 2022, pp. 10717–10724. Association for Computational Linguistics (2022). https://aclanthology.org/2022.emnlp-main.735
49. Zhong, Y., Zhang, Z., Zhang, W., Zhu, J.: BERT-KG: a short text classification model based on knowledge graph and deep semantics. In: Wang, L., Feng, Y., Hong, Y., He, R. (eds.) Natural Language Processing and Chinese Computing - 10th CCF International Conference, NLPCC 2021, Qingdao, China, 13–17 October 2021, Proceedings, Part I. Lecture Notes in Computer Science, vol. 13028, pp. 721–733. Springer, Cham (2021). https://doi.org/10.1007/978-3-030-88480-2_58
50. Zhuang, H., Qin, Z., Han, S., Wang, X., Bendersky, M., Najork, M.: Ensemble distillation for BERT-based ranking models. In: Proceedings of the 2021 ACM SIGIR International Conference on Theory of Information Retrieval, pp. 131–136. ICTIR 2021, Association for Computing Machinery, New York, NY, USA (2021). https://doi.org/10.1145/3471158.3472238

Reinforcement Learning with Temporal-Logic-Based Causal Diagrams

Yash Paliwal[1], Rajarshi Roy[2(✉)], Jean-Raphaël Gaglione[3], Nasim Baharisangari[1], Daniel Neider[4,5], Xiaoming Duan[6], Ufuk Topcu[3], and Zhe Xu[1]

[1] Arizona State University, Tempe, AZ, USA
xzhe1@asu.edu
[2] Max Planck Institute for Software Systems, Kaiserslautern, Germany
rajarshi008@gmail.com
[3] University of Texas at Austin, Austin, TX, USA
[4] TU Dortmund University, Dortmund, Germany
[5] Center for Trustworthy Data Science and Security, University Alliance Ruhr, Dortmund, Germany
[6] Department of Automation, Shanghai Jiao Tong University, Shanghai, China

Abstract. We study a class of reinforcement learning (RL) tasks where the objective of the agent is to accomplish temporally extended goals. In this setting, a common approach is to represent the tasks as deterministic finite automata (DFA) and integrate them into the state-space for RL algorithms. However, while these machines model the reward function, they often overlook the causal knowledge about the environment. To address this limitation, we propose the Temporal-Logic-based Causal Diagram (TL-CD) in RL, which captures the temporal causal relationships between different properties of the environment. We exploit the TL-CD to devise an RL algorithm in which an agent requires significantly less exploration of the environment. To this end, based on a TL-CD and a task DFA, we identify configurations where the agent can determine the expected rewards early during an exploration. Through a series of case studies, we demonstrate the benefits of using TL-CDs, particularly the faster convergence of the algorithm to an optimal policy due to reduced exploration of the environment.

Keywords: Reinforcement Learning · Causal Inference · Neuro-Symbolic AI

1 Introduction

In many reinforcement learning (RL) tasks, the objective of the agent is to accomplish temporally extended goals that require multiple actions to achieve.

The first three authors contributed equally.

Published by Springer Nature Switzerland AG 2023
A. Holzinger et al. (Eds.): CD-MAKE 2023, LNCS 14065, pp. 123–140, 2023.
https://doi.org/10.1007/978-3-031-40837-3_8

One common approach to modeling these goals is to use finite state machines. However, these machines only model the reward function and do not take into account the causal knowledge of the underlying environment, which can limit the effectiveness of the RL algorithms [2,3,6,17,20,25–28].

Moreover, online RL, including in the non-Markovian setting, often requires extensive interactions with the environment. This impedes the adoption of RL algorithms in real-world applications due to the impracticality of expensive and/or unsafe data collection during the exploration phase.

To address these limitations, in this paper we propose Temporal-Logic-based Causal Diagrams (TL-CDs) which can capture the temporal causal relationships between different properties of the environment, allowing the agent to make more informed decisions and require less exploration of the environment. TL-CDs combine temporal logic, which allows for reasoning about events over time, with causal diagrams, which represent the causal relationships between variables. By using TL-CDs, the RL algorithm can exploit the causal knowledge of the environment to identify configurations where the agent can determine the expected rewards early during an exploration, leading to faster convergence to an optimal policy.

We introduce an RL algorithm that leverages TL-CDs to achieve temporally extended goals. We show that our algorithm requires significantly less exploration of the environment than traditional RL algorithms that use finite state machines to model goals. By using TL-CDs, our algorithm identifies configurations where the agent can determine the expected rewards early during exploration, reducing the number of steps required to achieve the goal.

2 Motivating Example

Let us take a running example to illustrate the concept. There is a farmer who possesses a unique seed and his objective is to obtain a tree. There are two potential ways to achieve this goal. First, the farmer can plant the seed (p) and wait for the tree to grow (g). Alternatively, the farmer can sell the seed (s) and use the money to purchase a tree (b). The set of four propositions can thus be represented as $\mathcal{P} = \{p, g, s, b\}$. Figure 1a illustrates the corresponding task DFA $\mathcal{T}$ (note that, because $\mathcal{T}$ is deterministic, the transition from $v_0^{\mathcal{T}}$ to $v_2^{\mathcal{T}}$ also fires when $p \wedge s$ is true, but we assume that the agent cannot take both actions at once). Additional causal information is provided with the TL-CD $\mathfrak{C}$ (Fig. 1b), interpreted as follows: $p \twoheadrightarrow \mathbf{X}\, g$ expresses that planting a tree will result in a tree growing in the next time-step (e.g., year), and $s \twoheadrightarrow \mathbf{G} \neg \mathbf{X}\, b$ expresses that selling the seed leads to never being able to buy a tree (as the farmer will never find an offer for a tree that is cheaper than a seed). This TL-CD is equivalent to the *causal* DFA $\mathcal{C}$ illustrated in Fig. 1c (details are provided later).

3 Related Work

Causal inference answers questions about the mechanism by which manipulating one or a set of variables affects another variable or a set of variables [22]. In other

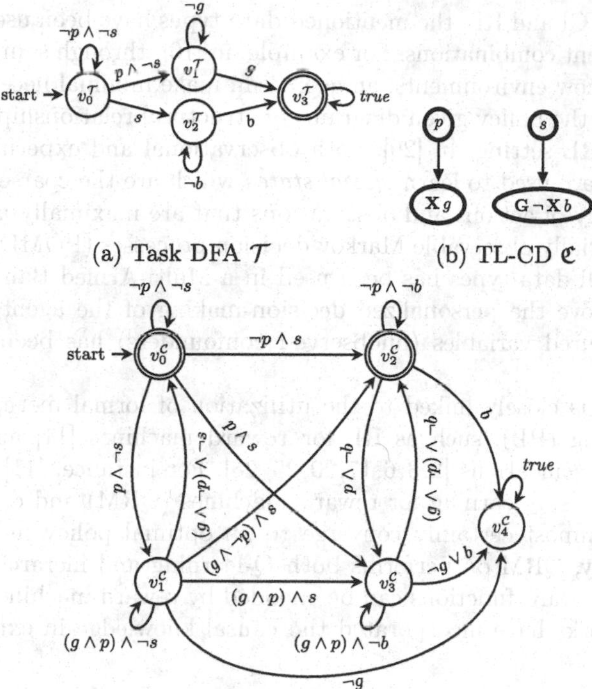

(a) Task DFA $\mathcal{T}$ (b) TL-CD $\mathfrak{C}$

(c) DFA $\mathcal{C}$ derived from $\mathfrak{C}$, and equivalent to the LTL$_f$ formula $\varphi^{\mathfrak{C}} = \mathbf{G}((p) \rightarrow (\mathbf{X}\, g)) \wedge \mathbf{G}((s) \rightarrow (\mathbf{G} \neg \mathbf{X}\, b))$

Fig. 1. The seed environment. The four propositions are p (the agent plants the seed), g (a tree grows), s (the agent sells the seed) and b (the agent buys a tree).

words, through causal inference, we infer the cause and effect relationships among the variables from observational data, experimental data, or a combination of both [16].

Recently, the inherent capabilities of reinforcement learning (RL) and causal inference (CI) have simultaneously been used for better decision-making including both interventional reasoning [13,14,24,30] and counterfactual reasoning [4,5,9,12] in different settings [15]. In other words, in an RL setting, harnessing casual knowledge including causal relationships between the actions, rewards, and intrinsic properties of the domain where the agent is deployed can improve the decision-making abilities of the agent [15].

Usually, incorporating CI in an RL setting can be done using three types of data including observational data, experimental data, and counterfactual data accompanied by the causal diagram of the RL setting, if available. An agent can have access to observational data by observing another agent, observing the environment, offline learning, acquiring prior knowledge about the underlying setting, etc. Experimental data can be acquired by actively interacting (intervening) with the environment. Counterfactual data can be generated using a specified model, estimated through active learning empirically [5,18,21].

In connecting CI and RL, the mentioned data types have been used by themselves or in different combinations. For example, in [10], through sampling observational data in new environments, an agent can make minimal necessary adaptions to optimize the policy given diagrams of structural relationship among the variables of the RL setting. In [29], both observational and experimental data (empirical data) are used to learn *causal states* which are the coarsest partition of the joint history of actions and observations that are maximally predictive of the future in partially observable Markov decision processes (POMDP). In [5], a combination of all data types has been used in a Multi-Armed Bandit problem in order to improve the personalized decision-making of the agent, where the effect of unmeasured variables (unobserved confounders) has been taken into consideration.

Our research is closely linked to the utilization of formal methods in reinforcement learning (RL), such as RL for reward machines [11] and RL with temporal logic specifications [2,3,6,17,20,25–28]. For instance, [11] proposed a technique known as Q-learning for reward machines (QRM) and demonstrated that QRM can almost certainly converge to an optimal policy in the tabular case. Additionally, QRM outperforms both Q-learning and hierarchical RL for tasks where the reward functions can be encoded by reward machines. However, none of these works have incorporated the causal knowledge in expediting the RL process.

4 Preliminaries

As typically done in RL problems, we rely on Markov Decision Processes (MDP) [23] to model the effects of sequential decisions of an RL agent. We, however, deviate slightly from the standard definition of MDPs. This is to be able to capture temporally extended goals for the agent and thus, want the reward to be non-Markovian. To capture non-Markovian rewards, we rely on simple finite state machines—deterministic finite automaton (DFA). Further, to express causal relationships in the environment, we rely on the de facto temporal logic, Linear Temporal Logic (LTL). We introduce all the necessary concepts formally in this section.

Labeled Markov Decision Process. A *labeled Markov decision process* [11] is a tuple $\mathcal{M} = \langle S, s_I, A, p, r, \mathcal{P}, L \rangle$ consisting of a finite set of states S, an agent's initial state $s_I \in S$, a finite set of actions A, a transition probability function $p : S \times A \mapsto \Delta(S)$, a non-Markovian reward function $r : (S \times A)^* \times S \mapsto \mathbb{R}$, a set of relevant propositions $\mathcal{P}$, and a labeling function $L : S \times A \times S \mapsto 2^{\mathcal{P}}$. Here $\Delta(S)$ denotes the set of all probability distributions over S. We denote by $p(s'|s, a)$ the probability of transitioning to state s' from state s under action a. Additionally, we include a set of propositions $\mathcal{P}$ that track the relevant information that the agent senses in the environment. We integrate the propositions in the labeled MDP using the labeling function L.

We define a *trajectory* to be the realization of the stochastic process defined by a labeled MDP. Formally, a trajectory is a sequence of states and actions $t = s_0 a_1 s_1 \cdots a_k s_k$ with $s_0 = s_I$. Further, we define the corresponding *label sequence* of t as $t^L := l_0 l_1 l_2 \cdots l_k$ where $l_i = L(s_i, a_{i+1}, s_{i+1})$ for each $0 \leq i < k$.

A stationary *policy* $\pi : S \to \Delta(A)$ maps states to probability distributions over the set of actions. In particular, if an agent is in state $s_t \in S$ at time step t and is following policy π, then $\pi(a_t|s_t)$ denotes its probability of taking action $a_t \in A$.

Deterministic Finite Automaton. A *deterministic finite automaton* (DFA) is a finite state machine described using tuple $\mathcal{A} = (V, 2^{\mathcal{P}}, \delta, v_I, F)$ where V is a finite set of states, $2^{\mathcal{P}}$ is the alphabet, $v_I \in V$ is the initial state, $F \subseteq V$ is the set of final states, and $\delta\colon V \times 2^{\mathcal{P}} \mapsto V$ is the deterministic transition function. We define the size $|\mathcal{A}|$ of a DFA as its number of states $|V|$.

A run of a DFA $\mathcal{A}$ on a label sequence $t^L = l_0 l_1 \ldots l_k \in (2^{\mathcal{P}})^*$, denoted using $\mathcal{A}\colon v_0 \xrightarrow{t^L} v_{k+1}$, is simply a sequence of states and labels $v_0 l_0 v_1 l_1 \cdots l_k v_{k+1}$, such that $v_0 = v_I$ and for each $0 \leq i \leq k$, $v_{i+1} = \delta(v_i, l_i)$. An accepted run is a run that ends in a final state $v_{k+1} \in F$. Finally, we define the language of $\mathcal{A}$ as $\mathcal{L}(\mathcal{A}) = \{t^L \in (2^{\mathcal{P}})^* \mid t^L \text{ is accepted by } \mathcal{A}\}$.

We define the parallel composition of two DFAs $\mathcal{A}^1 = (V^1, 2^{\mathcal{P}}, \delta^1, v_I^1, F^1)$ and $\mathcal{A}^2 = (V^2, 2^{\mathcal{P}}, \delta^2, v_I^2, F^2)$ to be the cross-product $\mathcal{A}^1 \times \mathcal{A}^2 = (V, 2^{\mathcal{P}}, \delta, v_I, F^2)$, where $V = V^1 \times V^2$, $\delta((s^1, s^2), l_i) = (\delta^1(s^1, l_i), \delta^2(s^2, l_i))$, $v_I = (v_I^1, v_I^2)$, and $F = F^1 \times F^2$. Using such a definition for parallel composition, it is not hard to verify that language $\mathcal{L}(\mathcal{A}^1 \times \mathcal{A}^2)$ is simply $\mathcal{L}(\mathcal{A}^1) \cap \mathcal{L}(\mathcal{A}^2)$.

Task DFA. Following some recent works [1,19], we rely on so-called *task DFA* $\mathcal{T} = \langle V^{\mathcal{T}}, 2^{\mathcal{P}}, \delta^{\mathcal{T}}, v_I^{\mathcal{T}}, F^{\mathcal{T}} \rangle$ to represent the structure of a non-Markovian reward function. We say a trajectory t has a positive reward if and only if the run of $\mathcal{T}$ on the label sequence t^L, $\mathcal{T}\colon v_0^{\mathcal{T}} \xrightarrow{t^L} v_{k+1}^{\mathcal{T}}$, ends in a final state $v_{k+1}^{\mathcal{T}} \in F^{\mathcal{T}}$.

Linear Temporal Logic. *Linear temporal logic* (over finite traces) (LTL$_f$) is a logic that expresses temporal properties using temporal modalities. Formally, we define LTL$_f$ formulas—denoted by Greek small letters—inductively as:

- each proposition $p \in \mathcal{P}$ is an LTL$_f$ formula; and
- if ψ and φ are LTL$_f$ formulas, so are $\neg\psi$, $\psi \vee \varphi$, $\mathbf{X}\,\psi$ ("neXt"), and $\psi \mathbf{U}\, \varphi$ ("Until").

As syntactic sugar, we allow Boolean constants *true* and *false*, and formulas $\psi \wedge \varphi := \neg(\neg\psi \vee \neg\varphi)$ and $\psi \to \varphi := \neg\psi \vee \varphi$. Moreover, we additionally allow commonly used temporal formulas $\mathbf{F}\,\psi := \mathit{true}\mathbf{U}\,\psi$ ("finally") and $\mathbf{G} := \neg\mathbf{F}\,\neg\varphi$ ("globally").

To interpret LTL$_f$ formulas over (finite) trajectories, we follow the semantics proposed by Giacomo and Vardi [7]. Given a label sequence t^L, we define recursively when an LTL$_f$ formula holds at position i, i.e., $t^L, i \models \varphi$, as follows:

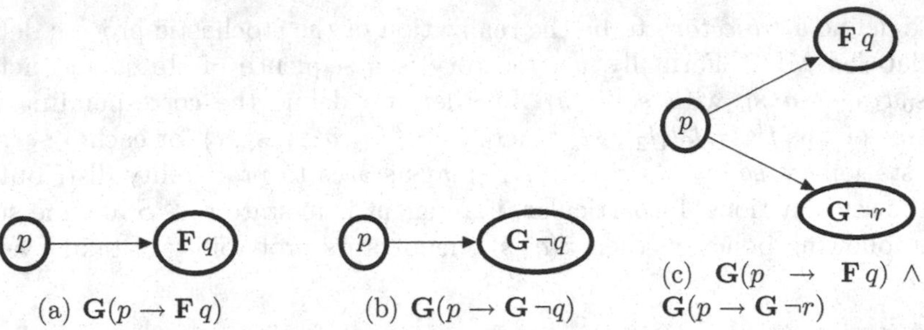

(a) $\mathbf{G}(p \rightarrow \mathbf{F}\,q)$ (b) $\mathbf{G}(p \rightarrow \mathbf{G}\,\neg q)$ (c) $\mathbf{G}(p \rightarrow \mathbf{F}\,q) \wedge \mathbf{G}(p \rightarrow \mathbf{G}\,\neg r)$

Fig. 2. Examples of TL-CDs with their corresponding description in LTL

$$
\begin{aligned}
&t^L, i \models p \text{ if and only if } p \in t^L[i] \\
&t^L, i \models \neg\varphi \text{ if and only if } t^L, i \not\models \varphi \\
&t^L, i \models \varphi \vee \psi \text{ if and only if } t^L, i \models \varphi \text{ or } t^L, i \models \psi \\
&t^L, i \models \mathbf{X}\,\varphi \text{ if and only if } i < |t^L| \text{ and } t^L, i+1 \models \varphi \\
&t^L, i \models \varphi \mathbf{U}\,\psi \text{ if and only if } t^L, j \models \psi \text{ for some} \\
&\qquad i \leq j \leq |t^L| \text{ and } t^L, i' \models \varphi \text{ for all } i \leq i' < j
\end{aligned}
$$

We say t^L *satisfies* φ if $t^L \models \varphi$, which, in short, is written as $t^L \models \varphi$.

Any LTL_f formula φ can be translated to an equivalent DFA $\mathcal{A}^\varphi$, that is, for any $t^L \in (2^\mathcal{P})^*$, $t^L \models \varphi$ if and only if $t^L \in \mathcal{L}(\mathcal{A}^\varphi)$ [7,31].

Deterministic Causal Diagrams. A causal diagram where every edge represents a cause leading to an effect with probability 1 is called a deterministic causal diagram. In a deterministic causal diagram, the occurrence of the cause will always result in the occurrence of the effect.

5 Temporal-Logic-Based Causal Diagrams

We now formalize causality in RL using (deterministic) Causal Diagrams [8], a concept that is widely used in the field of Causal Inference. We here augment Causal Diagrams with temporal logic since we like to express temporally extended relations. We call such Causal Diagrams as Temporal-Logic-based Causal Diagrams or TL-CDs in short. While, in principle, TL-CDs can be conceived for several temporal logics, we consider LTL_f due to its popularity in AI applications [7].

Structurally, for a given set of propositions $\mathcal{P}$, a *Temporal-Logic-based Causal Diagram* (TL-CD) is a directed acyclic graph $\mathfrak{C}$ where

- each node represents an LTL_f formula over propositions $\mathcal{P}$, and
- each edge ($\twoheadrightarrow$) represents a causal link between two nodes.

Examples of TL-CDs are illustrated in Fig. 2, where, in the causal relation $\psi \twoheadrightarrow \varphi$; ψ is considered to be the cause and φ to be the effect. The TL-CD in Fig. 2a describes that whenever the cause p happens, the effect q eventually (i.e., $\mathbf{F}\,q$) occurs. The TL-CD in Fig. 2b describes that whenever the cause p happens, the effect q never (i.e., $\mathbf{G}\neg q$) occurs. The TL-CD in Fig. 2c describes that whenever the cause p happens, effects q eventually (i.e., $\mathbf{F}\,q$) occurs and r never (i.e., $\mathbf{G}\neg r$) occurs.

For a TL-CD to be practically relevant, we must impose that the occurrence of the cause ψ must precede that of the effect φ. To do so, we introduce concepts that track the time of occurrence of an event such as the worst-case satisfaction $w_s(\varphi)$, the worst-case violation $w_v(\varphi)$, the best-case satisfaction $b_s(\varphi)$ and the best-case violation of a formula φ. Intuitively, the worst-case satisfaction $w_s(\varphi)$ (resp., the best-case satisfaction $b_s(\varphi)$) tracks the last (resp., the first) possible time point that a formula can get satisfied by a trajectory. Likewise, the worst-case violation $w_v(\varphi)$ (resp., the best-case violation $b_v(\varphi)$) tracks the last (resp., the first) possible time point that a formula can get violated by a trajectory. We introduce all the concepts formally in the following definition.

Definition 1. *For an LTL formula φ, we define the worst-case satisfaction time $w_s(\varphi)$, best-case satisfaction time $b_s(\varphi)$, worst-case violation time $w_v(\varphi)$ inductively on the structure of φ as follows:*

$$b_s(p) = w_s(p) = b_v(p) = w_v(p) = 0,$$

$$\neg\varphi : \begin{cases} b_s(\neg\varphi) = b_v(\varphi), \\ w_s(\neg\varphi) = w_v(\varphi), \\ b_v(\neg\varphi) = b_s(\varphi), \\ w_v(\neg\varphi) = w_s(\varphi); \end{cases}$$

$$\varphi_1 \wedge \varphi_2 : \begin{cases} b_s(\varphi_1 \wedge \varphi_2) = \max\{b_s(\varphi_1), b_s(\varphi_2)\}, \\ w_s(\varphi_1 \wedge \varphi_2) = \max\{w_s(\varphi_1), w_s(\varphi_2)\}, \\ b_v(\varphi_1 \wedge \varphi_2) = \min\{b_v(\varphi_1), b_v(\varphi_2)\}, \\ w_v(\varphi_1 \wedge \varphi_2) = \max\{w_v(\varphi_1), w_v(\varphi_2)\}; \end{cases}$$

$$\varphi_1 \vee \varphi_2 : \begin{cases} b_s(\varphi_1 \vee \varphi_2) = \min\{b_s(\varphi_1), b_s(\varphi_2)\}, \\ w_s(\varphi_1 \vee \varphi_2) = \max\{w_s(\varphi_1), w_s(\varphi_2)\}, \\ b_v(\varphi_1 \vee \varphi_2) = \max\{b_v(\varphi_1), b_v(\varphi_2)\}, \\ w_v(\varphi_1 \vee \varphi_2) = \max\{w_v(\varphi_1), w_v(\varphi_2)\}; \end{cases}$$

$$\mathbf{G}\varphi : \begin{cases} b_s(\mathbf{G}\varphi) = w_s(\mathbf{G}\varphi) = w_v(\mathbf{G}\varphi) = \infty, \\ b_v(\mathbf{G}\varphi) = b_v(\varphi); \end{cases}$$

$$\mathbf{F}\varphi : \begin{cases} b_s(\mathbf{F}\varphi) = b_s(\varphi), \\ w_s(\mathbf{F}\varphi) = w_v(\mathbf{F}\varphi) = b_v(\mathbf{F}\varphi) = \infty; \end{cases}$$

$$\mathbf{X}\varphi : \begin{cases} b_s(\mathbf{X}\varphi) = b_s(\varphi) + 1, \\ w_s(\mathbf{X}\varphi) = w_s(\varphi) + 1, \\ w_v(\mathbf{X}\varphi) = w_v(\varphi) + 1, \\ b_v(\mathbf{X}\varphi) = b_v(\varphi) + 1; \end{cases}$$

$$\varphi_1\mathbf{U}\varphi_2 : \begin{cases} b_s(\varphi_1\mathbf{U}\varphi_2) = b_s(\varphi_2), \\ w_s(\varphi_1\mathbf{U}\varphi_2) = w_v(\varphi_1\mathbf{U}\varphi_2) = \infty, \\ b_v(\varphi_1\mathbf{U}\varphi_2) = b_v(\varphi_1). \end{cases}$$

For each causal relation $\psi \twoheadrightarrow \varphi$ in a causal diagram $\mathfrak{C}$, we impose the constraints that $w_s(\psi) \leq \min\{b_s(\varphi), b_v(\varphi)\}$ and $w_v(\psi) \leq \min\{b_s(\varphi), b_v(\varphi)\}$. Such a constraint is designed to make sure that the cause ψ, even in the worst-time scenario, occurs before the event φ, in the best-time scenario. Based on the constraint, we rule out causal relations in which the cause occurs after the effect such as $\mathbf{X}p \twoheadrightarrow q$, where $w_s(\mathbf{X}p) = 1$ is greater than $b_s(q) = 0$.

To express the meaning of TL-CD in formal logic, we turn to its description in LTL. A causal relation $\psi \twoheadrightarrow \varphi$ can be described using the LTL$_f$ formula $\mathbf{G}(\psi \rightarrow \varphi)$, which expresses that whenever ψ occurs, φ should also occur. Further, an entire TL-CD $\mathfrak{C}$ can be described using the LTL$_f$ formula $\varphi^{\mathfrak{C}} := \bigwedge_{(\psi \twoheadrightarrow \varphi)} \mathbf{G}(\psi \rightarrow \varphi)$ which is simply the conjunction of the LTL$_f$ formulas corresponding to each causal relation in $\mathfrak{C}$.

Based on its description in LTL$_f$ $\varphi^{\mathfrak{C}}$, we can now define when a trajectory π satisfies a TL-CD $\mathfrak{C}$. Precisely, t satisfies $\mathfrak{C}$ if and only if its label sequence t^L satisfies $\varphi^{\mathfrak{C}}$.

In the subsequent sections, we also rely on a representation of a TL-CD as a deterministic finite automaton (DFA). In particular, for a TL-CD $\mathfrak{C}$, we can construct a DFA $\mathcal{C}^{\mathfrak{C}} = \langle V^{\mathcal{C}}, 2^{\mathcal{P}}, \delta^{\mathcal{C}}, v_I^{\mathcal{C}}, F^{\mathcal{C}} \rangle$ from its description in LTL$_f$ $\varphi^{\mathfrak{C}}$. We call such a DFA a *causal DFA*. When the TL-CD is clear from the context, we simply represent a causal DFA as $\mathcal{C}$, dropping its superscript.

In the motivating example (Sect. 2), the TL-CD $\mathfrak{C}$ pictured in Fig. 1b translates into the LTL$_f$ formula $\varphi^{\mathfrak{C}} = \mathbf{G}(p \rightarrow \mathbf{X}g) \wedge \mathbf{G}(s \rightarrow \mathbf{G}\neg\mathbf{X}b)$, which is equivalent to the causal DFA $\mathcal{C}$ pictured in Fig. 1c.

6 Reinforcement Learning with Causal Diagrams

We now aim to utilize the information provided in a Temporal-Logic-based Causal Diagram (TL-CD) to enhance the process of reinforcement learning in a non-Markovian setting. However, in our setting, we assume a TL-CD to be a ground truth about the causal relations in the underlying environment. As a result, we must ensure that a TL-CD is compatible with a labeled MDP.

Intuitively, a TL-CD $\mathfrak{C}$ is compatible with a labeled MDP $\mathcal{M}$ if all possible trajectories of $\mathcal{M}$ respect (i.e., do not violate) the TL-CD $\mathfrak{C}$. To define compatibility formally, we rely on the cross-product $\overline{\mathcal{M}} \times \mathcal{C}$, where $\overline{\mathcal{M}}$ is a (non-deterministic) finite state machine representation of $\mathcal{M}$ with states S, alphabet $2^{\mathcal{P}}$, transition $\delta(s, l) = \{s' \in S \mid L(s, a, s') = l \text{ for some } a \in A\}$, initial state s_I and final states S, and $\mathcal{C}$ is the causal DFA.

Formally, we say that a TL-CD $\mathfrak{C}$ is *compatible* with an MDP $\mathcal{M}$ if from any reachable state $(s, q) \in \overline{\mathcal{M}} \times \mathcal{C}$, one can always reach a state $(s', q') \in \overline{\mathcal{M}} \times \mathcal{C}$ where q' is a final state in the causal DFA $\mathcal{C}$. The above formal definition ensures that any trajectory of $\mathcal{M}$ can be continued to satisfy the causal relations defined by $\mathfrak{C}$.

We are now ready to state the central problem of the paper.

Problem 1 (Non-Markovian Reinforcement learning with Causal Diagrams). Let $\mathcal{M}$ be a labeled MDP, $\mathcal{T}$ be a task DFA, and $\mathfrak{C}$ be a Temporal Logic based Causal Diagram (TL-CD) such that $\mathfrak{C}$ is compatible with $\mathcal{M}$. Given $\mathcal{M}$, $\mathcal{T}$ and $\mathfrak{C}$, learn a policy that achieves a maximal reward in the environment.

We view TL-CDs as a concise representation of the causal knowledge in an environment. In our next subsection, we develop an algorithm that exploits this causal knowledge to alleviate the issues of extensive interaction in an online RL setting.

6.1 Q-Learning with Early Stopping

Our RL algorithm is an adaptation of QRM [11], which is a Q-learning algorithm [23] that is typically used when rewards are specified as finite state machines. On a high level, QRM explores the product space $\mathcal{M} \times \mathcal{T}$ in many episodes in the search for an optimal policy. We modify QRM by stopping its exploration early based on the causal knowledge from a TL-CD $\mathfrak{C}$. Before we describe the algorithm in detail, we must introduce some concepts that aid the early stopping.

For early stopping to work, the learning agent must keep track of whether a trajectory can lead to a reward. We do this by keeping track of the current configuration in the synchronized run of a trajectory on the product $\mathcal{T} \times \mathcal{C}$ of the task DFA and the causal DFA. We here identify two particular configurations that can be useful for early stopping: *causally accepting* and *causally rejecting.* Intuitively, a trajectory reaches a causally accepting configuration if all continuations of the trajectory from the current configuration that satisfy the TL-CD $\mathfrak{C}$ receive a reward of 1 (or a positive reward). On the other hand, a trajectory reaches a causally rejecting configuration if all continuations of the trajectory from the current configuration that satisfy the TL-CD $\mathfrak{C}$, do not receive a reward.

We formalize the notion of causally accepting configurations and causal rejecting configurations in the following two definitions. We use the terminology $\mathcal{A}_q$ to describe a DFA that is structurally identical to DFA $\mathcal{A}$, except that its initial state is q.

Definition 2 (Causally accepting). *We say $(v^{\mathcal{T}}, v^{\mathcal{C}}) \in V^{\mathcal{T}} \times V^{\mathcal{C}}$ is causally accepting if for each $t^L \in (2^{\mathcal{P}})^*$ for which the run $\mathcal{C} : v^{\mathcal{C}} \xrightarrow{t^L} v_f^{\mathcal{C}}$ ends in some final state $v_f^{\mathcal{C}} \in F^{\mathcal{C}}$, the run $\mathcal{T} : v^{\mathcal{T}} \xrightarrow{t^L} v_f^{\mathcal{T}}$ must also end in some final state $v_f^{\mathcal{T}} \in ^{\mathcal{T}}$. Equivalently, we say that $(v^{\mathcal{T}}, v^{\mathcal{C}}) \in V^{\mathcal{T}} \times V^{\mathcal{C}}$ is causally accepting if $\mathcal{L}(\mathcal{C}_{v^{\mathcal{C}}}) \subseteq \mathcal{L}(\mathcal{T}_{v^{\mathcal{T}}})$.*

Definition 3 (Causally rejecting). *We say $(v^{\mathcal{T}}, v^{\mathcal{C}}) \in V^{\mathcal{T}} \times V^{\mathcal{C}}$ is causally rejecting if for each $t^L \in (2^{\mathcal{P}})^*$ for which the run $\mathcal{C} : v^{\mathcal{C}} \xrightarrow{t^L} v_f^{\mathcal{C}}$ ends in some final state $v_f^{\mathcal{C}} \in F^{\mathcal{C}}$, the run $\mathcal{T} : v^{\mathcal{T}} \xrightarrow{t^L} v_f^{\mathcal{T}}$ must not end in any final state in $F^{\mathcal{T}}$. Equivalently, we say that $(v^{\mathcal{T}}, v^{\mathcal{C}}) \in V^{\mathcal{T}} \times V^{\mathcal{C}}$ is causally rejecting if $\mathcal{L}(\mathcal{C}_{v^{\mathcal{C}}}) \cap \mathcal{L}(\mathcal{T}_{v^{\mathcal{T}}}) = \emptyset$.*

Remark 1. A configuration $(v^{\mathcal{T}}, v^{\mathcal{C}})$ may be neither causally accepting nor causally rejecting.

To illustrate these concepts, we consider the motivating example introduced in Sect. 2, where $\mathcal{T}$ and $\mathcal{C}$ are depicted in Figs. 1a and 1c. The initial state $(v_0^{\mathcal{T}}, v_0^{\mathcal{C}})$ is neither causally accepting nor causally rejecting. If the agent decides to plant the seed, it encounters a label p and reaches $(v_1^{\mathcal{T}}, v_1^{\mathcal{C}})$, which is causally accepting since the only reachable configurations where $\mathcal{C}$ is accepting $\{(v_3^{\mathcal{T}}, v_0^{\mathcal{C}}), (v_3^{\mathcal{T}}, v_2^{\mathcal{C}})\}$ are accepting for $\mathcal{T}$. If the agent decides to sell the seed instead, it encounters a label s and reaches $(v_2^{\mathcal{T}}, v_2^{\mathcal{C}})$, which is causally rejecting since the only reachable configurations where $\mathcal{C}$ is accepting $\{(v_2^{\mathcal{T}}, v_2^{\mathcal{C}})\}$ are rejecting for $\mathcal{T}$.

Algorithm 1: causally accepting/rejecting state detection

Input: Task DFA $\mathcal{T}$, Causal DFA $\mathcal{C}$, a pair of states $(v^{\mathcal{T}}, v^{\mathcal{C}}) \in V^{\mathcal{T}} \times V^{\mathcal{C}}$.
$VC^{\mathcal{T}} \leftarrow \emptyset$ // set of reachable "causal states" of $\mathcal{T}$
$\mathcal{AP} \leftarrow \mathcal{T} \times \mathcal{C}$ // the parallel composition of $\mathcal{T}$ and $\mathcal{C}$
foreach *state* $(v_r^{\mathcal{T}}, v_r^{\mathcal{C}})$ *of* $\mathcal{AP}$ *reachable from* $(v^{\mathcal{T}}, v^{\mathcal{C}})$ **do**
  if $v_r^{\mathcal{C}} \in F^{\mathcal{C}}$ **then**
    $VC^{\mathcal{T}} \leftarrow VC^{\mathcal{T}} \cup \{v_r^{\mathcal{T}}\}$
// $(v^{\mathcal{T}}, v^{\mathcal{C}})$ is causally accepting if $VC^{\mathcal{T}} \subseteq F^{\mathcal{T}}$
// $(v^{\mathcal{T}}, v^{\mathcal{C}})$ is causally rejecting if $VC^{\mathcal{T}} \cap F^{\mathcal{T}} = \emptyset$
return $VC^{\mathcal{T}}$

We now present the pseudo-code of the algorithm used for detecting the causally accepting and causally rejecting configurations in Algorithm 1. Intuitively, the algorithm relies on a breadth-first search on the cross-product DFA $\mathcal{T} \times \mathcal{C}$ of the task DFA and the causal DFA.

Remark 2. The worst-case runtime of Algorithm 1 is $\mathcal{O}(|2^{\mathcal{P}}| \cdot |\mathcal{T}| \cdot |\mathcal{C}|)$ since the number of edges in the parallel composition $\mathcal{T} \times \mathcal{C}$ can be atmost $|2^{\mathcal{P}}| \cdot |\mathcal{T}| \cdot |\mathcal{C}|$.

Algorithm 2: Q-learning with TL-CD

Input: Labeled MDP $\mathcal{M}$, Task DFA $\mathcal{T}$, Causal diagram $\mathfrak{C}$.
Convert $\mathfrak{C}$ to causal DFA $\mathcal{C}$
Detect causally accepting and rejecting states of $\mathcal{T} \times \mathcal{C}$
foreach *training episode* **do**
  run QTLCD_episode

We now expand on our adaptation of the QRM algorithm. The pseudo-code of the algorithm is sketched in Algorithm 2. In the algorithm, we first compute the set of causally accepting or causally rejecting configurations as described in

Algorithm 1. Next, like a typical Q-learning algorithm, we perform explorations of the environment in several episodes to estimate the Q-values of the state-action pairs. However, during an episode, we additionally keep track of the configuration of the product $\mathcal{T} \times \mathcal{C}$. If during the episode, we encounter a causally accepting or causally rejecting configuration, we terminate the episode and update the Q-values accordingly.

Algorithm 3: QTLCD_episode

```
Hyperparameter: Q-learning parameters, episode length eplength.
Input: labeled MDP M, task DFA T, causal DFA C, learning rate α.
Output: the updated set of q-functions Q
s ← s_I; v^T ← v_I^T; v^C ← v_I^C                       // initialise states
R ← 0                                       // initialise cumulative reward
for 0 ≤ t < eplength do
    a ← GetEpsilonGreedyAction(q^{v^T}, s)           // get action from policy
    s' ← ExecuteAction(p(s, a))             // based on distribution p(s, a)
    v^T' ← δ^T(v^T, L(s, a, s'))                          // synchronize T
    v^C' ← δ^C(v^C, L(s, a, s'))                          // synchronize C
    R' ← 1_{F^T}(v^T')                          // compute cumulative reward
    // override reward based on causal analysis:
    if (v^T', v^C') causally accepting then R' ← 1
    if (v^T', v^C') causally rejecting then R' ← 0
    update q^{v^T}(s, a) using reward r = R' − R             // Bellman update
    if v^T' ∈ F^T then return Q
                                                          // end of episode
    if (v^T', v^C') causally accepting or rejecting then return Q
                                                // interrupt episode early
    s ← s'; v^T ← v^T'; v^C ← v^C'; R ← R'
return Q
```

The Q-learning with TL-CD algorithms consists of a loop of several episodes. The pseudo-code of one episode is sketched in Algorithm 3. The instant reward r is computed such that the cumulative reward R is 1 if and only if the Task DFA is accepting. The cumulative reward is then overridden if it is possible to predict the future cumulative reward, based on if the current configuration is causally accepting or causally rejecting. If the reward could be predicted, the episode is interrupted right after updating the q-functions, using that predicted reward. Note that the notion causally accepting and rejecting configurations is defined on unbounded episodes, and might predict a different reward than if the episode were to time out.

The above algorithm follows the exact steps of the QRM algorithm and thus, inherits all its advantages, including termination and optimality. The only notable difference is the early stopping based on the causally accepting and causally rejecting states. However, when these configurations are reached, based on their definition, all continuations are guaranteed to return positive and no reward, respectively. Thus, early stopping helps to determine the future reward and update the estimates of Q-value earlier. We next demonstrate the advantages of the algorithm experimentally.

7 Case Studies

In this section, we implement the proposed Q-learning with TL-CD (`QTLCD`) algorithm in comparison with a baseline algorithm in three different case studies.

In each case study, we compare the performance of the following two algorithms:

- Q-learning with TL-CD (`QTLCD`): the proposed algorithm, including early stopping of the episodes when a causally accepting/rejecting state is reached
- Q-learning with Reward Machines (`QRM`): the algorithm from [11], with the same MDP and RM but no causal diagram.

7.1 Case Study I: Small Office World Domain

We consider a small officeworld scenario in a 17×9 grid. The agent's objective is to first reach the location of either one key k_1 or k_2 and then exit the grid by reaching either e_1 or e_2. The agent navigates on the grid with walls, keys, and one-way doors. The set of actions is $A = \{S, N, E, W\}$. The action S, N, E, W correspond to moving in the four cardinal directions. The one-way doors are shown in Fig. 3a with green arrows. We specify the complex task of the RL agent in a maze as a *deterministic finite automaton (DFA)* $\mathcal{T}$ (see Fig. 3), as both event sequences $a - k_1 - e_1$ (open door a, pick up key k_1, exit at e_1) and $b - k_2 - e_2$ (open door b, pick up key k_2, exit at e_2) lead to completion of the task of exiting the maze (receiving reward 1). The agent starts at position o. If TL-CD in Fig. 3d is true, then the RL agent should never go along the sequence $b - k_2 - e_2$ as $k_2 \blacktriangleright \mathbf{G} \neg e_2$ means if the agent picks up key k_2 then it can never exit at e_2 (as c is a one-way door, so the agent can never get outside Room 3 once it enters c to pick up key k_2).

Results: Figure 3b presents the performance comparison of the RL agent with TL-CD and without TL-CD. It shows that the accumulated reward of the RL agent can converge to its optimal value around 1.5 times faster if the agent knows the TL-CD and learns never to open door b.

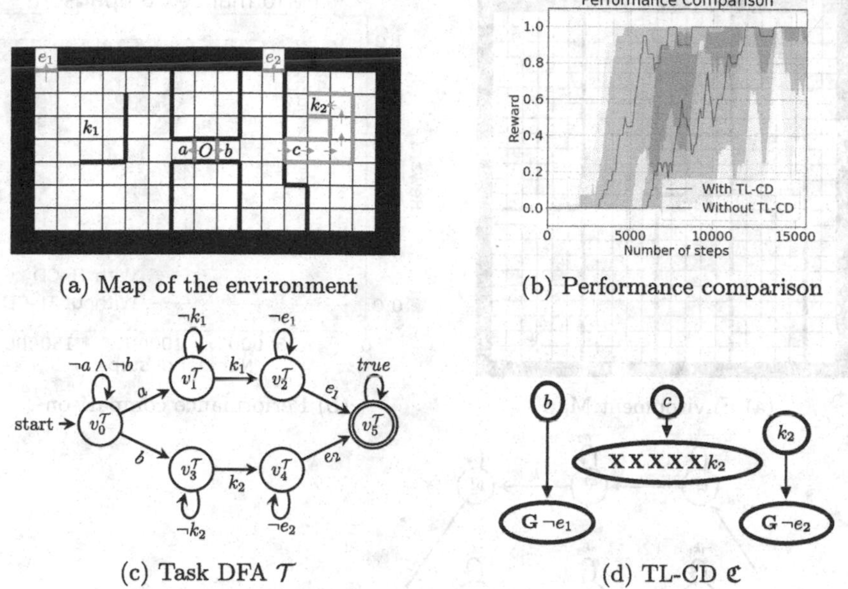

(a) Map of the environment

(b) Performance comparison

(c) Task DFA $\mathcal{T}$

(d) TL-CD $\mathfrak{C}$

Fig. 3. Case study I: small office world. The rewards attained in 10 independent simulation runs averaged for every 10 training steps.

7.2 Case Study II: Large Office World Domain

We consider a large office world scenario in a 25×25 grid. The objective of the agent is to collect both the keys, k_1 and k_2, and then exit the grid from e_1 or e_2 (since we assume that there are two locks in both exits, e_1 and e_2, which needs both keys to unlock). To achieve this objective, the agent needs to collect key k_1 first and then key k_2 since if it collects key k_2 first then it cannot be able to collect key k_1 in the map. The set of actions is $A = \{S, N, E, W\}$. The action S, N, E, W correspond to moving in the four cardinal directions. The one-way doors are shown in Fig. 4a with green arrows. The motivation behind this example is to observe the effect of increasing causally rejecting states on RL agents' performance. The task DFA and the TL-CD are depicted in Sect. 7.2.

Results: Figure 4b presents the performance comparison of the RL agent in a large office world scenario with TL-CD (`QTLCD`) and without TL-CD (`QRM`). It shows that the RL agent can converge to its optimal value around 3 times faster if the agent knows the TL-CD.

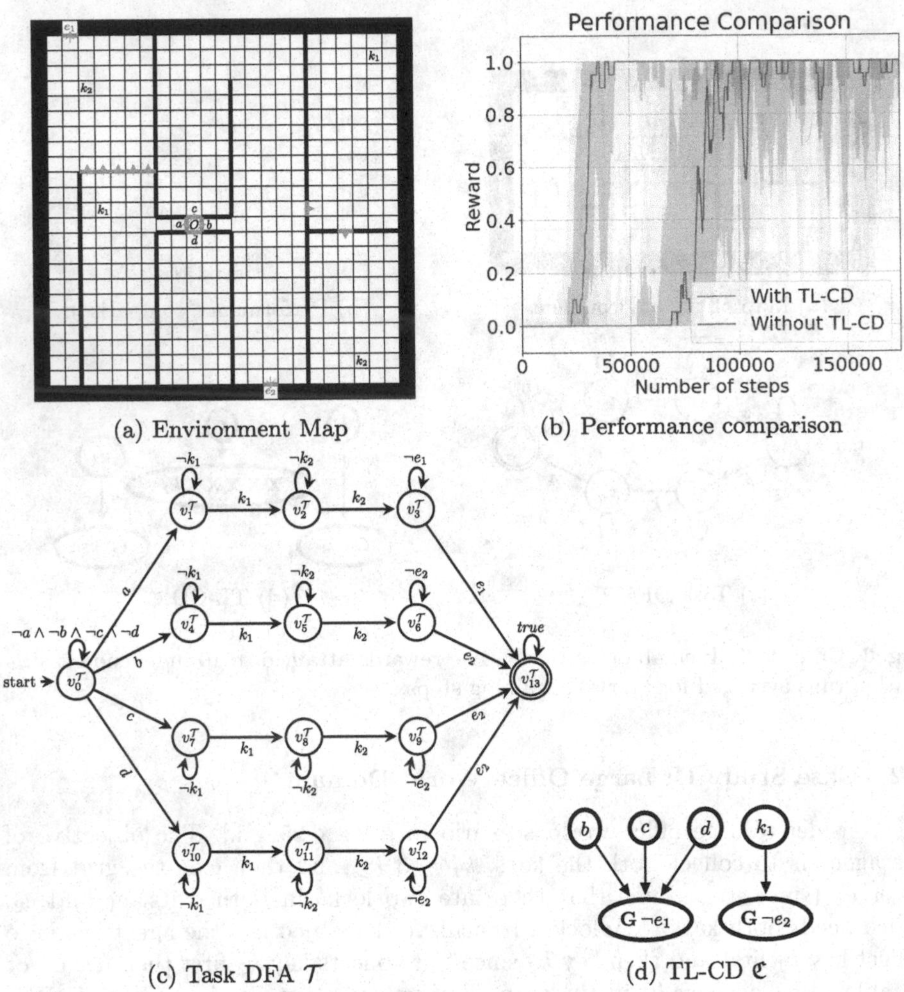

(a) Environment Map (b) Performance comparison

(c) Task DFA $\mathcal{T}$ (d) TL-CD $\mathfrak{C}$

Fig. 4. Case study II: large office world environment. The rewards attained in 10 independent simulation runs averaged for every 10 training steps.

7.3 Case Study III: Crossroad Domain

This experiment is inspired by the real-world example of crossing the road at a traffic signal. The agent's objective is to reach the other side of the road. The agent navigates on a grid with walls, crossroad, button, and light signal. The agent starts from a random location in the grid. The set of actions is $A = \{S, N, E, W, PressButton, Wait\}$. The action $PressButton$ presses the button at the crossroad to indicate it wants to cross the road. After pressing the button, at some later time, the pedestrian crossing light will be turned ON. The action $Wait$ will let the agent stay at a location. The actions S, N, E, W correspond to moving in the four cardinal directions.

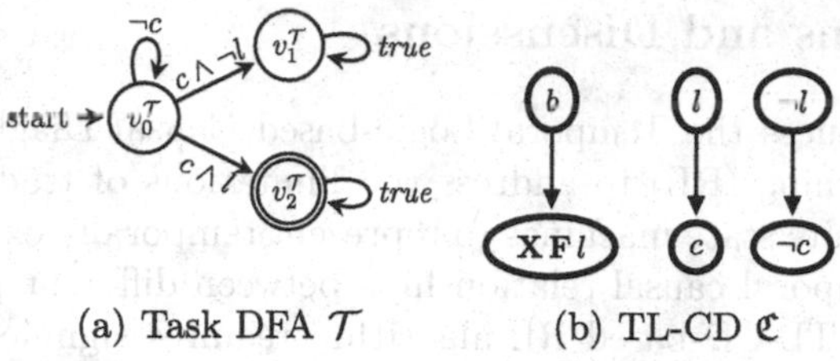

(a) Task DFA $\mathcal{T}$ (b) TL-CD $\mathfrak{C}$

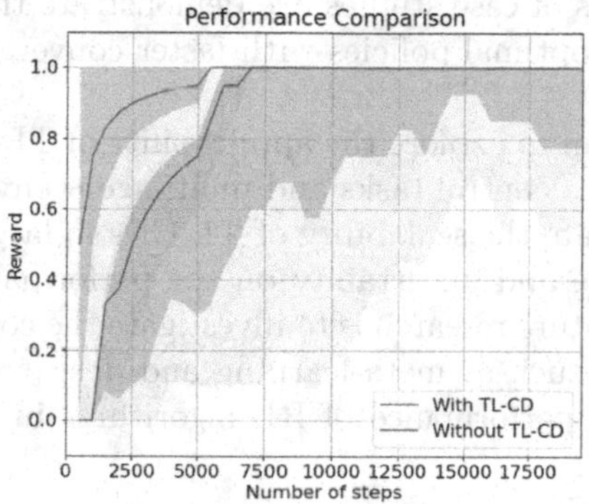

(c) Performance comparison

Fig. 5. Case study III: crossroad domain. The rewards attained in 10 independent simulation runs averaged for every 10 training steps.

To simplify the problem, we make the following assumptions. (1) The agent starts from a fixed location in the left half (one side) of the grid. (2) The agent knows to cross when the light is ON as after pressing the button the agent has only one valid action. That is, when the light is OFF, it can only wait; and when the light is ON, it will cross. After pressing the button at the crossroad, the crossing light will turn ON N steps later, where N is a random variable following a geometric distribution of success probability 0.01. Thus, the underlying MDP has five observable variables: x, y, the discrete coordinates of the agent on the grid, and b, p, l, three Boolean flags that indicate respectively that the button is currently pressed, that the button has been pressed and the light is still OFF, and that the light is turned ON. We specify that task is completed if the agent successfully crosses the road when the light signal is ON (see Fig. 5a).

We consider the causal LTL specification $\mathbf{G}(b \rightarrow \mathbf{F}\,\mathbf{X}\,l) \wedge \mathbf{G}(l \leftrightarrow c)$ (equivalent to the TL-CD in Fig. 5b), where the first part of the conjunction represents the knowledge that the pedestrian light has to turn ON some time later, and the second part represents the policy of the agent, because we suppose that the agent already knows to cross if and only if the light is on. Under these conditions, pressing the button leads to a causally accepting state.

Results: Figure 5c presents the performance comparison of the RL agent in the crossroad domain with TL-CD (`QTLCD`) and without TL-CD (`QRM`). It shows the RL agent will converge faster on average if it knows the TL-CD.

8 Conclusions and Discussions

This paper introduces the Temporal-Logic-based Causal Diagram (TL-CD) in reinforcement learning (RL) to address the limitations of traditional RL algorithms that use finite state machines to represent temporally extended goals. By capturing the temporal causal relationships between different properties of the environment, our TL-CD-based RL algorithm requires significantly less exploration of the environment and can identify expected rewards early during exploration. Through a series of case studies, we demonstrate the effectiveness of our algorithm in achieving optimal policies with faster convergence than traditional RL algorithms.

In the future, we plan to explore the applicability of TL-CDs in other RL settings, such as continuous control tasks and multi-agent environments. Additionally, we aim to investigate the scalability of TL-CDs in large-scale environments and the impact of noise and uncertainty on the performance of the algorithm. Another direction for future research is to investigate the combination of TL-CDs with other techniques, such as meta-learning and deep reinforcement learning, to further improve the performance of RL algorithms in achieving temporally extended goals.

Acknowledgements. This work has been supported by NSF CNS 2304863, ONR N00014-23-1-2505 and DFG grant number 434592664, ONR N00014-22-1-2254, NSF CNS-1836900

References

1. Abate, A., Almulla, Y., Fox, J., Hyland, D., Wooldridge, M.J.: Learning task automata for reinforcement learning using hidden Markov models. CoRR abs/2208.11838 (2022)
2. Aksaray, D., Jones, A., Kong, Z., Schwager, M., Belta, C.: Q-learning for robust satisfaction of signal temporal logic specifications. In: IEEE CDC 2016, pp. 6565–6570 (2016). https://doi.org/10.1109/CDC.2016.7799279
3. Alshiekh, M., Bloem, R., Ehlers, R., Könighofer, B., Niekum, S., Topcu, U.: Safe reinforcement learning via shielding. In: AAAI 2018 (2018)
4. Bareinboim, E., Forney, A., Pearl, J.: Bandits with unobserved confounders: a causal approach. In: Cortes, C., Lawrence, N., Lee, D., Sugiyama, M., Garnett, R. (eds.) Advances in Neural Information Processing Systems, vol. 28. Curran Associates, Inc. (2015). https://proceedings.neurips.cc/paper/2015/file/795c7a7a5ec6b460ec00c5841019b9e9-Paper.pdf
5. Forney, A., Pearl, J., Bareinboim, E.: Counterfactual data-fusion for online reinforcement learners. In: Precup, D., Teh, Y.W. (eds.) Proceedings of the 34th International Conference on Machine Learning. Proceedings of Machine Learning Research, vol. 70, pp. 1156–1164. PMLR (2017). https://proceedings.mlr.press/v70/forney17a.html
6. Fu, J., Topcu, U.: Probably approximately correct MDP learning and control with temporal logic constraints. Robotics: Science and Systems abs/1404.7073 (2014)
7. Giacomo, G.D., Vardi, M.Y.: Linear temporal logic and linear dynamic logic on finite traces. In: IJCAI, pp. 854–860. IJCAI/AAAI (2013)

8. Greenland, S., Pearl, J.: Causal diagrams. In: Lovric, M. (ed.) International Encyclopedia of Statistical Science, pp. 208–216. Springer, Cham (2011). https://doi.org/10.1007/978-3-642-04898-2_162
9. Howard, R., Matheson, J.: Influence diagrams. Decis. Anal. **2**, 127–143 (2005). https://doi.org/10.1287/deca.1050.0020
10. Huang, B., Feng, F., Lu, C., Magliacane, S., Zhang, K.: AdaRL: What, where, and how to adapt in transfer reinforcement learning. ArXiv: abs/2107.02729 (2021)
11. Icarte, R.T., Klassen, T.Q., Valenzano, R.A., McIlraith, S.A.: Using reward machines for high-level task specification and decomposition in reinforcement learning. In: ICML Proceedings of Machine Learning Research, vol. 80, pp. 2112–2121. PMLR (2018)
12. Koller, D., Milch, B.: Multi-agent influence diagrams for representing and solving games. Games Econ. Behav. **45**(1), 181–221 (2003). https://doi.org/10.1016/S0899-8256(02)00544-4, https://www.sciencedirect.com/science/article/pii/S0899825602005444. First World Congress of the Game Theory Society
13. Lattimore, F., Lattimore, T., Reid, M.D.: Causal bandits: learning good interventions via causal inference (2016). https://doi.org/10.48550/ARXIV.1606.03203, https://arxiv.org/abs/1606.03203
14. Lee, S., Bareinboim, E.: Structural causal bandits with non-manipulable variables. In: Proceedings of the AAAI Conference on Artificial Intelligence, vol. 33, no. 01, pp. 4164–4172 (2019). https://doi.org/10.1609/aaai.v33i01.33014164, https://ojs.aaai.org/index.php/AAAI/article/view/4320
15. Lee, S., Bareinboim, E.: Characterizing optimal mixed policies: where to intervene and what to observe. In: Advances in Neural Information Processing Systems. vol. 33, pp. 8565–8576. Curran Associates, Inc. (2020). https://proceedings.neurips.cc/paper/2020/file/61a10e6abb1149ad9d08f303267f9bc4-Paper.pdf
16. Lee, S., Correa, J.D., Bareinboim, E.: General identifiability with arbitrary surrogate experiments. In: Proceedings of The 35th Uncertainty in Artificial Intelligence Conference. Proceedings of Machine Learning Research, vol. 115, pp. 389–398. PMLR (2020). https://proceedings.mlr.press/v115/lee20b.html
17. Li, X., Vasile, C.I., Belta, C.: Reinforcement learning with temporal logic rewards. In: Proceedings of the IEEE/RSJ International Conference Intelligent Robots and Systems, pp. 3834–3839 (2017). https://doi.org/10.1109/IROS.2017.8206234
18. Lu, C., Huang, B., Wang, K., Hernández-Lobato, J.M., Zhang, K., Schölkopf, B.: Sample-efficient reinforcement learning via counterfactual-based data augmentation (2020). https://doi.org/10.48550/ARXIV.2012.09092, https://arxiv.org/abs/2012.09092
19. Memarian, F., Xu, Z., Wu, B., Wen, M., Topcu, U.: Active task-inference-guided deep inverse reinforcement learning. In: CDC, pp. 1932–1938. IEEE (2020)
20. Neider, D., Gaglione, J.R., Gavran, I., Topcu, U., Wu, B., Xu, Z.: Advice-guided reinforcement learning in a non-Markovian environment. In: Proceedings of the AAAI Conference on Artificial Intelligence, vol. 35, pp. 9073–9080 (2021)
21. Pitis, S., Creager, E., Garg, A.: Counterfactual data augmentation using locally factored dynamics. In: Proceedings of the 34th International Conference on Neural Information Processing Systems, NIPS 2020, Red Hook, NY, USA (2020)
22. Spirtes, P.: Introduction to causal inference. J. Mach. Learn. Res. **11**(54), 1643–1662 (2010). http://jmlr.org/papers/v11/spirtes10a.html
23. Sutton, R.S., Barto, A.G.: Reinforcement Learning - An Introduction. MIT Press, Cambridge (1998)
24. Tennenholtz, G., Mannor, S., Shalit, U.: Off-policy evaluation in partially observable environments. ArXiv: abs/1909.03739 (2019)

25. Wen, M., Papusha, I., Topcu, U.: Learning from demonstrations with high-level side information. In: Proceedings of the IJCAI 2017, pp. 3055–3061 (2017). https://doi.org/10.24963/ijcai.2017/426, https://doi.org/10.24963/ijcai.2017/426
26. Xu, Z., et al.: Joint inference of reward machines and policies for reinforcement learning. In: Proceedings of the International Conference on Automated Planning and Scheduling (ICAPS), Special Track on Planning and Learning (2020)
27. Xu, Z., Topcu, U.: Transfer of temporal logic formulas in reinforcement learning. In: Proceedings of the IJCAI 2019, pp. 4010–4018 (2019). https://doi.org/10.24963/ijcai.2019/557
28. Xu, Z., Wu, B., Ojha, A., Neider, D., Topcu, U.: Active finite reward automaton inference and reinforcement learning using queries and counterexamples. In: Holzinger, A., Kieseberg, P., Tjoa, A.M., Weippl, E. (eds.) CD-MAKE 2021. LNCS, vol. 12844, pp. 115–135. Springer, Cham (2021). https://doi.org/10.1007/978-3-030-84060-0_8
29. Zhang, A., et al.: Learning causal state representations of partially observable environments. ArXiv: abs/1906.10437 (2019)
30. Zhang, J., Bareinboim, E.: Transfer learning in multi-armed bandits: a causal approach. In: Proceedings of the 26th International Joint Conference on Artificial Intelligence, IJCAI 2017, pp. 1340–1346. AAAI Press (2017)
31. Zhu, S., Tabajara, L.M., Li, J., Pu, G., Vardi, M.Y.: Symbolic LTLF synthesis. In: IJCAI, pp. 1362–1369. https://www.ijcai.org/ (2017)

Using Machine Learning to Generate a Dictionary for Environmental Issues

Daniel E. O'Leary[1(✉)] and Yangin Yoon[2]

[1] University of Southern California, Los Angeles, CA 90089, USA
oleary@usc.edu

[2] Seoul National University of Science and Technology, Seoul, Korea
ben.yangin.yoon@seoultech.ac.kr

Abstract. The purpose of this paper is to investigate the use of machine learning approaches to build a dictionary of terms to analyze text for ESG content using a bag of words approach, where ESG stands for "environment, social and governance." Specifically, the paper reviews some experiments performed to develop a dictionary for information about the environment, for "carbon footprint". We investigate using Word2Vec based on Form 10K text and from Earnings Calls, and queries of ChatGPT and compare the results. As part of the development of our dictionaries we find that bigrams and trigrams are more likely to be found when using ChatGPT, suggesting that bigrams and trigrams provide a "better" approach for the dictionaries developed with Word2Vec. We also find that terms provided by ChatGPT were not as likely to appear in Form 10Ks or other business disclosures, as were those terms generated using Word2Vec. In addition, we explored different question approaches to ChatGPT to find different perspectives on carbon footprint, such as "reducing carbon footprint" or "negative effects of carbon footprint." We then discuss combining the findings from each of these approaches, to build a dictionary that could be used alone or with other ESG concept dictionaries.

Keywords: ESG · Carbon Footprint · Environment · Word2Vec · ChatGPT · Dictionary · Bag of Words · Ontology · Concept · Form 10K · Reducing Carbon Footprint · Hybrid Approach

1 Introduction

This paper investigates the generation of dictionaries for use in bag of words models to analyze text disclosure in accounting, finance, and economics. Typically, in a bag of words setting, the number of occurrences from dictionaries of different concepts are treated as independent variables and then the frequency of their occurrence is compared against some dependent variable, such as firm value to study the impact of the text on that variable. The relationships are then investigated using statistical analysis to try and understand the relationships captured in the regression equation. As an example, Allen et al. (2021) used a dictionary of tax terms to study the relationship between the occurrences of those terms in firm 10K documents (corporate disclosures) and the effective tax rates of those organizations.

Published by Springer Nature Switzerland AG 2023
A. Holzinger et al. (Eds.): CD-MAKE 2023, LNCS 14065, pp. 141–154, 2023.
https://doi.org/10.1007/978-3-031-40837-3_9

We generate wordlists using two different approaches, Word2Vec (Mikolov et al. 2013a, b) and ChatGPT, resulting in a hybrid approach. We focus on dictionaries for so-called ESG (**E**nvironment, **S**ocial and **G**overnance) issues, and particularly on a dictionary for the environment. We develop ESG dictionaries because there is substantial real-world interest in understanding the impact of firm ESG text disclosures on the value of the firm (e.g., Gordon and Bell 2022).

1.1 Findings

As we analyze those dictionaries that we develop, we have several findings. First, although singleton words are used extensively in psychology bag of words applications, we find that singleton words are not as descriptive of our environmental seed concept of "carbon footprint" as bigrams and other multi-grams. When using Word2Vec the generated singleton words generally are not descriptive of the concept. However, when we move to bigrams, the resulting words are more descriptive of the seed concept. As further evidence, generating three different lists using ChatGPT we find that out of 75 phrases for "carbon footprint", all were at least 2 words long and a few were 4 words long. None were a single word. We use the theory of requisite variety to help explain the result. Second, we find that we can generate dictionaries that are more specific to Form 10K and Earnings Call Disclosures, using Word2Vec based on text generated from firm disclosures in the Form 10K's and Earnings Conference calls, than with the more general and broadly-based ChatGPT. There was limited overlap among the words generated by ChatGPT and Word2Vec. Third, we find that we can use ChatGPT and Word2Vec to study characteristics of wordlists for the environment. For example, the human in the loop can analyze our word lists to find events and characteristics that would have positive and negative effects on organizations. As a result, in order to use our word lists, we need partition the words into different lists for different purposes. Finally, as a result, the paper suggests using a hybrid approach, based on both ChatGPT and Word2Vec, to generate dictionaries for ESG purposes.

1.2 This Paper

This paper proceeds in the following manner. Section 2 provides some background information on the basic nature of Word2Vec, ChatGPT, two key sources of business text used in the analysis of ESG text disclosures, and some roles of humans in the loop. Section 3 investigates the notion of bags of words in more detail, reviewing concepts in general, providing an example and briefly analyzing the concept of interest in this paper. Section 4 analyzes a list of single "similar" words to "Carbon Footprint" generated using Word2Vec. Section 5 analyzes a set of Word2Vec generated bigrams and compares those words from a business source of text (corpus) to a set of words generated by ChatGPT. Section 6 generates a set of sub-dictionaries and Sect. 7 investigates some additional queries of ChatGPT to study different partitions of the notion of carbon footprint. Section 8 briefly summarizes the paper, the paper's contributions and provides some potential extensions.

2 Background

The purpose of this section is to provide a brief background into the systems used to create word lists for our analysis.

2.1 Word2Vec – CBOW and Skip-Gram

Word2Vec (Mikolov et al. 2013a, b) uses two different neural network architectures to find the words that are "similar" to a seed word: CBOW and Skip-Gram. Together the two approaches provide an ensemble, from which a single list of similar words for the dictionary can be created. As seen below in the examples, there can be overlap between the two approaches, but the two approaches capture different words.

As discussed in Mikolov et al. (2013a, b) Word2Vec generates words that are "similar" to a seed word. The algorithm can generate words that have both syntactic and semantic similarities. For example, as noted in Mikolov (2013a, b, p. 5) the "… word big is similar to bigger in the same sense that small is similar to smaller." Since Word2Vec generates lists of words that are similar to some seed word, it is particularly useful in the development of bags of words used to represent a concept.

2.2 ChatGPT

ChatGPT (https://openai.com/blog/chatgpt) is a massive language system that was built using a wide range of Internet text available as of 2021. ChatGPT is a robust system that has been shown to be able to discuss a wide range of issues and answer questions about many different issues. As an example, it has been used by a judge in Columbia to assist in a court ruling (Zoppo 2023).

ChatGPT offers a free and widely available source of artificial intelligence that can perform a number of tasks, including generate word lists. However, unlike Word2Vec and other approaches, it is based on a very broad-based set of text resources. As a result, any lists provided by ChatGPT are likely to differ from lists generated from using more specialized text sources. Ultimately, our bag of word analysis will be used to investigate specific text, so that can be a drawback of using ChatGPT. As we will see, the concepts found by ChatGPT will be a bit different than those found by Word2Vec, because of the difference in the text on which they draw (e.g., O'Leary 2022 and 2023).

2.3 Business Text – Form 10K and Earning Call Conferences

Bag of words in accounting, finance and economics, typically are aimed at the analysis of business text such as the Form 10K and Earnings Conference Calls. These text sources are directly developed by specific companies and capture the text that those particular companies have disclosed about specific concerns and issues. In the Form 10K, firms disclose accounting information, risk information and other types of information. Unlike Wikipedia and other sources, generally the text in the Form 10K is structured to include a promulgated set of information and the accounting portions contain substantial amounts of numbers. In the Earnings Conference Calls, firms discuss the impact of events (such as the pandemic) on the firm and their corresponding expectations for earnings. The text disclosures are less structured than the Form 10K, but based on scripted verbal disclosures, questions and answers.

2.4 Role of Human in the Loop

Word2Vec and ChatGPT do not function independent of people. Their use typically requires people to perform several tasks. The activities of humans in the loop, have been investigated by a range of researchers, including O'Leary (2003), Holzinger (2016), Crootof et al. (2022) and others. However, the roles of the human in the loop in building dictionaries is not well-established. When using Word2Vec, the human needs to choose the seed words, and which approaches to use, whether CBOW or Skip-Gram. For ChatGPT, the human needs to choose which questions to ask. For both, when given the lists, the human needs to choose which words best meet their needs and which of the words should be used in the project to best capture the concepts of interest.

3 Bag of Words

The bag of words approach to text analysis is a simple model of processing text, based solely on the number of occurrences of different words in some sample set of text. The approach does not use the part of speech or the location of the word in the text, it simply counts the number of occurrences of dictionary words in some text. However, the approach is quite robust and has the advantage of facilitating a statistical analysis of the text occurrences as compared to some dependent variable.

3.1 Concepts as Ontologies Represented as Bag of Words

Gruber (1993, p. 199) defined an ontology as "… an explicit specification of a conceptualization." In the same sense, bag of words approaches use independent variables, representing ontology "concepts" to study how text content affects some dependent variable. Each concept is represented as a set of "similar" words that capture different dimensions of the concept.

In the case of dictionaries, the explicit specification of an independent variable is a list of words, also called a dictionary. Typically, a concept is represented by a single variable and that dictionary of terms is used to define that variable. The number of occurrences of that variable in some text will determine the value of that independent variable for that specific text.

3.2 Example System - LIWC

Perhaps the best-known bag of words system is the well-known LIWC (Linguistic Inquiry and Word Count). LIWC provides several related dictionaries for analysis of text content of psychology concepts. For example, as noted by LIWC (https://www.liwc.app/) the different dictionaries can help "… analyze others' language can help you understand their thoughts, feelings, personality, and the ways they connect with others."

LIWC includes different types of concepts. For example, there are multiple "Psychological Processes," including the concept of "Drives," where there are three sub concepts of "Affiliation," "Achievement," and "Power," where there are dictionaries for each of affiliation, achievement and power, and then a variable representing those three for drives. Using this approach allows LIWC to provide over a hundred different concept and sub-concept dictionaries. We use the same design approach here.

However, LIWC uses exclusively single words, never bigrams or trigrams to capture a concept. Although a single word approach may be appropriate for a system designed for investigating psychological concepts, it is arguable that single words are appropriate for all such settings, including one designed to investigate ESG issues.

3.3 Concept of Interest – Carbon Footprint

In this paper we investigate the concept of "carbon footprint." That concept captures both an environmental element and contaminant "carbon" and a state suggesting there is a corresponding amount, "footprint." As a result, we expect that for a term to be "similar" it would need to be able to also capture both aspects of that term. Accordingly, whatever words are to be judged as "similar" are likely to require the capture of similar capabilities.

There is theory that suggests that to capture concepts "similar" to such bigrams will require multiple types of information, and thus can require multi-gram words. In particular, in cybernetics and in uncertainty theory, there is a "law" that has been stated in similar ways. For example, Ashby's Law of requisite variety (1956) states, "it takes variety to destroy variety." Similarly, as noted by Karl Weick (1969), "it takes equivocality to remove equivocality." Accordingly, we expect that a concept that requires multiple capabilities or words, will also require multiple words. In some cases, this might require a new word, a meaningful abbreviation or possibly a word from some other language.

4 Word2Vec – Single Words

In this section, we limit the word list developed by Word2Vec to a list of single words, i.e., singletons, in order to test our hypothesis that a multigram is necessary to fully capture the meaning associated with each of the words in "carbon footprint."

4.1 Data for Word2Vec

Our text data that we use with Word2Vec comes from two corporate sources: Form 10K's from 2020 and 2021, and Earnings Conference Calls from 2021.

4.2 Approach

In this section we review the set of single words found using Word2Vec when the environmental seed word is "Carbon Footprint." We used both CBOW and Skip-gram that are a part of Word2Vec, and we used them on text from both Firm 10Ks and Earnings Conference Calls. We limited our search to twenty-five words using each approach, however that can be easily extended to a larger number of words.

4.3 Findings

The results are summarized in Table 1. Although the words appear broadly related to our seed words, the singleton nature of the words makes many of them appear not relevant as relating to the seed word. These results led us to focus more on bigrams, as seen in the next section.

Some of our single words are clearly important to our dictionary, such as "emissions," however, most of the words would not be strong signals of carbon footprint. For example, "footprint" could refer to any type of footprint. Other words, such as "cleaner" and "shrimp," also would seldom be related to carbon footprint.

Interestingly, the abbreviations that we found can have a broader meaning than the words. For example, "ghg," found in the top six of each of the approaches and text sources, means "green-house gases" and is an important related concept to carbon footprint, and could be an important part of our dictionary.

Table 1. Word2Vec Single Words – Seed Word "Carbon Footprint"

2020 and 2021 Method #1	2020 and 2021 Method #2	Earnings call_2021 Method #1	Earnings call_2021 Method #2
sulfur	emissions	emissions	greenhouse
footprints	footprints	greenhouse	emissions
nitrogen	greenhouse	methane	footprints
pyrolytic	ghg	emission	ghg
ghg	dioxide	nox	emission
emissions	emission	ghg	methane
nox	sulfur	footprints	tailpipe
presence	hydrogen	competitiveness	scope
viscosity	nitrogen	fugitive	roofline
calgon	fuels	sugar	neutrality
density	efficiency	presence	gases
uniformity	efficiencies	scope	dioxide
sulphur	scale	ci	hydrogen
pitch	cleaner	fiber	decarbonization
blacks	gases	gases	particulates
efficiency	oxide	density	renewable

(continued)

Table 1. (*continued*)

2020 and 2021 Method #1	2020 and 2021 Method #2	Earnings call_2021 Method #1	Earnings call_2021 Method #2
ambient	renewable	network	sequestering
shrimp	fuel	recyclability	gigatons
opacity	initiative	capacity	calpeco
hid	cement	emitting	gigaton
xingheyongle	steelmaking	cost	terravault
tetrachloride	eaf	co	decarbonizing
friction	electrification	flaring	nanotubes
mercury	mercury	emitters	carbonizing
biodiesel	sustainable	electricity	monoxide

4.4 Implications

Although the list of words captures some key words for the dictionary, such as emissions, methane, nitrogen and others, it also captures some that may not necessarily provide precision at identifying issues of environmental concern, such as "footprints" and "greenhouse." Other words such as "density," "shrimp," "hid," and others are not likely to be useful in analysis of environment text. However, the occurrence of abbreviations provides powerful symbolic capture of concepts.

4.5 Human in the Loop

Based on these lists, it is easy to see the need for a "human in the loop" to choose which words to include in the development of a dictionary for ESG. Perhaps that human would be an expert or even a committee to choose which words should be chosen to model the concepts of interest. As seen in the analysis of the abbreviations and terms, some expertise is necessary to determine which words are likely to be signals of the key concepts being considered.

5 Comparison of Word Lists Between Word2Vec and ChatGPT

5.1 Approach

Our approach was to generate three sets of lists of 25 words using Word2Vec bigrams using both CBOW and Skip-Gram, and ChatGPT. For Word2Vec our seed word was "carbon footprint." For ChatGPT, we asked the question, "Can you give me 25 words or bigrams for carbon footprint?".

5.2 Using ChatGPT to Facilitate List Analysis

As part of our analysis, we investigated the use of ChatGPT to help us in our use of Word2Vec. We used ChatGPT to determine the nature of phrases being used to characterize environmental issues associated with "carbon footprint." The lists from that analysis are in Table 2.

We used ChatGPT to find "25 words or bigrams for carbon footprint." In so doing we found that the list of words was entirely bigrams, trigrams and quad-grams, in contrast to psychology software, such as LIWC. While analyzing three such sets generated by ChatGPT, we found out of 75 phrases, that there were 51 bigrams, 22 trigrams and 2 quad-grams. No single words were generated as part of those queries. Accordingly, this suggests that any dictionary for the environment generally will incorporate multiple word phrases.

5.3 Findings

It is easy to see that the words in Tables 1 and 2 are substantially different. The words in Table 2 are all multi-grams. Further, there is only one abbreviation in Table 2.

In addition, we compared the lists generated using Word2Vec based on Form 10Ks compared to lists generated by ChatGPT as seen in Table 2. Interestingly we found only one word from the ChatGPT list appeared in the lists from the focused Word2Vec approach. These results provide one measure of the potential use of ChatGPT words to investigate text drawn from business financial statements. The more general setting provided by ChatGPT provides limited insight into the words actually disclosed by firms in the Form 10K and Earnings Calls.

5.4 Implications

There are a number of implications of the lack of overlap between the different approaches. If a dictionary, based on ChatGPT was used to investigate the impact of the concept of "Carbon Footprint", while analyzing Form 10K's as the source of text, then it appears that the ChatGPT list would "under" represent the number of similar concepts that actually appear in the Form 10K and Earnings Calls, as only one term from ChatGPT appears on the Word2Vec lists from those sources.

Table 2. Word2Vec Bigrams vs. ChatGPT[1]

Word2Vec - Bi-gram 2020 and 2021 10K - CBOW	Word2Vec - Bi-gram 2020 and 2021 Skip-Gram	ChatGPT - Can you give me 25 words or bigrams for carbon footprint? (#3)
environmental footprint	environmental footprint	Greenhouse gas emissions
carbon emissions	carbon intensity	Fossil fuel use
carbon intensity	energy consumption++	Energy consumption++
energy consumption++	carbon emissions	Transportation impact
friction and	efficiency and	Renewable energy sources
fuel consumption	reduce carbon	Carbon offsetting
sulfur emissions	lower carbon	Climate change impact
fuel efficiency	consumption and*	Sustainable living
mercury emissions	water consumption	Environmental impact
carbon footprints	while reducing	Ecological footprint
viscosity	significantly reducing	Energy efficiency
dependency on	fuel consumption	Carbon offset credits
power consumption	operational efficiency	Low-carbon economy
water consumption	fuel efficiency	Carbon neutral
its carbon	reducing costs	Carbon capture and storage
energy usage	electricity consumption	Emissions reduction
overall costs	while helping	Carbon sequestration
carbon and	sustainability goals	Energy conservation
emissions intensity	efficiency measures	Sustainable transportation
castability	while keeping	Green lifestyle
carbon economy	improve efficiency	Carbon accounting
funding costs	while improving	Carbon pricing
sulfur content	streamline operations	Carbon tax
consumption and	marketing spend	Life cycle assessment
ghg emitting	goal of	Environmental sustainability

6 Building a Dictionary

If we were to build a dictionary to represent the concept of "carbon footprint," we would draw on each of the different lists created as part of the development of this paper, including words from Word2Vec and ChatGPT in Tables 1 and 2. However, an analysis of Table 2 suggests that there are multiple perspectives (sub-dictionaries) on the words

[1] ++ is the only ChatGPT term to appear in both 10K Word2Vec bigrams.

related to carbon footprint and that those perspectives vary based on the whether the Word2Vec or ChatGPT are used. Thus, we likely would construct multiple dictionaries under the carbon footprint construct. Further, based on the source text, we can expect different word sets.

6.1 Carbon Footprint

The purpose of this section is to outline an initial dictionary for the concept carbon footprint, based on Tables 1 and 2. As with LIWC, we can imagine that dictionary would have a number of sub-dictionaries associated with it. As we analyze the word groupings, it is clear that the groupings are typically generated using one of the tools.

First, the single words in Table 1, appear to relate to a variety of types of *emissions*. For example, the following are included there: nitrogen, sulfur, ghg, emissions, tetrachloride, mercury and monoxide. Second, in the bigrams, Word2Vec also identifies different types of *emissions*: carbon emissions, sulfur emissions, mercury emissions, emissions intensity, and ghg emitting, In the short list requested of ChatGPT, only greenhouse gas emissions was proposed. Third, one clear set of words relates to the *carbon economy*, and those words come largely from ChatGPT. While Word2Vec using CBOW identified one term related to the "carbon economy," ChatGPT lists a number of terms related to the carbon economy, including carbon offsetting, carbon offset credits, low-carbon economy, carbon accounting, carbon tax, carbon pricing, carbon capture and storage. Fourth, although Word2Vec found the term *sustainability goals*, ChatGPT, focused more on sustainability. ChatGPT included sustainable living, sustainable transportation, green lifestyle, energy conservation, suggesting less content about sustainability in corporate disclosures. Fifth, the category of *consumption* could provide some important concepts. Word2Vec found energy consumption, water consumption, electricity consumption, power consumption and energy usage. ChatGPT found only energy consumption.

This discussion illustrates that the words generated using the seed word of carbon footprint appear to vary quite a bit based on the text sources, in spite of the broad generality of ChatGPT. Accordingly, a hybrid approach can be used to develop a robust dictionary with several carbon footprint-based concepts.

6.2 Other Environmental Dictionaries

In addition, Tables 1 and 2 could provide us with additional potential concepts to create dictionaries related to other environmental issues and sustainability. For example, mercury emissions and water consumption could be part of other environment dictionaries, aimed at issues beyond implications of just a carbon footprint.

7 Positive, Negative or Action Dictionaries

Dictionaries can be developed that capture positive effects, negative consequences or just the generic use of words. To illustrate those issues, Table 3 provides the results of three questions to ChatGPT:

- Can you provide a list of some things that organizations can do to reduce their carbon footprint?
- Can you provide a list of words that capture the negative consequences of carbon footprint?
- Can you give me 25 words or bigrams for carbon footprint?

These questions generate some interesting and different responses than our previous analyses. As an example, the question about negative consequences generates ocean acidification, deforestation, desertification, health risks, melting glaciers and other concerns. Although these are real issues, it is unlikely that they will appear in corporate financial disclosures and likely not useful for analysis of Form 10Ks or Earnings Calls disclosures. However, the question about organizations reducing their carbon footprint are feasible terms that may well be included in corporate disclosures. For example, the sub-concept of *organization efforts* might include recycling, waste reduction, green procurement, renewable energy and others, as identified on these lists. In any case each of these sets of disclosures could provide additional lists to the overall concept of carbon footprint.

Table 3. Carbon Footprint: Alternative Questions for Dictionaries

Question: "… some things that organizations can do to **reduce their "carbon footprint"**?"	Question: "A list of words that capture the **negative consequences of "carbon footprint"**?"	Question: "Can you give me 25 words or bigrams for carbon footprint?"
Energy efficiency	Global warming	Greenhouse gas emissions
Renewable energy	Climate change	Fossil fuel use
Carbon offsetting	Greenhouse effect	Energy consumption
Sustainable transportation	Air pollution	Transportation impact
Waste reduction	Ocean acidification	Renewable energy sources
Recycling	Deforestation	Carbon offsetting
Green procurement	Melting glaciers	Climate change impact
Energy conservation	Rising sea levels	Sustainable living
Carbon accounting	Extreme weather	Environmental impact
Eco-friendly practices	Drought	Ecological footprint
Green building	Heatwaves	Energy efficiency
Telecommuting	Desertification	Carbon offset credits
Paperless operations	Species extinction	Low-carbon economy

(*continued*)

Table 3. (*continued*)

Question: "… some things that organizations can do to **reduce their "carbon footprint"**?"	Question: "A list of words that capture the **negative consequences of "carbon footprint"**?"	Question: "Can you give me 25 words or bigrams for carbon footprint?"
Sustainable supply chain	Health risks	Carbon neutral
Employee education	Respiratory problems	Carbon capture and storage
Sustainable packaging	Environmental degradation	Emissions reduction
Water conservation	Loss of biodiversity	Carbon sequestration
Emission reduction targets	Coral bleaching	Energy conservation
Efficient heating and cooling systems	Disrupted ecosystems	Sustainable transportation
Green IT solutions	Food scarcity	Green lifestyle
Product life cycle assessment	Water scarcity	Carbon accounting
Eco-friendly commuting options	Natural disasters	Carbon pricing
Renewable energy procurement	Energy insecurity	Carbon tax
Behavior change campaigns	Displacement	Life cycle assessment
Carbon footprint monitoring	Social inequality	Environmental sustainability

8 Summary, Contributions and Extensions

This paper has examined the process of building a dictionary for use in the analysis of environment bag of words as related to the impact of text disclosures on firm value or other dependent variable measures. We built our dictionaries using two different neural net-based approaches: Word2Vec and ChatGPT. Ultimately, a dictionary for our concept of carbon footprint, used words from each source. Although the Word2Vec approach is more focused on terms specific to business documents, we can build a broader use dictionary by incorporating words from both sources. Perhaps this approach may also "anticipate" additional word lists in corporate disclosures.

8.1 Contributions

This research has several contributions. First, this research has produced a set of words that could be used in a bag of word model of environmental concerns, with a focus on the environment investigating firms' text context related to "carbon footprint." Second, we find that for the seed concept of "carbon footprint," it appears that the primary approach should employ multiple word grams, rather than single words. Single words do not appear to capture the concept as well as multiword grams. However, the Word2Vec analysis did isolate some single word concepts indicative of emissions. Third, we also find that Word2Vec's ability to focus on specific corpuses, such as the Form 10K, allows us to generate a more focused word set than words generated from a broader source, as has been done with ChatGPT. We expect that to be the case for other ESG concepts. Fourth, as a result, it is clear that a "hybrid" approach can use aspects associated with both Word2Vec and ChatGPT. We use Word2Vec to generate words from a specific set of documents (Form 10K's and Earnings Calls) to capture the specific context. We use ChatGPT to determine the likely number of words in the phrases that we seek, say 1, 2 or 3 words. In addition, we use ChatGPT to generate other words that may not have been captured from specific business corpuses. This allows the generation of a more generic dictionary that can be used in other settings. Finally, we can use ChatGPT to establish a list of words with a particular sentiment or purpose. For example, we can ask ChatGPT to provide a list of words associated with the "negative" or "positive" consequences of concerns such as "Carbon Footprint".

8.2 Extensions

There are several potential extensions to the research. First, we have only examined a dictionary for environmental concepts, and have not considered other aspects, including social and governance concerns. Second, we have only considered one particular aspect of the environment, the issue of "carbon footprint." As a result, other environmental issues could be considered, and dictionaries developed for them. Accordingly, the development of bag of word dictionaries for ESG concern would likely include the development of multiple concept dictionaries, potentially with multiple dictionaries for each environment, social and governance concerns. Third, our analysis was for a two-word concept "carbon footprint" and we found that two-word and three-word descriptors appear to provide better descriptiveness than single words. Perhaps multiple word concepts are typically better described using multiple word descriptors, rather than single word descriptors. Single word descriptors of psychology concepts may be appropriate, but it does not seem to be sufficient with the concept of "carbon footprint." Fourth, we could expand our analysis to a larger number of items from both Word2Vec and from ChatGPT. In the current version, we used word lists of twenty-five items for feasibility of presentation. Fifth, we could expand our analysis to a broader range of concepts, developing an ESG model analogous to LIWC's psychological model, with dictionaries for many different concepts and sub-concepts.

References

Allen, E., O'Leary, D.E., Qu, H., Swenson, C.W.: Tax specific versus generic accounting-based textual analysis and the relationship with effective tax rates: building context. J. Inf. Syst. **35**(2), 115–147 (2021)

Ashby, W.R.: An Introduction to Cybernetics. Chapman & Hall, London (1956)

Boyd, R.L., Ashokkumar, A., Seraj, S., Pennebaker, J.W.: The development and psychometric properties of LIWC-22, pp. 1–47. University of Texas at Austin, Austin (2022)

Crootof, R., Kaminski, M.E., Price II, W.N.: Humans in the loop. Vanderbilt Law Rev. **76**(2), 22-10 (2022)

Earth.Org: 14 Biggest Environmental Problems of 2023. https://earth.org/the-biggest-environmental-problems-of-our-lifetime/#:~:text=One%20of%20the%20biggest%20environmental%20problems%20today%20is%20outdoor%20air,contains%20high%20levels%20of%20pollutants. Accessed 20 June 2023

Gordon, T., Bell, M.: The EY Global Corporate Reporting and Institutional Investor Survey finds a significant reporting disconnect with investors on ESG disclosures (2022). https://www.ey.com/en_us/assurance/how-can-corporate-reporting-bridge-the-esg-trust-gap?WT.mc_id=10821007&AA.tsrc=paidsearch&gad=1&gclid=Cj0KCQjwmN2iBhCrARIsAG_G2i5nQyuMeduz7JpaITIJnSygQT5Fldf4GaW-1AU6XhM44H0Kum7jkskaApOpEALw_wcB. Accessed 20 June 2023

Grimmer, J., Stewart, B.M.: Text as data: the promise and pitfalls of automatic content analysis methods for political texts. Polit. Anal. **21**(3), 267–297 (2013)

Gruber, T.R.: A translation approach to portable ontology specifications. Knowl. Acquis. **5**(2), 199–220 (1993)

Holzinger, A.: Interactive machine learning for health informatics: when do we need the human-in-the-loop? Brain Inform. **3**(2), 119–131 (2016). https://doi.org/10.1007/s40708-016-0042-6

Mikolov, T., Chen, K., Corrado, G., Dean, J.: Efficient estimation of word representations in vector space. arXiv preprint arXiv:1301.3781 (2013a)

Mikolov, T., Sutskever, I., Chen, K., Corrado, G.S., Dean, J.: Distributed representations of words and phrases and their compositionality. In: Advances in Neural Information Processing Systems, pp. 3111–3119 (2013b)

O'Leary, D.E.: Auditor environmental assessments. Int. J. Account. Inf. Syst. **4**(4), 275–294 (2003)

O'Leary, D.E.: Massive data language models and conversational artificial intelligence: emerging issues. Intell. Syst. Account. Finance Manag. **29**(3), 182–198 (2022)

O'Leary, D.E.: An analysis of three chatbots: BlenderBot, ChatGPT and LaMDA. Intell. Syst. Account. Finance Manag. **30**(1), 41–54 (2023)

Pennebaker, J.W., Booth, R.J., Boyd, R.L., Francis, M.E.: Linguistic Inquiry and Word Count: LIWC2015 (2015). https://www.researchgate.net/profile/Ryan-Boyd-8/publication/337731895_Linguistic_Inquiry_and_Word_Count_LIWC2015/links/6088fef0907dcf667bcadc17/Linguistic-Inquiry-and-Word-Count-LIWC2015.pdf. Accessed 20 June 2023

Pennington, J., Socher, R., Manning, C.: GloVe: global vectors for word representation. In: Moschitti, A., Pang, B., Daelemans, W. (eds.) Proceedings of the 2014 Conference on Empirical Methods in Natural Language Processing (EMNLP), pp. 1532–1543. Association for Computational Linguistics, Stroudsburg (2015)

Weick, K.E.: The Social Psychology of Organizing. Addison-Wesley, Reading (1969)

Zoppo, A.: ChatGPT Helped Write a Court Ruling in Colombia. Here's What Judges Say About Its Use in Decision Making. https://www.law.com/nationallawjournal/2023/03/13/chatgpt-helped-write-a-court-ruling-in-colombia-heres-what-judges-say-about-its-use-in-decision-making/?slreturn=20230408153845#:~:text=A%20Colombian%20judge%20last%20month,tools%20in%20judicial%20decision%2Dmaking. Accessed 20 June 2023

Let Me Think! Investigating the Effect of Explanations Feeding Doubts About the AI Advice

Federico Cabitza[1,2(✉)], Andrea Campagner[2], Lorenzo Famiglini[1], Chiara Natali[1], Valerio Caccavella[3], and Enrico Gallazzi[3]

[1] Universitá degli Studi di Milano-Bicocca, Milan, Italy
federico.cabitza@unimib.it
[2] IRCCS Istituto Ortopedico Galeazzi Milano, Milan, Italy
[3] UOC Patologia Vertebrale e Scoliosi, ASST G. Pini - CTO, Milan, Italy

Abstract. Augmented Intelligence (AuI) refers to the use of artificial intelligence (AI) to amplify certain cognitive tasks performed by human decision-makers. However, there are concerns that AI's increasing capability and alignment with human values may undermine user agency, autonomy, and responsible decision-making. To address these concerns, we conducted a user study in the field of orthopedic radiology diagnosis, introducing a reflective XAI (explainable AI) support that aimed to stimulate human reflection, and we evaluated its impact of in terms of decision performance, decision confidence and perceived utility. Specifically, the reflective XAI support system prompted users to reflect on the dependability of AI-generated advice by presenting evidence both in favor of and against its recommendation. This evidence was presented via two cases that closely resembled a given base case, along with pixel attribution maps. These cases were associated with the same AI advice for the base case, but one case was accurate while the other was erroneous with respect to the ground truth. While the introduction of this support system did not significantly enhance diagnostic accuracy, it was highly valued by more experienced users. Based on the findings of this study, we advocate for further research to validate the potential of reflective XAI in fostering more informed and responsible decision-making, ultimately preserving human agency.

Keywords: eXplainable AI · Medical machine learning · reflective AI · similarity metrics

1 Motivations and Background

In the circles concerned with AI developments, it is commonly believed that Artificial Intelligence (AI) and Augmented intelligence (AuI) are closely related phenomena [4,41]. On the one hand, it is thought that AI, by automating menial tasks and repetitive activities, can free human intelligence to apply itself to

Published by Springer Nature Switzerland AG 2023
A. Holzinger et al. (Eds.): CD-MAKE 2023, LNCS 14065, pp. 155–169, 2023.
https://doi.org/10.1007/978-3-031-40837-3_10

more creative and intellectually nobler tasks, such as supervisory or relational duties [25]. On the other hand, it is assumed that AuI can be realized precisely by virtue of the functions of AI that amplify the scope or speed of certain human cognitive actions. However, there is a growing number of voices that express concern regarding an opposite and seemingly paradoxical effect: the more accurate, capable, and aligned with human values and expectations AI becomes, the more its users may perceive their agency as increasingly eroded, and lose their autonomy in favor of the technology, along with their willingness to take risks, self-confidence, and ability to take responsibility (e.g. [10,21]). Moreover, ever-improving decision supports may affect even more intensely the processes of learning skills, abilities, and mindsets that enable correct and responsible decision making [26,35].

In this second strand of research, which is still in its infancy, the role of the AI system designer in determining the process of users' appropriation of the system or the degree to which they will rely on it is well recognized [14]. For example, some authors have proposed to purposefully develop cognitive disruption or "sources of attrition" [21] that would require users to continue exercising some degree of will, judgment, preference, and responsibility, even at the expense of efficiency, ease of use, or cognitive comfort [21]. For instance, Buçinca et al. [6] propose to design and evaluate "cognitive-forcing functions"; Dai and Fishbach [18] speak of "deliberation-promoting nudges", aimed at encouraging people to give a kind of sober second thought [37]; also Cabitza [10] reported the results from some studies regarding so-called "programmed inefficiencies", inspired by the work by Ohm and Frankle [32] and Chalmers [15]. Some authors have focused on how AI could foster analogical reasoning [27], for instance by presenting users with the most similar cases (to the case at hand) that are already annotated according to the available ground truth [2]. Similarly, counterfactuals have been proposed as to help users in achieving an intuitive understanding of cause-and-effect relationships that involve the factors considered by the AI system in output generation [7], although they require to analyze multiple scenarios of different plausibility. In Prabhudesai et al. [33], the authors "found that communicating uncertainty about ML predictions forced participants to *slow down and think analytically* about their decisions" (our emphasis). Other authors have proposed studying and developing AI systems that would push their users into an active role of reflection [17] by directly prompting and questioning them, and by fostering their evaluation [31] by presenting contrasting evidence and opposite arguments [38] even at the expense of the efficiency, and even against the suggestions they receive from the machine.

In this work we report about a user study that we designed and conducted to investigate the impact of a particular kind of XAI support on user decision performance and perceptions. We define this XAI support as "reflective", as its main aim is to make the user think and reflect before they make the final decision. This reflective XAI support can be considered a sort of "post-hoc-explanations-by-example" support [29], where the explanations are *pixel attribution maps* [16] and the examples are the cases found by the systems to be the most similar ones available to the case at hand that the system both correctly classified and misclassified with respect to the ground truth (see Fig. 1). Therefore, this support was conceived to make users reflect on the reliability of the system advice on a given case, by showing how well (or badly) the system performed on the most similar cases. This acts as a proxy of the AI skill to correctly classify cases like the one at hand, and hence a proxy of its trustworthiness.

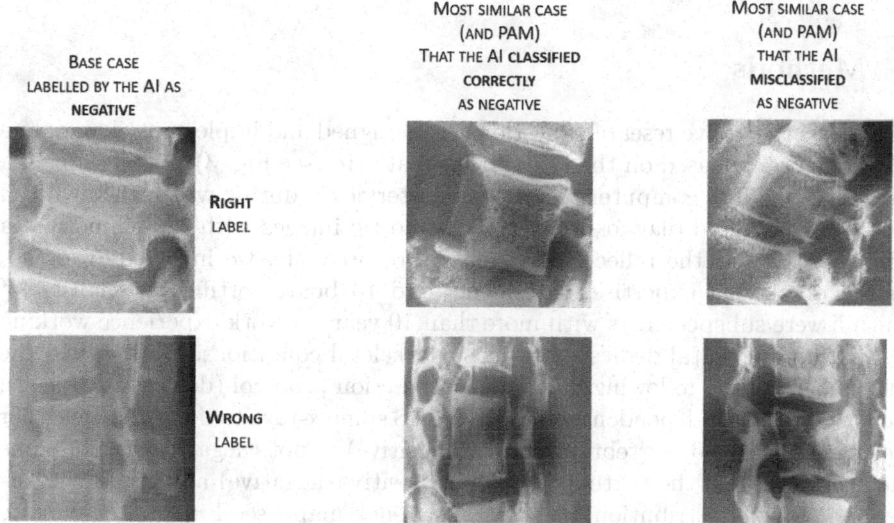

Fig. 1. Two cases shown to the participants in the user study: on the left, the base cases associated with an advice of no fractures (negative); in the top case the label was right; in the bottom case, the label was wrong; on the right, the two most similar cases with the corresponding pixel attribution maps (PAMs) associated with each base case; they indicate, in the middle, the case correctly identified by the AI and, on the right, the case incorrectly indicated by the AI as positive (to the presence of fractures). This means that in the first base case, the middle XAI case should have reinforced the idea that the AI was right; in the second base case, conversely, the misclassified case on the right should have prompted users to be cautious of the AI's advice.

In particular, our user study aims at investigating the following research questions:

- **RQ1**: Does the reflective support described above affect decision-making performance? (i.e., does it have raters change their minds?). Addressing this RQ is equivalent to verifying whether the reflective support impacts accuracy more than just AI support, and whether this support can lead to the phenomenon of technology dominance [8] in either a positive or negative sense, i.e., whether it makes people avoid mistakes (change for the better) or, conversely, misleads them.

- **RQ2**: Does the reflective support impact final diagnostic confidence? This is equivalent to verifying whether the support can instill doubts, or conversely, make raters more cautious.
- **RQ3**: Is the reflective support found to be useful by its users? That is, does it help users get an idea of the local, instance-related, reliability of the system? And connected, **RQ3b**: is there any difference in perceived usefulness between residents (in-training physicians) and specialists? And between experts and less experienced raters?

2 Methods

To address the above research questions, we designed and implemented an online questionnaire, released on the Limesurvey platform (see Fig. 3), and used it in a series of individual computer-assisted web interviews, during which the involved physicians provided diagnoses a set of diagnostic images with the support of a simulated AI and the reflective system. More precisely, we involved 16 orthopedists of varying expertise: 6 residents and 10 board-certified specialists, of which 5 were subspecialists with more than 10 years of work experience working in a teaching hospital dedicated to musculoskeletal conditions. We involved the orthopedists in the following human-AI interaction protocol (depicted in Fig. 2): each orthopedist independently interpreted 18 spine x-rays to determine whether they were positive to vertebral fractures (positive) or not (negative); in the same page they received the textual AI advice (positive/negative) and consult a corresponding pixel attribution map (i.e., a saliency map - see Fig. 3). To generate these Pixel Attribution Maps (PAMs), we used a process called Grad-CAM (Gradient-weighted Class Activation Mapping) [34] with a finetuned ResNext model [40] to detect the presence/absence of fractures in X-ray. These maps highlight the regions of the input image that were key to the model's decision, providing a visual explanation of the AI's advice.

In the technical jargon introduced in Cabitza et al. [12], the envisioned protocol is an AI-first (or second-opinion) protocol with the XAI (reflective) support shown after the more traditional (categorical and visual) AI support.

The 18 cases had been chosen from a previous study [22] to be representative of cases that were medium-to-hard to diagnose: 9 of those x-rays presented fractures, while the other 9 were negative (they presented no vertebral fracture). The AI support was simulated (on the basis of the model developed for [22]) and provided 14 right diagnoses out of 18 (i.e., 78% accurate[1]). Also its sensitivity and specificity, as well as predictive values and F1 score, were .78.

After providing their diagnosis, the participants expressed the degree of confidence on their decision, on a 4-value ordinal scale (a semantic differential from 1 - not at all certain to 4 - practically certain). Subsequently, the system presented a new page with XAI reflective support. This included two other pixel

[1] This performance was set since in the previous study described above, this was found to be the average accuracy of the expert x-rays readers therein involved. Human average accuracy was then 78% (N = 16), SD 7%, median 80%, max 89%, min 67%.

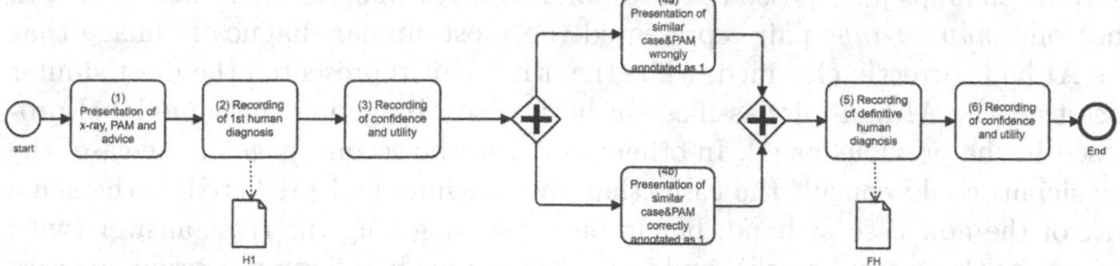

Fig. 2. Human-AI Interaction Protocol, represented in BPMN notation, adopted in the user study, and the main data collected. Steps 1–3 are performed on the first page of the questionnaire; Steps 4–6 are in the second page of the questionnaire. The protocol is repeated for each base case. The similar cases are displayed in the same page, at the same moment (step 4).

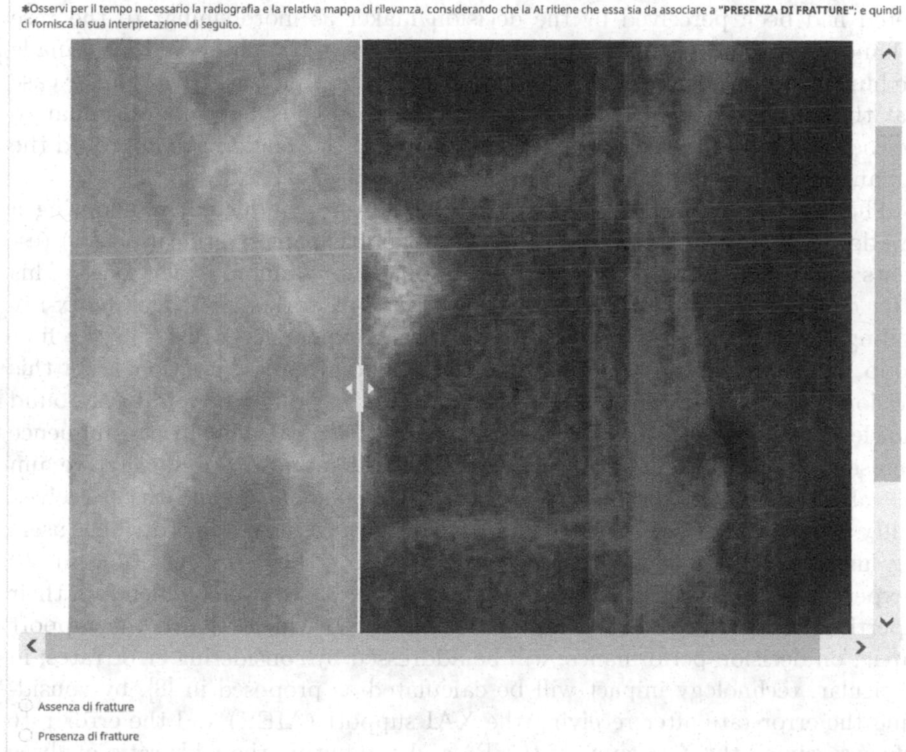

*Osservi per il tempo necessario la radiografia e la relativa mappa di rilevanza, considerando che la AI ritiene che essa sia da associare a **"PRESENZA DI FRATTURE"**; e quindi ci fornisca la sua interpretazione di seguito.

○ Assenza di fratture

○ Presenza di fratture

Fig. 3. Screenshot of a question from the questionnaire provided via the Limesurvey platform. The question presented, translated from Italian, is: "Observe for as long as necessary the radiography and its relevance map, considering that AI believes it is to be associated with PRESENCE OF FRACTURES; and then give us your interpretation below." The two response options are "Absence of fractures" and "Presence of fractures"

attribution maps juxtaposed to corresponding x-ray images. Physicians were told that one *map—x-rays* pair represented the most similar diagnostic image that the AI had correctly classified, while the other pair represented the most similar case that the AI had misclassified (in both cases, by giving the same label proposed in the previous page). In other words, in the second page of the case, the physicians could consult the cases that the machine had predicted in the same way of the new case at hand, but in one case by giving the right answer (with respect to the ground truth), and in the other case by giving the wrong answer. We instructed the physicians by saying that the case that had been classified correctly could suggest how the AI reasoned to provide its advice for the case at hand, as if it had compared the case at hand with that previous successful prediction; on the other hand, we suggested that the wrong case could give them an indication of the reliability of the current advice, as that case was the most similar one in the training set to the case at hand that the machine had actually misinterpreted. Our idea was that, if the case that the machine had misinterpreted had been perceived by the decision maker as more similar to the case at hand than the case in which the machine got it right, this could have made the human suspicious of the previous advice. On the contrary, if the similar case that the machine had correctly identified had been considered more similar to the case at hand than the other misclassified case, this could have improved the human's trust on the machine advice about the case at hand.

The analyses reported in the next sections were conducted by adopting a confidence level of 95%. We will consider two participant stratifications: 1) residents vs specialists; and 2) less expert respondents vs more expert ones. This latter dichotomization will be based on the years of work experience (and x-ray reading): raters with less than 5 years of work experience will go in the first group, those with more than 5 years in the second group. The rationale for this division is rooted in our interest in assessing whether differences in background knowledge and real-world experience in the radiological domain can influence the users' varying benefits from the reflective XAI support. Specifically, we aim to evaluate its impact on decision performance, confidence, and the perceived utility of the XAI advice. This includes investigating the extent to which users may uncritically accept the AI advice, disregarding their own intuition due to inexperience, or alternatively, reject the AI advice due to overconfidence in their expertise. In regard to the research questions: RQ1, which regards the support impact on decision performance, will be addressed by considering error rates; in particular, technology impact will be calculated as proposed in [8], by considering the error rate after receiving the XAI support (AIER) and the error rate before receiving the XAI support (CER) and computing the odds ratio of these rates:

$$\frac{CER}{1-CER}\frac{1-AIER}{AIER}$$

RQ2, which regards impact on confidence, will be addressed by considering the distribution of the responses given in the ordinal scale adopted to operationalize this psychometric construct; finally, RQ3, which regards perceived usefulness of the support, will be addressed by considering both the ordinal response distribution and the proportions of responses in the lower/higher side of the ordinal scale adopted to operationalize this concept.

2.1 Statistical Analysis

The study utilizes several statistical tests and measures to assess the impact of the XAI reflective support on decision performance, decision confidence, and perceived utility: Two-proportion Z-test is employed to compare the proportions of two groups [39]. We also employed the Mann-Whitney U Test, a non-parametric test that compares differences between two independent groups when the dependent variable is ordinal or continuous but not normally distributed [5]. Welch's T-test, a variant of the independent samples t-test, is used when the variances of the two groups being compared are not assumed to be equal [19]. Finally, One-Proportion Test is a statistical procedure we used to test whether a sample proportion significantly differs from a specific value [42]. In addition to these tests, effect sizes are calculated to understand the magnitude of the differences observed. Effect size is a quantitative measure of the strength of a phenomenon or effect [28]. An effect size can be a very useful tool to understand how much of an impact the experimental manipulation (in this case, the introduction of XAI reflective support) has had on the outcomes. Different measures of effect size will be employed in accordance with the corresponding statistical tests.

3 Results

In what follows we organize the section by considering each research questions sequentially, in the order of presentation above.

3.1 RQ1: Impact on Decision Performance

The XAI reflective support had an observable impact on 4.2% of decisions: not surprisingly, on those for which raters expressed a lower confidence after the AI-only support (2.58 vs 2.97). However, the impact was detrimental. Indeed, as we can see from the benefit diagram depicted in Fig. 4, while differences in accuracy were very small among all less expert users, some participants were negatively impacted by considering the two additional maps and x-rays: one expert rater even worsened their performance by almost a fifth, while another expert, who did not commit any diagnostic error when presented with the first AI support, committed some errors only after receiving the additional reflective support. With reference to Fig. 5, we can see that the technology impact of the reflective XAI is significantly negative (notably, the confidence interval does not cross the 1 value line), that is the reflective XAI support misled users more than it aided them in avoiding mistakes.

More in particular, presenting the similar cases induced 12 changes (out of 288 decisions): 8 (67%) of these involved the more expert raters; interestingly,

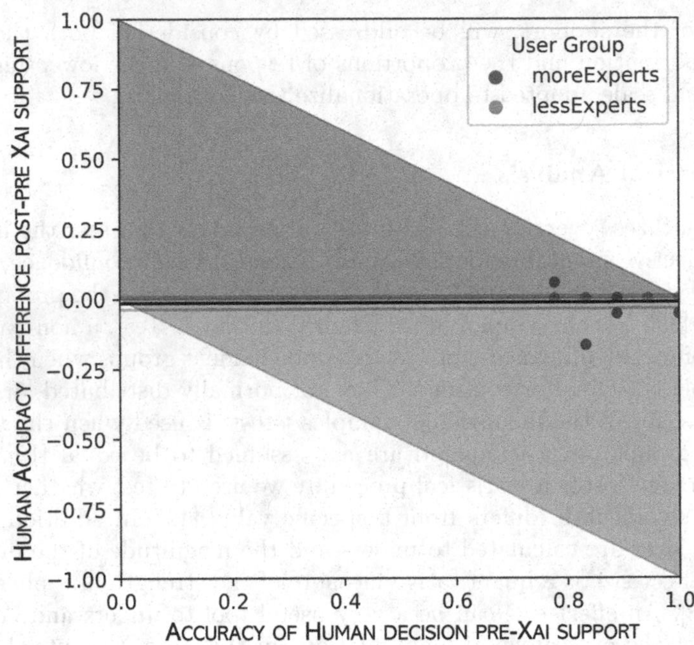

Fig. 4. Benefit diagram of the introduction of the XAI support (after the AI support). Each point corresponds to a single participant in the study, the solid black line represents the average post-pre XAI support accuracy difference, while the shaded grey lines represent the corresponding 95% confidence interval. The red region of the diagram denotes a worsening, in terms of accuracy, between post- and pre-XAI support; while the blue region denotes an improvement in terms of accuracy. The confidence interval of the aggregate effect contains the zero line, although the average line is slightly below it, so no significant difference in accuracy could be detected in the user study. Generated with https://haiiassessment.pythonanywhere.com/ (Color figure online)

two thirds of the changes were for the worse, that is, the support did likely mislead the raters more often than it made them avoid a mistake (see Fig. 5). This happened in two thirds of the cases (67% vs 33%). Moreover, and quite unexpectedly, the expert raters were misled more than the less expert ones: the three quarters of the induced mistakes were made by them.

This is also reflected by the decrease in accuracy that we observed in the responses: pre-support accuracy was 88%, post-support accuracy 87%. While, obviously, the difference is not statistically significant (P-value = .64, Z = 0.47 two proportion test) and the effect size is negligible (0.04), it would nevertheless be difficult to assert that presenting similar cases that the AI either misclassified or classified correctly would actually improve the accuracy of the decision makers.

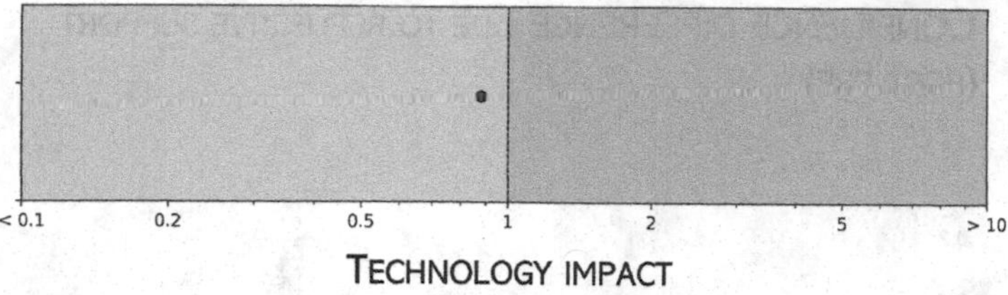

Fig. 5. Technology Impact Diagram. This diagram, generated by https://haiiassessment.pythonanywhere.com/, represents the odds ratios of the errors made after seeing the reflective XAI support and the errors made with the AI support only. We refer the reader to [8] for additional detail on the computation of the represented odds ratio.

3.2 RQ2: Impact on Decision Confidence

No significant difference was found in confidence between the diagnosis formulated before receiving the XAI reflective support and after having received it: pre- and post-support average confidence differed only very slightly (2.95 vs. 2.98, p = .81, Z = −0.2418, U = 41021, Mann-Whitney test), thus we cannot assert that the support made raters more cautious, nor more confident. However, there was a significant difference in confidence difference between the less expert raters and the more expert raters: the former even saw their confidence in the last decision decrease, while the latter reported a higher confidence (−0.05 vs 0.10, p-value = .23, T = −2.28, Welch's T-test, effect size = 0.27) in their final decision (see Fig. 6).

3.3 RQ3: Impact on Perceived Utility

The reflective support was generally found useful by the physicians involved, although not significantly so (149 vs 139 values of perceived utility were above 2, p-value = .6, one proportion test), and no significant difference was found in the positive proportion of responses (i.e., above 2) between residents and specialists (53/108 vs 96/180, p-value = .48, two proportions test), nor in the ordinal values (p-value = .72; Z = −0.36, U = 9486, Mann-Whitney test); however, a significant difference was found between less expert raters (i.e. those with less than 5 years after specialization) and more expert ones (i.e. those with more than 5 years after specialization): 60/144 vs 89/144, p-value = .000, effect size = 0.41, which was confirmed also with respect to ordinal values (p-value = .000, Z = −3.667, U = 7894.5, Mann Whitney test, effect size = 0.22), see Fig. 7. The more expert raters also found the XAI support useful in absolute terms (89 values higher than 2 out of 144, p-value = .006, one proportion test, effect size = 0.24).

In short, the experts considered the support useful, and significantly more useful than the less expert raters did (see Fig. 7). Experts felt more confident of their diagnosis after being exposed to the support (see Fig. 6), but their performance did not improve (if at all).

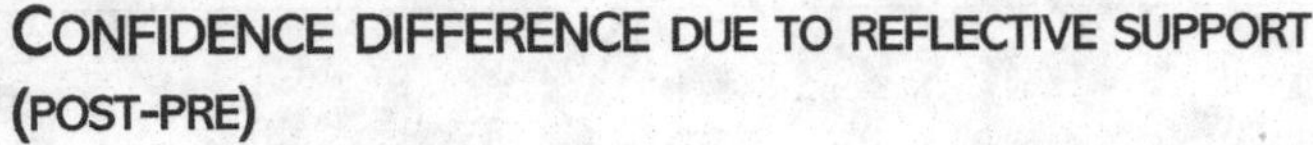

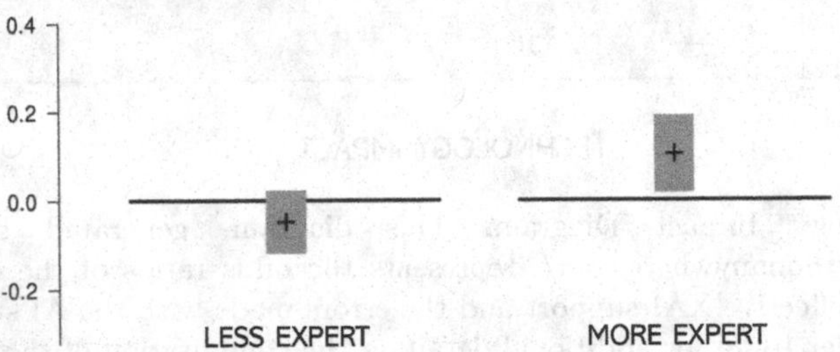

Fig. 6. Plot of the difference in perceived confidence between the decision made before and after seeing the similar cases among the less expert raters and the more expert ones. The solid black cross represents the average confidence difference, while the surrounding shaded gray area represents the corresponding 95% confidence interval.

4 Discussion

We start from acknowledging the main limitation of this study: this latter was set in an experimental setting with obvious time constraints and the risk that some "laboratory effect" [24] could have affected the doctors' performance. However, residents and specialists exhibited the same accuracy (after receiving the AI support) and this suggests that the above effect should be small, if any (since novices and students may have more recent experience with lab-based simulations in a teaching setting, as opposed to senior radiologists who have been found in the work by Gur et al. [24] to experience a worsening in performance *in laboratory* compared to *in vivo*). Moreover, the constrained structure of the questionnaire (currently consisting of the selection of self-reported numbers, as in the case of confidence) may not have captured more subtle and less quantifiable insights on user experience: in future research, the questionnaire could evolve as to integrate other soft metrics, for example by allowing for open text field input, or the recording of think-aloud protocols [20].

While we made efforts to ensure the selected cases would be representative of a wide range of cases of medium-to-high complexity, we recognize that 18 cases is a small sample, only partially mitigated by the relatively large number of experts involved (16) and thus of diagnostic decisions considered (288) with a perfect completion rate (we recall that the questionnaire was filled out by physicians in the presence of some of the authors who were available for clarification and technical assistance.). While this limitation could be addressed in future work, we believe that its impact on our study was limited: the primary objective of our study was not to identify statistically significant differences, but rather to identify possible effects of interest, and to help future studies identify the most appropriate and efficient sample size and power of testing procedures to be employed for achieving statistical significance. To this latter regard, although the interpretation of effect sizes is an open issue that cannot be easily generalized, we believe that for the research questions related to confidence and perceived usefulness (i.e., RQ2 and RQ3) the two effect sizes found (.27 and .22, respectively) are not pragmatically small (it only takes approximately 500 decisions to observe a significant effect) and indeed suggest that more research should be done in this direction.

In light of the above cautious words, we can then summarize the main points that we can derive from our user study: Evaluative or reflective AI does not necessarily bring higher accuracy (indeed, by impacting trust on accurate AI it could even decrease it, even if almost negligibly), or higher confidence in human decisions (which was observed in this study with a very small effect size). However, physicians can appreciate this kind of inconclusive support, and the more so the more expert they are. This is an interesting finding about which we can conjecture several explanations, but none is conclusive. The explanation that we tentatively make here relates this preference to the fact that experts might appreciate the often implicit value of leveraging knowledge on past cases, have more developed capabilities to assess the reliability of a colleague or a teammate (as the AI can be seen) from its past decisions, and have more refined abilities to judge the complexity of a case and therefore have precise expectations as to what an expert may or may not get wrong; in this latter regard, the assessment of the difficulty of similar cases may have played an important role in the experts' appreciation of the support. This may also explain the contrast between experts and users with less expertise in terms of confidence in the final decision: while the first displayed a confidence increase, confidence decreased for the latter. While less experienced users may have experienced a decrease of confidence due to the (intentional) similarity of both positive- and negative-cases to the one at hand, experts may have been more capable in leveraging the implicit information provided by the two similar cases.

Also the seemingly paradoxical effect that we observed in the experts' decrease of accuracy (or higher negative dominance), although associated with a very small effect size (0.067), can be tentatively explained with the low representativeness of the similar cases retrieved from the training set (which was observed in other studies [9]): the experts could have misjudged the AI reliability (both for the better and the worse) in virtue of the false assumption that past performance (right or wrong predictions) on these cases (whose complexity was accurately assessed) was representative of the performance on the current case. If similarity scores used to retrieve these cases were not associated with really semantically similar cases, the conceived support is much less effective than desired. However, it should be clear that this kind of support is not proposed to improve decision accuracy in the here-and-now decision (or is it unrealistic to think it can actually do it). Rather, its value should be considered in the long run, as a remedy to technology over-reliance, agency erosion, loss of skill and self-confidence, because it requires users of DSS keep reflecting on the technology reliability for the specific case at hand and exert full responsibility on their final decisions or diagnoses.

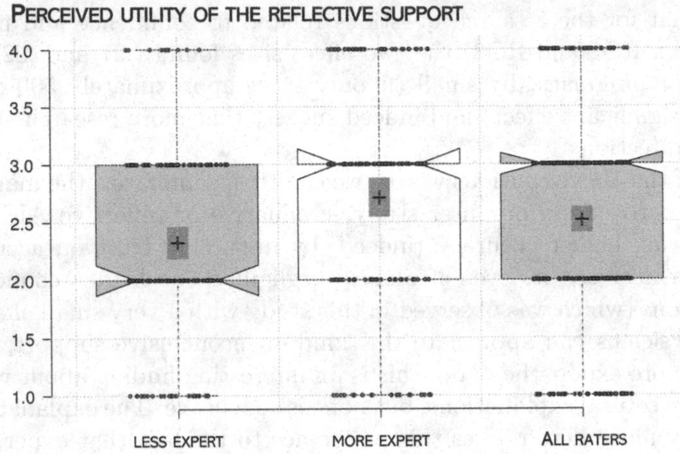

Fig. 7. Box plot of the perceived usefulness of having to consider the similar cases that the AI classified either rightly or wrongly in the validation set. The solid cross represents the average perceived usefulness, while the surrounding gray area represents the corresponding 95% confidence interval.

5 Conclusions

A highly popular book from the early 2000s [30], which sold over 100,000 copies on the niche subject of usability of interactive systems, boldly argued that a truly usable system's interface should not require users to think. In fact, the title conveyed this demand from a user's perspective: "Don't make me think!" Through heuristics and guidelines that remain relevant today, the book emphasized the importance of intuitive and user-friendly systems that minimize cognitive friction in supporting tasks. However, some of these suggestions may have inadvertently led to designs aimed at eliciting immediate and thoughtless reactions from users. Examples of such designs include dark patterns [23] or choice architectures [1] used on websites to manipulate users into actions that primarily benefit the implementers, often at the expense of the users themselves.

In this study, we explored the impact of a support system designed to serve the opposite purpose: encouraging users to think *before* making a particular (legally relevant) decision. Specifically, we focused on a feature resembling a form of XAI explanation (i.e., supplementary information related to machine-generated advice [11]) intended to assist users in gauging the local and instance-specific dependability of their decision support. This was achieved by presenting users with a pair of cases deemed highly similar to the one being evaluated, and informing them of the machine's performance in those instances (i.e., whether it misclassified the case or not).

This feature resembles posing the question, "Are you sure?" and appears to focus more on scrutinizing the machine's capability rather than affirming it (akin to stating, "In a similar case, the machine was, however, incorrect in identifying the presence/absence of a fracture"). In essence, we tested a feature aimed at decelerating the decision-making process and potentially increasing uncertainty (although we observed the contrary outcome), as a feature to de-bias the decision support, especially with regard to the priming and anchoring effect of presenting the AI advice before thoroughly considering the case [3].

As a result, what could seem a counterproductive or, at best, an inconsequential decision support system is actually more useful than allegedly more efficient systems. In fact, we believe, with other observers (e.g., Tim Miller [31] and Ben Shneiderman [36]), that we need to move from an oracular [13], answer-providing artificial intelligence (aimed at increasing the accuracy of the human-machine hybrid or human+AI team), to one instrumental to helping framing the problem, asking the right questions, enhancing the human being's ability to judge situations: a decision support that lets users affirm their autonomy, and that recognizes the need for them to feel the decision as totally their own, after gathering evidence both for and against the machine's suggestions. In this line of research, in a parallel study that we are conducting on the same set of diagnostic images [9] we are also experimenting the impact and acceptance of a support that does not give advice, neither in categorical nor in probabilistic form, but it precisely provides only elements *for* and *against* a certain decision: We have termed this kind of AI *judicial AI* specifically to highlight how systems within this class of application generate outputs that adhere to the dialectical and contrasting format prescribed by the adversarial procedure in the legal domain for court trials.

We argue that systems developed over centuries, which leverage refined skills, extensive training, and protective measures against abuse and injustice, still have the potential to enhance human decision-making also in settings that employ machine learning, data mining, and information retrieval techniques. By leveraging extensive historical data and (supposedly) correct decisions, we believe these systems can facilitate more informed and responsible decision-making while preserving human agency and skill. This study aims to explore and validate this notion.

References

1. Arnott, D., Gao, S.: Behavioral economics for decision support systems researchers. Decis. Support Syst. **122**, 113063 (2019)
2. Baselli, G., Codari, M., Sardanelli, F.: Opening the *black box* of machine learning in radiology: can the proximity of annotated cases be a way? European Radiology Experimental **4**, 1–7 (2020). https://doi.org/10.1186/s41747-020-00159-0
3. Bertrand, A., Belloum, R., Eagan, J.R., Maxwell, W.: How cognitive biases affect XAI-assisted decision-making: a systematic review. In: Proceedings of the 2022 AAAI/ACM Conference on AI, Ethics, and Society, pp. 78–91 (2022)
4. Bhandari, M., Reddiboina, M.: Augmented intelligence: a synergy between man and the machine. Indian J. Urol. IJU: J. Urol. Soc. India **35**(2), 89 (2019)
5. Bhattacharya, C., et al.: An application of the Mann-Whitney "U" test. Technical report. Indian Institute of Management Ahmedabad, Research and Publication Department (1981)

6. Buçinca, Z., Malaya, M.B., Gajos, K.Z.: To trust or to think: cognitive forcing functions can reduce overreliance on AI in AI-assisted decision-making. Proc. ACM Hum.-Comput. Interact. **5**(CSCW1), 1–21 (2021)
7. Byrne, R.M.: Counterfactuals in explainable artificial intelligence (XAI): evidence from human reasoning. In: IJCAI, pp. 6276–6282 (2019)
8. Cabitza, F., Campagner, A., Angius, R., Natali, C., Reverberi, C.: AI shall have no dominion: on how to measure technology dominance in AI-supported human decision-making. In: Proceedings of the 2023 CHI Conference on Human Factors in Computing Systems (CHI 2023). Association for Computing Machinery, New York, USA, Article 354, pp. 1–20 (2023). https://doi.org/10.1145/3544548.3581095
9. Cabitza, F., Campagner, A., Natali, C., Famiglini, L., Caccavella, V., Gallazzi, E.: Never tell me the odds. Investigating the concept of similarity and its use in pro-hoc explanations in radiological AI settings (2023, Submitted)
10. Cabitza, F.: Cobra AI: exploring some unintended consequences of our most powerful technology. In: Machines We Trust: Perspectives on Dependable AI, MIT Press (2021). ISBN 978-0262542098
11. Cabitza, F., et al.: Quod erat demonstrandum?-Towards a typology of the concept of explanation for the design of explainable AI. Expert Syst. Appl. **213**, 118888 (2023)
12. Cabitza, F., et al.: Rams, hounds and white boxes: investigating human-AI collaboration protocols in medical diagnosis. Artif. Intell. Med. **138**, 102506 (2023)
13. Cabitza, F., Campagner, A., Simone, C.: The need to move away from agential-AI: empirical investigations, useful concepts and open issues. Int. J. Hum Comput Stud. **155**, 102696 (2021)
14. Carroll, J.: Completing design in use: closing the appropriation cycle. In: ECIS 2004 Proceedings, p. 44 (2004)
15. Chalmers, M.: Seamful design and ubicomp infrastructure. In: Proceedings of Ubicomp 2003 Workshop at the Crossroads: The Interaction of HCI and Systems Issues in Ubicomp, pp. 577–584 (2003)
16. Chen, H., Gomez, C., Huang, C.M., Unberath, M.: Explainable medical imaging AI needs human-centered design: guidelines and evidence from a systematic review. NPJ Digit. Med. **5**(1), 156 (2022)
17. Cornelissen, N.A.J., van Eerdt, R.J.M., Schraffenberger, H.K., Haselager, W.F.G.: Reflection machines: increasing meaningful human control over Decision Support Systems. Ethics Inf. Technol. **24** (2022). Article number: 19. https://doi.org/10.1007/s10676-022-09645-y
18. Dai, X., Fishbach, A.: When waiting to choose increases patience. Organ. Behav. Hum. Decis. Process. **121**(2), 256–266 (2013)
19. Delacre, M., Lakens, D., Leys, C.: Why Psychologists Should By Default Use Welch's t-test instead of Student's t-test. Int. Rev. Soc. Psychol. **30**(1), 92–101 (2017)
20. Fonteyn, M.E., Kuipers, B., Grobe, S.J.: A description of think aloud method and protocol analysis. Qual. Health Res. **3**(4), 430–441 (1993)
21. Frischmann, B., Selinger, E.: Re-Engineering Humanity. Cambridge University Press, Cambridge (2018)
22. Gallazzi, E., Famiglini, L., La Maida, G., Giorgi, P., Misaggi, B., Cabitza, F.: Coloured shadows: understanding the value of visual aided diagnosis through AI-generated saliency maps. In: Orthopaedic Proceedings, vol. 104. The British Editorial Society of Bone & Joint Surgery (2022)

23. Gray, C.M., Kou, Y., Battles, B., Hoggatt, J., Toombs, A.L.: The dark (patterns) side of UX design. In: Proceedings of the 2018 CHI Conference on Human Factors in Computing Systems, pp. 1–14 (2018)
24. Gur, D., et al.: The "laboratory" effect: comparing radiologists' performance and variability during prospective clinical and laboratory mammography interpretations. Radiology **249**(1), 47–53 (2008)
25. Jaiswal, A., Arun, C.J., Varma, A.: Rebooting employees: upskilling for artificial intelligence in multinational corporations. Int. J. Hum. Resour. Manag. **33**(6), 1179–1208 (2022)
26. Karlsen, T.K., Oppen, M.: Professional knowledge and the limits of automation in administrations. In: Göranzon, B., Josefson, I. (eds.) Knowledge, Skill and Artificial Intelligence. HCS, pp. 139–149. Springer, London (1988). https://doi.org/10.1007/978-1-4471-1632-5_13
27. Keane, M.: Analogical mechanisms. Artif. Intell. Rev. **2**(4), 229–251 (1988). https://doi.org/10.1007/BF00138817
28. Kelley, K., Preacher, K.J.: On effect size. Psychol. Methods **17**(2), 137 (2012)
29. Kenny, E.M., Ford, C., Quinn, M., Keane, M.T.: Explaining black-box classifiers using post-hoc explanations-by-example: the effect of explanations and error-rates in XAI user studies. Artif. Intell. **294**, 103459 (2021)
30. Krug, S.: Don't Make Me Think!: A Common Sense Approach to Web Usability. Pearson Education India (2000)
31. Miller, T.: Explainable AI is dead, long live explainable AI! Hypothesis-driven decision support. arXiv preprint arXiv:2302.12389 (2023)
32. Ohm, P., Frankle, J.: Desirable inefficiency. Fla. L. Rev. **70**, 777 (2018)
33. Prabhudesai, S., Yang, L., Asthana, S., Huan, X., Liao, Q.V., Banovic, N.: Understanding uncertainty: how lay decision-makers perceive and interpret uncertainty in human-AI decision making. In: Proceedings of the 28th International Conference on Intelligent User Interfaces, pp. 379–396 (2023)
34. Selvaraju, R.R., Cogswell, M., Das, A., Vedantam, R., Parikh, D., Batra, D.: Grad-CAM: visual explanations from deep networks via gradient-based localization. In: Proceedings of the IEEE International Conference on Computer Vision, pp. 618–626 (2017)
35. Seo, K., Tang, J., Roll, I., Fels, S., Yoon, D.: The impact of artificial intelligence on learner-instructor interaction in online learning. Int. J. Educ. Technol. High. Educ. **18**, 1–23 (2021)
36. Shneiderman, B.: Human-Centered AI. Oxford University Press, Oxford (2022)
37. Sunstein, C.R.: Sludge audits. Behav. Public Policy **6**(4), 654–673 (2022)
38. Verma, S., Dickerson, J., Hines, K.: Counterfactual explanations for machine learning: a review. arXiv preprint arXiv:2010.10596 (2020)
39. Wagner, B.D., Robertson, C.E., Harris, J.K.: Application of two-part statistics for comparison of sequence variant counts. PLoS ONE **6**(5), e20296 (2011)
40. Xie, S., Girshick, R., Dollár, P., Tu, Z., He, K.: Aggregated residual transformations for deep neural networks. In: Proceedings of the IEEE Conference on Computer Vision and Pattern Recognition, pp. 1492–1500 (2017)
41. Yau, K.L.A., et al.: Augmented intelligence: surveys of literature and expert opinion to understand relations between human intelligence and artificial intelligence. IEEE Access **9**, 136744–136761 (2021)
42. Zou, K.H., Fielding, J.R., Silverman, S.G., Tempany, C.M.: Hypothesis testing I: proportions. Radiology **226**(3), 609–613 (2003)

Enhancing Trust in Machine Learning Systems by Formal Methods

With an Application to a Meteorological Problem

Christina Tavolato-Wötzl[1] and Paul Tavolato[2](✉)

[1] Austro Control Digital Services GmbH, Vienna, Austria
[2] Research Group Security and Privacy, University of Vienna, Vienna, Austria
paul.tavolato@univie.ac.at

Abstract. With the deployment of applications based on machine learning techniques the need for understandable explanations of these systems' results becomes evident. This paper clarifies the concept of an "explanation": the main goal of an explanation is to build trust in the recipient of the explanation. This can only be achieved by creating an understanding of the results of the AI systems in terms of the users' domain knowledge. In contrast to most of the approaches found in the literature, which base the explanation of the AI system's results on the model provided by the machine learning algorithm, this paper tries to find an explanation in the specific expert knowledge of the system's users. The domain knowledge is defined as a formal model derived from a set of if-then-rules provided by experts. The result from the AI system is represented as a proposition in a temporal logic. Now we attempt to formally prove this proposition within the domain model. We use model checking algorithms and tools for this purpose. If the proof is successful, the result of the AI system is consistent with the model of the domain knowledge. The model contains the rules it is based on and hence the path representing the proof can be translated back to the rules: this explains, why the proposition is consistent with the domain knowledge. The paper describes the application of this approach to a real world example from meteorology, the short-term forecasting of cloud coverage for particular locations.

Keywords: Explainable AI · Machine Learning · Formal Methods · Model Checking · Solar Radiation Forecast · Meteorology

1 Introduction

The increasing digitization in all areas of life is reaching meanwhile more and more critical domains. In the last few years, methods of artificial intelligence (AI), especially methods of machine learning (ML), have gained considerable influence in these areas. Based on the results of these systems effective decisions are taken automatically, paving the way to at least partially autonomous control of critical systems [19].

In such systems the necessary knowledge is not provided by human experts, but rather acquired autonomously by machine learning algorithms. These algorithms base their

Published by Springer Nature Switzerland AG 2023
A. Holzinger et al. (Eds.): CD-MAKE 2023, LNCS 14065, pp. 170–187, 2023.
https://doi.org/10.1007/978-3-031-40837-3_11

findings on historic or specifically prepared data (the so called training data) and generate a model representing the learned rules; this model is then applied to new data to come up with new information – predictions and/or operation instructions. Such systems raise a considerable number of ethical and legal questions. For example: How much trust in the system can (or must) a human operator have? To what extend can a system be held at least partially responsible for a (false) decision? Can a system be responsible at all (whatever "responsible" means)? How do we guarantee the transparency demanded by article 5 of the EU GDPR (European Union general data protection rules)[1]? We are clearly in a situation, where systems based on complicated machine learning algorithms provide suggestions and generate effective operation instructions. In a semi-automatic situation a human operator in the loop has to decide on the plausibility of these instructions. Does s/he have to merely believe what the system outputs or is there a possibility to understand, why the system's conclusions are plausible? Most systems in use nowadays are not transparent at all and give no explanations whatsoever to the users about their conclusions and why the users should or could trust these conclusions.

Such systems work in two phases:

1. In the first phase – the learning phase – the systems are confronted with a set of training data – mostly historic data, sometimes even augmented with artificial data that has not been observed in reality. The systems use machine learning algorithms, such as statistical methods or structural methods like neural networks to build a model, usually of considerable complexity. One of the most essential aspects of machine learning is the selection of relevant features (properties of the data), which will be used by the machine learning algorithm. For sake of clarity we will call the system that operates this first phase the "ML system".
2. In the second phase, this model is applied to data from a new (ongoing) situation that was not part of the training data set with the aim to generate forecasts and operating instructions. To distinguish it from the ML system we will call the system that operates this second phase the "AI system".

The advantage of this method is twofold: First, an information processing system can take properties of the data into account that might be of importance, but are not directly available to a human analyst (e.g. because they are derived from the raw data by more or less complex mathematical transformations). Second, the system – because of the mere speed of modern data processing – could use a very large number of training examples – much more than a human expert could ever see during his professional lifetime. The main disadvantage of present-day systems, however, is that they work like a black box: For the human (the responsible operator) it is not transparent and hence not comprehensible, how the system arrives at its result. S/he has to believe in the system's findings. And "belief" is not a scientific category. Moreover, the more complex the learning algorithm, the less comprehensible the result. On the other side, applying more complex learning algorithms may generate better results.

[1] In Chapter II, Article 5, Paragraph 71 of the introduction reads "… must guarantee … the right … to obtain an explanation of the decision reached after such assessment". https://www.privacy-regulation.eu/en/22.htm, last accessed 2023/03/01.

A main drawback of such systems from a technical point of view is the missing robustness of many of them: The systems heavily depend on the quality and amount of the training data used in the ML phase. Without a careful sensitivity analysis – which is not provided with most of the systems – one cannot guarantee that the resulting model of the ML system does not show significant changes induced by just minimal perturbations in the training data, as shown in [15].

The acceptance of such a system by the users is essential for a successful integration into the everyday-practice of critical systems. A main ingredient of acceptance is **trust**. As defined by the Merriam-Webster Dictionary trust is the "assured reliance on the character, ability, strength, or truth of someone or something". Therefore, the development of methods that are able to increase the users' trust in the results of a system based on machine learning methods is an urgent demand [24]. One such method to achieve this goal would be to verify the system's result in the domain specific knowledge of the user – which is the core point of this paper.

Before we can attack this challenge, we must clarify some fundamental concepts: What does "explainability" mean in the context of AI systems based on machine learning? How must such an explanation look like, so it is comprehensible and interpretable by users from the application domain? What is a practical balance between the complexity of the learning algorithm, and hence the quality of its results, and its ability to provide a comprehensible explanation of its mode of operation? This paper attempts to answer some of these questions.

After a discussion of the state of the art in Sect. 2, the following section tries to investigate the concept of "explanation" as used in a scientific context in a short historical perspective. In Sect. 4 we develop our notion of an explanation with respect to its possible meaning in the realm of machine learning tools. This notion is based on several ideas from the philosophy of science and uses the concept of "trust" as its central decisive point. Further on, we propose a practical method to implement this notion.

In order to bring the discussion close to practice we will demonstrate our ideas on a real-world example: meteorological nowcasting (short term predictions) of cloud coverage for specific locations. Machine learning using neural networks is used for this purpose. An archive of satellite images serves as training data to predict the upcoming cloud coverage. The goal of these short term predictions of solar radiance is the advance calculation of the daily power output of solar power plants. Knowing about the upcoming weather situation in advance for a specific location helps to calculate and pre-arrange for the expected power loss or gain each day. Thus, such a system can help to avoid financial loss when less power is harvested than expected due to overcast, frontal systems or convective cells.

2 State of the Art

The development of theory, methods and tools for Explainable Artificial Intelligence (XAI) is a new, but very active area of research. Two directions of research, sometimes colliding and sometimes complementary to each other, have evolved. The first is aimed at developing tools for increasing the transparency of automatically learned and defined prediction models, as for instance by deep learning or reinforcement learning algorithms.

The second focuses on anticipating the negative impacts of opaque models in order to regulate or control consequences of false predictions, especially in sensitive, safety-critical areas.

Early AI was not so complex – due among others to hardware limitations –, and hence retraceable, interpretable, thus understandable by and explainable to humans. Consequently, there is some historical research that uses abductive reasoning for scientific explanation of the results of AI systems. Examples are Pople in 1973 [23], later Poole et al. in 1986 [22], or Muggleton in 1991 [21], up to Evans et al. in 2018 [8]. Pople describes an algorithm and develops a model for abduction applied to medical diagnosis. Poole et al. provided a logic programming system in first order logic, which could subsume non-monotonic reasoning theories and implemented a proof-of-concept framework which uses explanatory hypothesis for any application domain. Muggleton proposed a further refinement referred to as inductive logic programming where hypotheses are identified by inductive constraints within any logic, including higher-order logics. Finally, these adoptions have been generalized to explanations based on inductive logic programming by Evans et al. [8]. This rather recent work connects with information theoretic ideas used to compare differences in how to learn probability distributions that are modelled by machine learning methods.

Many scholars have tried to review research works about explanations [2, 3, 9, 19, 29] and especially [4], which quotes more than 400 papers more or less dealing with the subject. These research papers cover a broad range of different methods for providing explanations depending on the wide spectrum of different applications. As an example in [5] the authors provide a classification framework based on comparing levels of explanation with autonomy-levels in driving.

In recent years, explainable AI methods mainly focus on tracing back the results of a black-box machine learning systems to the input data, highlighting input relevant parts via heat-mapping. The idea behind such procedures is to show the parts of the input, which contributed most significantly to the result. With models based on neural networks, this is a very challenging task involving advanced mathematical methods as shown in [6] for image analysis. Unfortunately, this solves only a part of the problem as there is a need to take into account a concept called causability, too (see [14] or [20]).

In [18] current concepts, applications, research challenges and visions of explainable artificial intelligence describe the big picture, ideas and their role in advancing the development of explainable AI systems and to emphasizes the biggest challenges to moving forward. A main issue remains: are – at least theoretically – all machine learning models explainable or are some models inherently unexplainable [25]. In the latter case, the requirement of explainability would restrict the choice of the machine learning model. Whether this leads to a lower quality of the outcome is still an open question.

Fundamental considerations on the topic can be found in a paper by Juan Durán [7]. Restricted to the application areas of medicine and healthcare he advocates for scientific explanations (as opposed to mere classifications): an explanation should be called scientific only if they “conform to a specific well-defined structure capable of advancing our understanding of the world”; he calls explanations following such a structure “bona fide” and claims that most of the studies published on explainable AI in the medical field fail to be bona fide.

All these approaches show that explainability in AI systems, especially in critical domains, is indispensable for bringing them to practical use. The scientific literature offers only first results in this new area of research called Explainable AI or short XAI. However, a lot of basic research still needs to be done.

3 What is an "Explanation"?

So far, some definitions of the term "explanation" with respect to the area of ML systems can be found in the literature. [10] for example gives the following definition: "Produce more explainable models, while maintaining a high level of learning performance (prediction accuracy); and enable human users to understand, appropriately trust, and effectively manage the emerging generation of artificially intelligent partners". In some publications the authors make a difference between "explainable" and "interpretable", see for example [4]. There, interpretability is defined as "the ability to explain or to provide the meaning in understandable terms to a human"; and explainability is "associated with the notion of explanation as an interface between humans and a decision maker that is, at the same time, both an accurate proxy of the decision maker and comprehensible to humans". All these definitions have one point in common: they are extremely informal and hence do not qualify for a stringent scientific treatment (not to mention the circular reference in the definition of [4]). Though our approach is closer to the concept of interpretability, we will nevertheless use the term explainability, as it is widely used in the literature and reflected in the generally accepted term "XAI" as an abbreviation for "Explainable AI".

3.1 A Brief Survey of the Term Explanation in the Philosophy of Science

In the light of such vague definitions it seems appropriate to review how the concept of explanation is treated in the philosophy of science. This discussion started in 1948 with a paper by Hempel and Oppenheim [12] introducing the HO model (Hempel-Oppenheim model). Based on the philosophical tradition of logical empiricism (Hempel was a member of the "Vienna Circle") they define scientific explanation as a deductive structure consisting of initial conditions and law-like generalizations that entail the truth of the event to be explained. In other words: the explanation of a proposition is defined as a logical derivation from some given conditions according to generally valid laws. Hempel and Oppenheim mention two possibilities how such laws can be obtained: either they are universal generalizations (they call that deductive-nomological) or they are of statistical nature (inductive-statistical); they strongly prefer deductive-nomological laws over inductive-statistical ones and accept the latter ones only as approximations of the former. There has been some criticism of the HO model, though. The main objections are:

- the problem of relevance of the initial conditions (irrelevant premisses could be part of an explanation), and
- the problem of asymmetry meaning that the discrimination between the fact to be explained and the initial conditions is not always evident (if for example the barometer is falling rapidly and this is followed by a storm, it is not clear whether the falling barometer is an explanation of the storm or the storm explains the falling barometer).

Due to this criticism some alternative approaches to define the concept of explanation were published during the last 60 years. These new definitions can roughly be classified into two groups: realistic and epistemic (non-realistic). An epistemic interpretation means that the task of a theory is to order our experience, while a realistic interpretation is based on the correspondence of the theory with the external reality. With regard to machine learning only epistemic approaches are relevant, because machine learning deals with data, which must be ordered in a way to produce a model that is in line with this and similar data. The HO model is clearly epistemic. Other epistemic models were published by van Fraassen [27], Achinstein [1], and Holland et al. [13].

Van Fraassen [27] defines an explanation as an answer to a why-question, so the pair question-answer replaces the premises and the conclusion of the HO model. A main point in this argument is that the validity of the relation between the question and the answer is defined by the interests of the person asking, which means that explanations are subjective with respect to the person demanding the explanation.

Achinstein [1], too, defines explanations in terms of questions and answers, but focuses on the process of explanation itself. The goal of an explanation is to generate understanding in the asking person by answering her/his questions. This requires to take into account the asking person's knowledge when constructing the answer. Therefore, Achinstein's suggestion is subjective, too, despite the fact that he tries to overcome some problems of van Fraassen's proposal by an attempt to clarify the concept of understanding and hence eliminate tautological answers as valid explanations. Nevertheless some essential concepts in his theory of explanation remain vague.

A third approach to defining explanations described in Holland et al. [13] refers to concepts from cognitive science combining findings from neuroscience and aspects from AI. The central idea is the use of a mental model, which is an internal representation of a hierarchical structure of if-then type rules. Generating an explanation means searching and finally finding a path through this structure. Due to the hierarchical structure of the search space the search algorithm must contain backtracking mechanisms. Moreover, the idea of the mental model must incorporate learning algorithms in order to improve its structure for future use.

A generally important aspect can be found in Habermas [11]: he does not talk literally of "explanations", but rather of "understanding". But he argues that understanding another person's point of view requires this other person to "explain" her/his point of view, thus understanding is just the other side of an explanation, seen from the recipient of the explanation. According to Habermas understanding is "intersubjective mutuality … shared knowledge, mutual trust, and accord with one another" [11]. An explanation, therefore, must have the capability of generating this intersubjective mutuality and mutual trust.

A model specifically designed for medical AI systems, but containing many generally applicable aspects, was proposed by Durán [7]. He demands that XAI must answer **why-questions** (as requested by van Fraassen [27] more than 40 year ago) as opposed to most conventional work in XAI that answers how-questions only. He distinguishes between the unit carrying out the explanation (called explanans), the unit that will be explained (called explanandum) and the objective relations of dependence between these (called the explanatory relation).

3.2 Defining Explanation for Systems Based on Machine Learning

In the following we will give a specific definition of the concept of explanation based on the philosophical discussion and fitting to the field of XAI. From the aforementioned various philosophical considerations we will take the following cornerstones into account for our definition: the concept of trust from Habermas, the concept of a logical derivation from the HO model; the consideration of the knowledge of the target person of the explanation; and the concept of a mental model as the basis for finding a path to the proposition to be explained. We claim that the overall goal of an explanation is to build up trust in the results of the AI system within the users. Moreover, we have to take into account three specific requirements:

- The definition must take into account that the machine learning algorithms generate a model of the (training) data, which is not per se accessible to the human user due to its potential complexity.
- The explanation, however, must be understandable to the human user and hence take the user's prior knowledge into consideration.
- The process must be apt for automatic processing implying that it be formal.

In contrast to the various philosophical considerations about epistemic definitions of explanations, the situation in XAI is characterized by the fact that the machine learning algorithms create a model that is supposed to describe the application area under consideration sufficiently accurate so that the AI system can generate predictions and operating suggestions for the application. In this situation two separate models of the application exist: the one generated by the machine learning algorithm and the application domain model inherent in the user's knowledge about the application. In general, these models are different (otherwise machine learning would not be necessary). Hence, there are two different ways of providing an explanation to create understanding for the AI system's predictions and suggestions:

1. either the explanation tries to instruct the user about the model generated by the machine learning algorithm,
2. or the explanation tries to find a plausible conclusion for the AI system's proposals within the user's mental model of the domain (his domain knowledge).

The first of these possibilities is the one almost exclusively found in the scientific literature about XAI: common XAI systems try to explain to the user the model generated by the machine learning algorithm by providing a relation between the features used by the machine learning algorithm and the outcome of the system. In terms of the HO model the explanation is given by describing the initial conditions that lead to the result. The mostly statistical generalizations are implicit in the generated model and usually not part of the explanation. The reason for this is that often this model – such as a neural network – is too complex to be easily accessible to the human user. We think that such an approach can only fulfil the claim to be a valid explanation in cases where either the machine learning model is rather simple (e.g. a not too large decision tree) or the user's domain model is evidently correlated with the result of the system (as for example in special cases of image recognition, where it is easy for the user to check whether the result of the system – say, the system recognized a cat in the image – is correct or not).

We suggest an alternative approach to explanation based on the second possibility, namely on the knowledge available to the user about the application domain. The explanation should be found in the user's mental model and thus provide an understanding for the result of the AI system. Helpful in this sense is the definition of "understanding" given by Habermas as mentioned above: understanding is "intersubjective mutuality … shared knowledge, mutual trust, and accord with one another" [11]. The crucial word here is "trust": it signifies the ultimate goal of the explanation: the user shall trust the result of the AI system as it is compliant with his understanding of the application domain. Trust can be built up either by understanding in detail how the system came to its solution (see the possibility 1 mentioned above) or by checking the plausibility of the solution by comparing it to a solution achieved in a different, trustful way. This second possibility – the one we ground our approach on – is like going for a second opinion: If this second opinion is equivalent to the first, our trust in the solution is enhanced. Otherwise we would remain sceptical.

These considerations lead to the following definition of the concept of an explanation: in the field of XAI an **explanation** – in contrast to a mere description – is **a formal proof of a proposition in a formal model of the relevant domain theory**. The **proposition** in this definition is the output of the AI system, which bases its results on models derived by a machine learning algorithm. And the explanation relates this output to the user's domain knowledge. **The relevant domain theory** in this definition is the common understanding of the rules governing the domain as collectively understood by the users of the AI system.

4 Generating an Explanation

The idea is the construction of an application domain model representing the users' knowledge in addition to the one generated by the ML system as shown in Fig. 1. Note that the model generated by the machine learning algorithm is not necessarily a subset of the application domain model! The AI system could base its answers on other information than a human operator would (or even could).

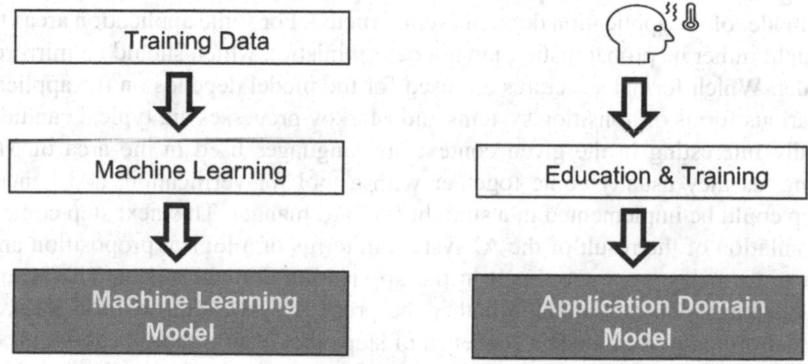

Fig. 1. The two models

To generate an explanation for the results, which the AI system produces for a new problem, we must try to find a way to prove this result in the users' application domain model (see Fig. 2). If we are successful, this will increase our trust in the result of the AI system and the steps on this way will give an explanation of the result in terms of the users' domain. To be valid in a strict sense this proof of the result in the application domain model must be a **formal proof** of the result of the AI system **in a formal definition of the application domain model**. The steps of the proof could then be used as an explanation of the result.

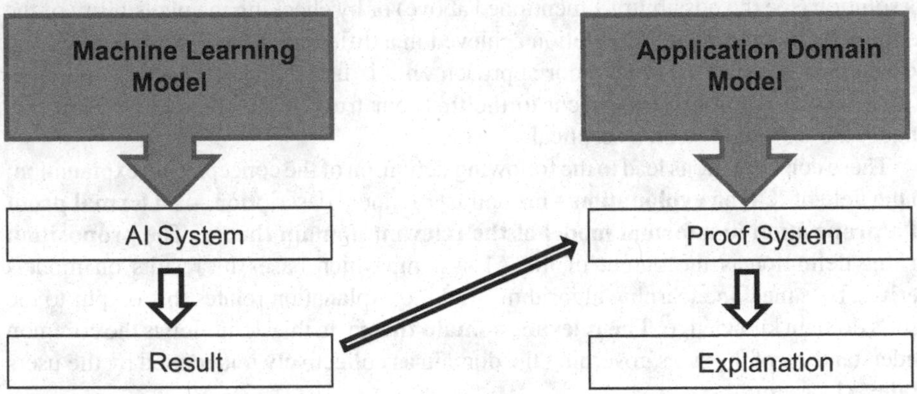

Fig. 2. Generating an explanation

The application domain model must represent the users' domain knowledge sufficiently precise to be a sound basis for solving the given problem. But, having in mind the ultimate goal, namely the generation of an explanation, the model must guarantee that a comprehensible description of the steps taken to come to a solution of the problem is possible. This is a balancing act between preciseness and simplicity.

Starting points for the construction of the application domain model are the set of if-then-rules, an expert user would follow in order to produce a result based on her/his knowledge and the available input data for the problem. Out of these prerequisites a formal model of the application domain is constructed. For some application areas these rules might rather be probabilistic (and not deterministic), which should be mirrored in the model. Which formal structures are used for the model depends on the application area. Various forms of transition systems and Markov processes are typical candidates. Especially interesting in the given context are languages used in the area of Model Checking, as they usually come together with a tool for verification, and hence the next step could be implemented in a straight forward manner. This next step comprises the formulation of the result of the AI system in terms of a logical proposition and an attempt to formally prove the result in the application domain model. The proofs in model checking tools (no matter whether the proof is successful or it fails) have the convenient property to produce a sequence of steps that lead to the successful proof or to a contradiction. If these steps are attributed with the rules of the application domain model, this sequence of rules is an explanation.

5 Applying the Method to a Meteorological Example

5.1 Description of the Meteorological Problem

The goal of the application is the prediction of cloud coverage or solar radiation for a specific limited area based on satellite data. This information represents an important parameter for calculating the expected output of a solar power plant and thus in advance helps to manage situations where less power is harvested than needed due to cloud coverage from frontal systems or convective cells. The earlier you know you have to purchase additional energy, the lower the price. The users of the system have meteorological knowledge and are capable of interpreting weather data such as satellite and radar images.

Within the last years this problem has gained considerable attention in the meteorological literature; see [26] for a recent review. Generally, one has to distinguish between ultra-short-term forecasts (in the range of up to 10 min) on the one hand and forecasts in the range of several hours up to one day. Ultra-short-term forecasts are necessary for dealing with the fluctuations of the power output of solar plants and the management of short-term energy storage systems; as this is not the main focus of the meteorological example described here, they will not be considered further here.

Forecasts in the range of several hours can be achieved by different methods: either by methods of machine learning (see for example [28]), where a wide range of algorithms is suggested spanning from statistical methods like KNN or SVM to various versions of neural networks or by analytical methods based on models of numerical weather prediction. It must be noted, though, that out of the more than one hundred and fifty papers referenced in [26] only three resort to satellite images as input.

In the example described here a machine learning approach using a neural network is used for solar radiation prediction based on satellite images and the goal is to enhance trust in this system's results by generating an explanation.

5.2 Machine Learning Approach

In the learning phase the prediction model is constructed with the help of a convolutional neural network with five layers. The algorithm is faced with an archive of satellite frames spanning approximately four and a half years. These frames are taken every 5 min and cover a rectangular area enclosing the territory of Austria. Each satellite frame comprises 4 channels: a high-resolution visibility channel (HRV), an infrared channel, and two water vapour channels[2]. Additionally, the solar angle for each measurement point and generally some topographical data about the area scanned is provided. For illustrative purposes Fig. 3 shows an example of an HRV image from Europe. One can see a weather front here over Great Britain (belonging to a low-pressure vortex over the Atlantic) and some convective cells, for example near Salzburg and in Slovenia.

Preprocessing. Preprocessing includes a discretization of the images. Each image is sectioned by a 320×256 grid and each grid point is assigned its topological height. All images are grey-scale; the grey-values are classified into 16 classes and for each grid

[2] For more information see for example https://en.allmetsat.com.

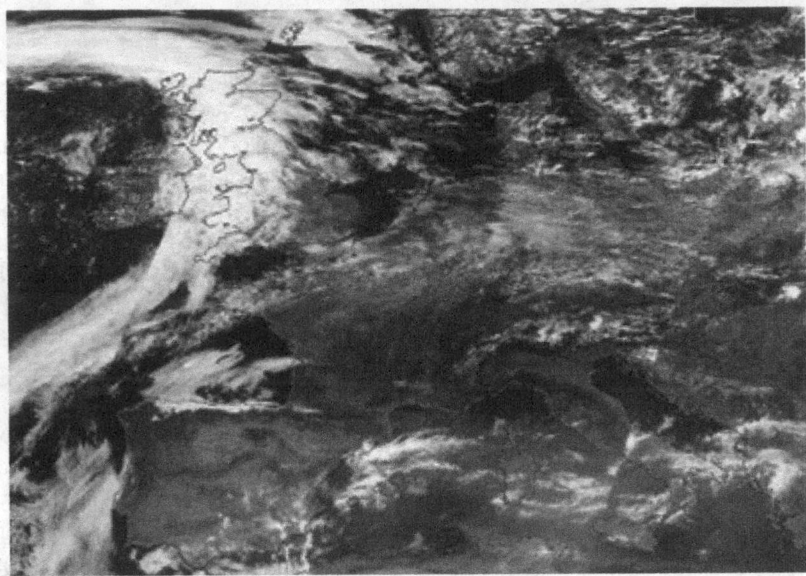

Fig. 3. HRV image of Europe from August 11, 2021 (Source: https://worldview.earthdata.nasa.gov/, last accessed 2023/03/08)

point in each image the respective class is assigned. Each image has its timestamp and the solar angle. Solar angles are used to facilitate the determination of day/night and the season with respect to the topological situation of the grid point.

Machine Learning. Out of an archive of such satellite images from the past four and a half years the learning algorithm uses a convolutional neural network to describe the movement, emergence and disappearance of sky cover for the grid points over time. The resulting ML model is highly non-linear and hence does not lend itself to easy interpretation or even understanding.

Prediction. In the analysis or prediction phase the 12 last frames (each including the four channels) taken at an interval of 5 min are input to the neural network (the machine learning model) generated in the learning phase together with the topology (height) and the solar angle of the location, for which the prediction should be calculated. The output of the model is a prediction of the cloud coverage for every grid point for the next 6 h in intervals of 15 min. Due to the complexity of the task and the restriction to the 12 last frames, there exist situations, were the same sequence of images can lead to different evolutions, which means that there can be more than one possible prediction. For each prediction a probability is calculated and the AI system chooses the prediction with the highest probability. For the intended purpose (prediction of the daily power output of solar power plants) we can limit the analysis to daytime for obvious reasons.

5.3 Constructing the Explanation

The task of constructing an explanation can be subdivided into subtasks:

1. Define the necessary knowledge for the application domain model: gather the knowledge from experts and available sources and formulate it as rules.
2. Develop a formal model of this knowledge: Discrete Time Markov Chains seem appropriate to represent the knowledge in this example.
3. Implement this model: formulate the model in the language of a model checking tool (for example PRISM [17]).
4. Generate the explanation: formulate the result of the AI system in the language of the tool and try to formally prove it within the formal model by model checking.
5. Present the explanation in the users' domain language: translate the path representing the proof (or the contradiction if the proof failed) that constitutes the explanation in terms of the rules used on this path.

Definition of the Application Domain Model. As mentioned above the users of the system have meteorological knowledge, especially concerning the interpretation of satellite data. This knowledge constitutes the users' domain model and for our intended purpose must be described in a formal way. We collected this knowledge from interviews with such persons and from common information sources such as [30] or [31]. We coded this knowledge as a set of textual rules, mainly in an if-then form (these rules will be used later for the formulation of the explanation). Starting point for the model are the topological information of the area under consideration and the 12 last satellite frames (each including the four channels) taken at an interval of 5 min, which are preprocessed in the way mentioned above. The next steps are:

(1) The detection of large-scale cloud patterns (weather fronts) and of small-scale cloud patterns (convective cells and fog) in an image, each identified by the sets of grid points they comprise. They can be extracted from the satellite data: from the HRV channel (during day only) and more details from the infrared channel (the more light-colored the colder and hence the higher up rises the cloud). This is especially relevant during winter where clouds and snow-covered underground resemble each other very much in the HRV image.

(2) The calculation of the direction and the velocity of the flow of these patterns from the 12 consecutive input frames of the last hour. Large scale structures are relatively inert: they neither emerge spontaneously nor do they vanish spontaneously and they do not change their direction of flow immediately. If a large structure arrives at a barrier like mountains (e.g. the Alps) the structure usually does not overstep this barrier (at least if it had not done so in the past). Small scale cloud structures, on the other hand, have a shorter life cycle. They can emerge spontaneously in specific weather situations; for example during summer a convection can develop before a front leading to a shower or thunderstorm. Small scale structures can grow rapidly and they may disappear rapidly (e.g. fog). So, their lifetime, and the conditions at the time of their appearance and disappearance are calculated, too (if possible). Moreover, the velocity of cloud structures depends on additional parameters including topographical properties of the subjacent landscape, the time of the day and the airmass involved. Depending on those parameters cloud patterns may change over time.

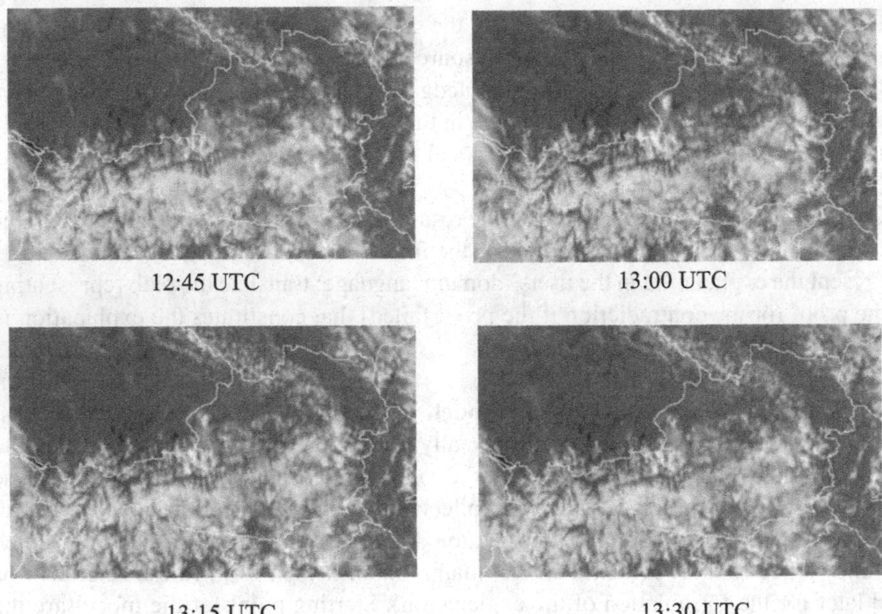

Fig. 4. Development of cloud patterns over Austria, 2023-04-21

Figure 4 shows such a development over one hour in steps of 15 min on April 21, 2023 over Austria (note the emergence of a local shower north of Salzburg)[3].

Formalization of the Application Domain Model. To this end we use Discrete Time Markov Chains (DTMC). The use of Markov Chains for cloud coverage predictions has for example been described in [16]. A Discrete Time Markov Chain (DTMC) consists of a set of states S and a transition matrix P = ($p_{i,j}$) with |S| rows and columns; $p_{i,j}$ is the probability that when the DTMC is in state i at time t, then it will be in state j at time t + Δ (remember that time steps for predictions are in intervals of $\Delta = 15$ min). P can be regarded as function P: $S \times S \rightarrow [0, 1]$. Furthermore, for each state a set of information is defined by a labelling function L: $S \rightarrow 2^{AP}$ that assigns a label to each state based on a set of atomic propositions AP. In addition we define labelling RL, which labels the transitions with the rule applied RL: $S \times S \rightarrow R$, where R is the set of rules. One state, called s_0, is defined as the starting state.

The states describe the weather situation at a certain time t consisting of the cloud structures together with the attributes valid at time t such as the direction and velocity of the structures observed so far and – for small scale structures – the time of appearance and disappearance. The transition matrix represents the probabilities that a weather situation represented by a state s at time t will evolve into a weather situation represented by state s' at time t + Δ. The labelling function L assigns to each state the cloud coverage for the area under consideration. The set AP of atomic propositions can be used to classify the cloud coverage into a limited set of different grades of coverage (say full coverage,

[3] Source: https://en.allmetsat.com/

75% coverage, 50% coverage, 25% coverage, no coverage). The labelling function RL assigns the rule that is used for a transition from a state s to a state s'.

For the application of the model in a specific situation it has to be initialized with the information belonging to this situation: The starting state s_0 is the cloud pattern of the last input frame and the time this frame was taken is t_0. It is labelled with the state information (velocity and direction are calculated from several frames taken before). For the first prediction for time $t + \Delta$ we use the domain knowledge rules. This can either lead to one new state (with transition probability 1) if everything is clear or to two (or sometimes even more) new states if there are ambiguities in the knowledge base. For the transition to each new state the knowledge base must include the transition probabilities (which must sum up to 1). New states must be labelled with the information generated by the prediction. Additionally the edge leading to the next state is labelled with the rules used to generate this state transition. This must be done for each time step. A prediction of 6 h will thus need 24 time steps.

Implementation of the Application Domain Model. This model is implemented in the model checking tool PRISM[4] [17]. This tool allows probabilistic model checking and hence is suited for the purpose. Furthermore it provides the possibility of directly implementing DTMCs. A detailed description of the application domain model and the DTMC fills many pages and therefore is not possible here.

Generation of the Explanation. To generate an explanation of a result from the AI system (which predicts the cloud coverage for a specific grid point for a specific point in time during the next 6 h) we formulate the result in terms of PRISM's property specification language (a syntax for formulating expressions of a temporal logic). After having correctly initialized the model we run the model checker on the property. As it is a probabilistic model checker, it yields the maximum and minimum probability of the situation formulated by the property. If we are content with these probabilities – if they increase our trust in the prediction of the AI system in a satisfactory way – we accept the result as an explanation. If the probability that the property is true is not sufficient for us, we can try to look for a property with better probabilities and compare, whether the two properties are close enough to each other giving us confidence in the result. If there is no such property we conclude that the result of the AI system cannot be explained by the application domain model.

As an example let's assume the situation from Fig. 4 and let's concentrate on Ried im Innkreis as the location of interest marked red (see Fig. 5).

It was sunny with 16 °C and a humidity of 57%. The cloud cover prediction for the next hour by the AI system was that it will still be sunny (no cloud coverage) with a probability of more than 0.95. This prediction is now formulated in the property specification language of PRISM:

$$P > 0.95\,[F = 4 \text{ cloudcover } = 0] \tag{1}$$

This formula in temporal logic means: The probability P is greater than 0.95 that after exactly 4 time steps the variable cloudcover will have a value of 0. F is the future

4 http://www.prismmodelchecker.org/

Fig. 5. Ried im Innkreis, 2024-04-21, 12:45

operator and F = 4 indicates "4 time steps ahead" (time steps are in intervals of 15 min). The variable cloudcover can only assume 5 different values (0 for no cloud cover, 1 for a 25% cloud cover, 2 for a 50% cloud cover 3 for a 75% cloud cover, and 4 for a complete cloud cover).

In the users' domain the situation can be described as follows: the weather front seen here was coming from the south and is stuck at the Alps which it cannot cross. Therefore, for the specific location no change in the cloud coverage is to be expected from this front at least in the next hour. There is still a chance that a convective cell could emerge, but the probability for this is very low as the humidity of 57% is not high enough, the temperature is relatively low and the topological situation of Ried im Innkreis does not enhance an immediate upstreaming of the air. The emergence of a convective cell needs an unstable state of the air masses, so warm and humid air can stream upwards and there meet some residual air from the front that succeeded in crossing the mountains. The clash of the air masses then results in the convective cell. The parameters (temperature, humidity, and the topological situation) are not in favor of the forming of a convective cell and so the probability for it is really low. This leads to the prediction of "no cloud cover in one hour in Ried im Innkreis with a probability greater than 0.95".

After having correctly initialized the model we run the model checker on the property. The result is whether the property is true or false. Alternatively we could have calculated the minimum and maximum probabilities for the property. In the property specification of PRISM this reads:

$$\text{Pmin} =? [\text{F} = 4 \ \text{cloudcover} = 0] \tag{2}$$

$$\text{Pmax} =? [\text{F} = 4 \ \text{cloudcover} = 0] \tag{3}$$

If we are content with these probabilities – if they increase our trust in the prediction of the AI system in a satisfactory way – we accept the result. In order to get an explanation in textual form, we follow the path from the starting point of the proof to its endpoint. The labelling function RL that labels all state transitions gives a sequence of rules followed by the model checking procedure when generating the proof in the DTMC. In the example this will read as follows:

1. The state at the location under consideration in the beginning: sunny, 16 °C, humidity 57%; a front approaching from the south.
2. As south of the location under consideration there are high mountains, the front will not cross the mountains, stop its movement and hence will not reach the location.
3. The emergence of a convective cell at the location has only a probability of 0.05 due to the values of the influencing parameters: temperature, humidity and topographical situation.
4. Therefore the overall prediction for the location Ried im Innkreis is cloud coverage of 0% (no clouds) with a probability of 0.95.

If the proposition representing the result of the AI system cannot be proved or the probability that the property is true is not sufficient for us, we conclude that the result of the AI system cannot be explained by the application domain model. This is then not enhancing our trust in the result of the AI system. We could now try to look for another property, which was not predicted by the AI system and calculate its probabilty. If this property has a considerably higher probability than the property representing the AI system's prediction we have to decide which prediction is giving us better confidence.

6 Conclusions

Building up trust in the results of an AI system based on machine learning is one of the most important scientific topics nowadays, as these methods are used in more and more different application areas, some of them critical domains. So far a strict and nevertheless usable definition of the concept of explanation is still in discussion. This paper adds a new view into this definition problem by trying to correlate the results of the machine learning system with the domain theory of the users. It tries to give a theoretically sound, nevertheless practically viable answer to the following question: How can we enhance trust in the results of a machine learning system, when this is an essential requirement? Without an AI system's ability to give understandable reasons for the plausibility of its results, the system will never cope with the requirements concerning safety and accountability and will never generate the necessary trust within the personnel. Systems cannot only use up-to-date machine learning algorithms, they must be able to show the plausibility of their results in a language understandable by the users, too. In other words they must be able to give an "explanation" of the results.

This paper assumes generally that the main goal of an "explanation" is to enhance trust in the results of the AI system. It suggests to define such an explanation as a formal proof of a proposition in a formal model of the relevant domain theory. As opposed to the majority of papers in XAI this definition does not mean that the internal operation of the machine learning algorithm should be explained to the user, but rather tries to explain the result of the AI system in terms of the users' domain knowledge, which need not even partially overlap with the model generated by the machine learning system.

The viability of this approach is shown with the help of an example taken from the field of meteorology: machine learning methods are used to generate a model for short term predictions of cloud coverage for specific locations from satellite images. To enhance the users' trust in the results the users' domain knowledge is described by means of rules. These rules are used to generate an application domain model in the form of a Discrete Time Markov Chain. The DTMC is implemented in a model checking tool (PRISM); the prediction of the AI system is formulated as a proposition in temporal logic, and finally an attempt is made to prove this proposition – the result of the AI system – (or at least a situation close enough to the prediction in terms of probability) in this model. The path in the DTMC (the application domain model), leading from the input data to the predicted result represents not only the proof, but it also is an explanation: it explains the result by giving consecutive steps that will lead to the predicted result together with the rules from the users' domain knowledge that induced the transitions from one step to the next. This information is directly related to the domain knowledge of the users and can therefore be understood by them. Thus it is an explanation in the sense of our definition, which can generate trust in the results of the machine learning system. In the meteorological example the sequence of rules could be augmented with the consecutive images showing the development of the cloud cover over time.

We are well aware that the details of this procedure are strongly influenced by the meteorological example chosen. Nevertheless, we think the approach is general enough to be used in other applications areas as well. Future work will encompass two directions: first, the refinement of the meteorological domain model and second it will centre around extending the method to other application areas.

Acknowledgement. This research was funded in whole, or in part, by the Austrian Science Fund (FWF) P 33656. For the purpose of open access, the author has applied a CC BY public copyright license to any Author Accepted Manuscript version arising from this submission.

References

1. Achinstein, P.: The Nature of Explanation. Oxford University Press, New York (1983)
2. Adadi, A., Berrada, M.: Peeking inside the black-box: a survey on explainable artificial intelligence (XAI). IEEE Access **6**, 52138–52160 (2018)
3. Arras, L., Osman, A., Müller, K.-R., Samek, W.: Evaluating recurrent neural network explanations. In: Proceedings of the 2019 ACL Workshop BlackboxNLP: Analyzing and Interpreting Neural Networks for NLP, pp. 113–126 (2019)
4. Arrieta, A.B., et al.: Explainable artificial intelligence (XAI): concepts, taxonomies, opportunities and challenges toward responsible AI. Inf. Fusion **58**, 82–115 (2020)
5. Atakishiyev, S., et al.: A multi-component framework for the analysis and design of explainable artificial intelligence. Mach. Learn. Knowl. Extr. **3**, 900–921 (2021)
6. Bach, S., et al.: On pixel-wise explanations for non-linear classifier decisions by layer-wise relevance propagation. PLoS ONE **10**, e0130140 (2015)
7. Durán, J.: Dissecting scientific explanation in AI (sXAI): a case for medicine and healthcare. Artif. Intell. **297**(C), 103498 (2021)
8. Evans, R., Grefenstette, E.: Learning explanatory rules from noisy data. J. Artif. Intell. Res. **61**, 1–64 (2018)

9. Guidotti, R., et al.: A survey of methods for explaining black box models. ACM Comput. Surv. (CSUR) **51**, 1–42 (2018)
10. Gunning, D.: DARPA's explainable artificial intelligence (XAI) program (2019)
11. Habermas, J.: Communication and the Evolution of Society. Toronto: Beacon Press. The book contains translations of 5 essays by Habermas. The quotation is taken from the first essay "What Is Universal Pragmatics", p. 3. The original German version „Was heißt Universalpragmatik?" was written 1976 and published by Suhrkamp 1984 in: Vorstudien und Ergänzungen zur Theorie des kommunikativen Handelns, pp. 353–440 (1979)
12. Hempel, C.G., Oppenheim, P.: Studies in the Logic of Explanation, 1948. In: Readings in the Philosophy of Science, pp. 8–38. Prentice Hall, Englewood Cliffs (1970)
13. Holland, J., Holyoak, K., Nisbett, R., Thagart, P.: Induction: Processes of Inference, Learning, and Discovery. MIT Press, Cambridge (1986)
14. Holzinger, A., et al.: Causability and explainability of artificial intelligence in medicine. Wiley Interdiscip. Rev. Data Min. Knowl. Discov. **9**, e1312 (2019)
15. Holzinger, A.: The next frontier: AI we can really trust. In: Kamp, M., et al. (eds.) Machine Learning and Principles and Practice of Knowledge Discovery in Databases. ECML PKDD 2021. CCIS, vol. 1524, pp. 427–440. Springer, Cham (2021). https://doi.org/10.1007/978-3-030-93736-2_33
16. Hou, X., Papachristopoulou, K., Saint-Drenan, Y., Kazadzis, S.: Solar radiation nowcasting using a Markov chain multi-model approach. Energies **15**(9), 2996 (2022)
17. Kwiatkowska, M., Norman, G., Parker, D.: Stochastic model checking. In: Bernardo, M., Hillston, J. (eds.) SFM 2007. LNCS, vol. 4486, pp. 220–270. Springer, Heidelberg (2007). https://doi.org/10.1007/978-3-540-72522-0_6
18. Longo, L., et al.: Explainable artificial intelligence: concepts, applications, research challenges and visions. In: International Cross-Domain Conference for Machine Learning and Knowledge Extraction, pp. 1–16 (2020)
19. Makridakis, S.: The forthcoming artificial intelligence (AI) revolution: its impact on society and firms. Futures **90**, 46–60 (2017)
20. Miller, T., Howe, P., Sonenberg, L.: Explainable AI: Beware of inmates running the asylum or: How I learnt to stop worrying and love the social and behavioural sciences (2017). arXiv preprint arXiv:1712.00547
21. Muggleton, S.: Inductive logic programming. N. Gener. Comput. **8**, 295–318 (1991)
22. Poole, D., Goebel, R., Aleliunas, R.: A Logical Reasoning System for Defaults and Diagnosis, University of Waterloo, Dep. of Computer Science. Research Rep. CS-86-06 (1986)
23. Pople, H.E.: On the mechanization of abductive logic. In: IJCAI'73: Proceedings of the 3rd International Joint Conference on Artificial Intelligence, pp. 147–152 (1973)
24. Preece, A.: Asking 'Why' in AI: explainability of intelligent systems–perspectives and challenges. Intell. Syst. Account. Financ. Manag. **25**, 63–72 (2018)
25. Rudin, C.: Stop explaining black box machine learning models for high stakes decisions and use interpretable models instead. Nat. Mach. Intell. **1**, 206–215 (2019)
26. Singla, P., Duhan, M., Saroha, S.: A comprehensive review and analysis of solar forecasting techniques. Front. Energy **16**, 187–223 (2022)
27. van Fraassen, B.C.: The Scientific Image. Clarendon Press, Oxford (1980)
28. Voyant, C., et al.: Machine learning methods for solar radiation forecasting: a review. Renew. Energy **105**, 569–582 (2017)
29. Zhang, Q.-S., Zhu, S.-C.: Visual interpretability for deep learning: a survey. Front. Inf. Technol. Electron. Eng. **19**(1), 27–39 (2018). https://doi.org/10.1631/FITEE.1700808
30. https://en.allmetsat.com. Accessed 01 Mar 2023
31. earthobservatory.nasa.gov/features/ColorImage?msclkid=21fe225da5ff11ec94190320202 8b5d1. Accessed 01 Mar 2023

Sustainability Effects of Robust and Resilient Artificial Intelligence

Torsten Priebe(✉), Peter Kieseberg, Alexander Adrowitzer, Oliver Eigner, and Fabian Kovac

Institute of IT Security Research, St. Pölten University of Applied Sciences, St. Pölten, Austria
{torsten.priebe,peter.kieseberg,alexander.adrowitzer, oliver.eigner,fabian.kovac}@fhstp.ac.at

Abstract. It is commonly understood that the resilience of critical information technology (IT) systems based on artificial intelligence (AI) must be ensured. In this regard, we consider resilience both in terms of IT security threats, such as cyberattacks, as well as the ability to robustly persist under uncertain and changing environmental conditions, such as climate change or economic crises. This paper explores the relationship between resilience and sustainability with regard to AI systems, develops fields of action for resilient AI, and elaborates direct and indirect influences on the achievement of the United Nations Sustainable Development Goals. Indirect in this case means that a sustainability effect is reached by taking resilience measures when applying AI in a sustainability-relevant application area, for example precision agriculture or smart health.

Keywords: artificial intelligence · machine learning · resilience · security · sustainability

1 Introduction

Artificial intelligence (AI) can be usefully applied to build resilience in many areas. However, this can simultaneously open up new threats in the area of IT security, as the use of technology creates the risk of failure, vulnerability, or misuse. This paper discusses, how these threats can be countered and initially demonstrates how the creation of robust and resilient AI can also have sustainability effects in line with the United Nations (UN) Sustainable Development Goals [35].

To this end, we must first define what resilient AI means. We base this on multiple definitions by the National Institute of Standards and Technology (NIST), which we consolidate as follows: Resilience is the ability of an information system

This research was funded in part by the German Federal Ministry for the Environment, Nature Conservation, Nuclear Safety and Consumer Protection (BMUV) by resolution of the German Bundestag through the cooperative project CO:DINA as part of the AI Lighthouse initiative.

A. Holzinger et al. (Eds.): CD-MAKE 2023, LNCS 14065, pp. 188–199, 2023.
https://doi.org/10.1007/978-3-031-40837-3_12

to mitigate the impact of known or unknown changes in the operating environment (including intentional attacks, accidents, and naturally occurring threats) by (a) anticipating and preparing for such events (e.g., through risk management, contingency, and business continuity planning), (b) the ability to withstand and adapt to attacks, adverse conditions, or other stresses and to continue operations (or rapidly regain the ability to do so), while maintaining essential and required operational capabilities, and (c) restoring full operational capability after such disruption in a time-frame consistent with mission requirements. Thus, it is also about "robustness" in the face of a changing environment. Großklaus [16] refers to this "ability to successfully drive the sustainable development of society under uncertain and changing conditions" as "transformative resilience".

We would like to point out that AI systems can never be considered as isolated, abstracted entities, but must be seen in their social context as sociotechnical systems; it must be understood that algorithms and especially AI are not just technical artifacts – in the sense of "physical, human-designed objects that have both a function and a plan of use" as defined by Vermaas et al. [36] – but complex systems characterized by collective and distributive action [30]. In our case, this means that the concept of resilient AI has links to issues of acceptance and trust. No AI system can be called resilient if its use is fraught with fundamental mistrust, accountability gaps, accusations of unfairness, or criticism of its black-box nature. Likewise, the EU's Joint Research Centre [19] distinguishes four dimensions of resilience: *societal*, *economic*, *organizational*, and *technological*. The focus of this work is on the *technological* side, touching *organizational* aspects where appropriate. The *societal* sense of resilience is in fact what we refer to as *sustainability*. This work is based on the assumption that the guiding principles of resilience (in a technical/organizational sense) and sustainability complement each other: In a crisis-ridden world, resilience becomes a basic requirement for the success of sustainability goals. To validate this assumption, this paper develops fields of action for resilient AI and examines their sustainability impacts – in general and in selected sustainability-relevant application areas. The underlying study is not yet complete; therefore, this paper represents initial considerations and results of an ongoing work-in-progress effort.

The rest of this paper is organized as follows: Sect. 2 outlines our research method and in particular the increasing focus in our stepwise approach. Section 3 introduces robust and resilient AI and develops a roadmap for fields of action. Section 4 discusses direct sustainability effects of resilient AI as well as indirect effects, if according measures are taken in sustainability-relevant application areas. Finally, Sect. 5 concludes the paper, given that we are presenting work-in-progress, with a focus on current and future work.

2 Research Method

As shown in Fig. 1, we proceed in three steps. In Eigner et al. [11], a survey of the scientific state-of-the-art as well as the identification of possible fields of action of robust and resilient AI was carried out; a summary can be found in

Sect. 3. This paper presents the continuation of this work. We have analyzed potential sustainability impacts of the identified action areas based on academic literature, project results and brainstorming with experts. In this second step we focused on sustainability-relevant application areas. The resulting overview of sustainability effects is provided in Sect. 4. Last not least, we are currently identifying recommended actions for public actors with a further increased focus on smart cities and regions, critical infrastructures and ecological sustainability goals using an exploratory scenario analysis [24].

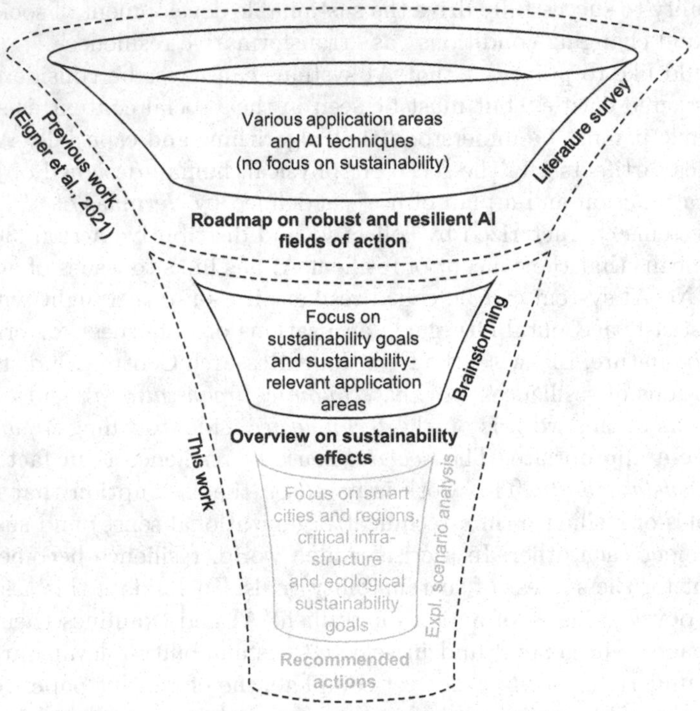

Fig. 1. Increasing focus and methods applied throughout our research

As shown in Fig. 2, different forms of dependencies between robust and resilient AI and sustainability emerge:

a. Direct sustainability effects of a field of action, e.g., the reduction of bias in AI algorithms has a direct positive impact on gender equality.
b. Indirect sustainability effects of robust and resilient AI in a specific application area, e.g., increasing the robustness of an AI in precision agriculture contributes to reducing hunger.
c. We are developing concrete recommendations for resilient AI in selected areas, which may bring up new impacts. E.g., a recommendation to address drift by frequently retraining machine learning (ML) models may increase energy consumption and therefore have a negative sustainability effect.

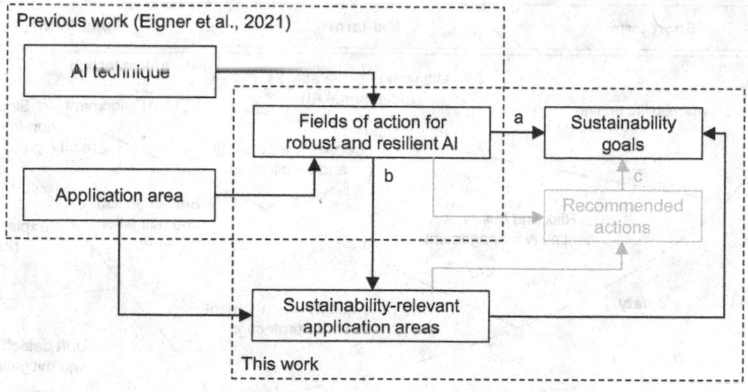

Fig. 2. Dependencies of the various concepts in our research

3 Robust and Resilient Artificial Intelligence

AI systems are becoming essential to our daily lives. Organizations should ensure their resilience as with any other critical asset. However, the black-box approach typically found in AI may make assessing and ensuring resilience different compared to traditional IT systems. In Eigner et al. [11], we provide an overview of the emerging field of resilient AI, both from the perspective of selected application areas and specific AI techniques. From this, we derive fields of action for robust and resilient AI. In Fig. 3 we structure these in the form of a roadmap, as some targets have already been or are being extensively addressed, while others will only reach practical applicability in the medium or long term. Following the European Union (EU) High-Level Expert Group on Artificial Intelligence [17], we divide the fields of action into *security*, *safety*, *accuracy* and *reliability*.

Security. Security incidents in AI can be distinguished by (a) the AI technology used, (b) the type of incident, or (c) the stage of the AI/ML pipeline in which the incident occurs. The type of disruption can be broadly divided into intentional disruptions, which include all types of hostile attacks on AI systems and unintentional disruptions, which can range from careless human interaction to rare special cases that systems may never have encountered before. With this in mind we see *plausibility tests*, i.e. rules that identify at least outliers or unexpected results of an AI algorithm, as a basic level of protection. More sophisticated methods of *mitigating hostile attacks* are knows as *adversarial AI* [23].

Penetration testing ("pentesting" for short) is a key technique for assessing the security and resilience of IT services and products. Since there are myriads of possible threats in the field of AI that can affect the proper operation of AI

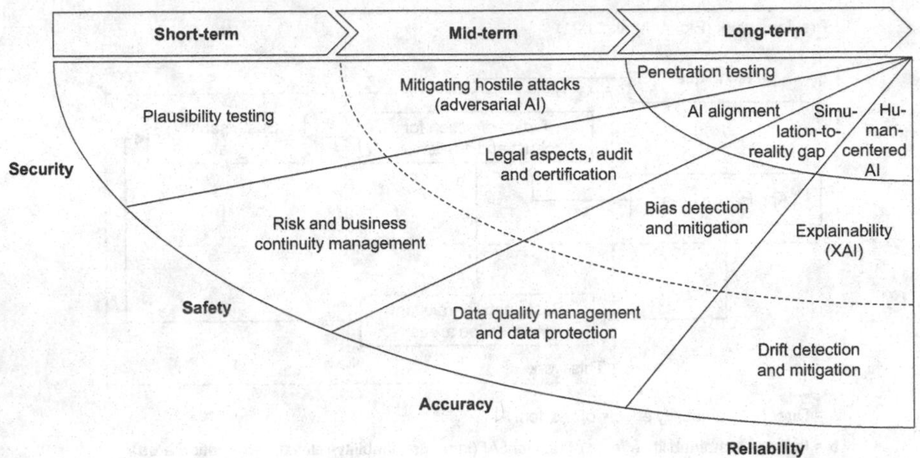

Fig. 3. Roadmap for robust and resilient AI fields of action

applications, pentesting of AI will certainly gain importance in the near future. Initial approaches have been proposed, e.g., by Das et al. [8] or Tjoa et al. [33], or are under development, still, currently no best practices methodologies exist.

Safety. Regarding the field of action *risk and continuity management*, KPMG highlights in its AI Risk and Controls Matrix various risks associated with AI [25]. Risks to be highlighted in this area include inadequate fallback solutions related to infrastructure, the AI solution itself, and business operations. In addition, the inability to restore service after an incident is highlighted as a specific AI risk, as the last good AI state may not be easily restored due to its complex and often black-box nature.

In terms of *legal aspects*, the United Kingdom Information Commissioner's Office [21] identifies three key areas: the legal status of algorithms, sector-specific standards, and the interdependencies of privacy and AI. In addition, monitoring robustness to (even non-adverse) changes in the environment opens up another area of research requiring a holistic view of *audit and certification* of AI systems [37]. In related terms, the upcoming EU AI-Act [12] differentiates AI systems into four risk categories, ranging from "unacceptable" risks to "high", "limited" and "minimal" risk levels. The categorization is not only depending on the AI technology in use, but also on the sensibility of data and application area.

AI alignment aims align AI systems with human preferences and ethical principles. AI systems, especially when using reinforcement learning, are based on specified objectives. As it can be difficult do define all desired and, in particular, undesired behaviors, AI systems may find loopholes to accomplish their objectives efficiently but in unintended, sometimes harmful ways [38].

Accuracy. Accuracy is one of the most important areas in the field of AI and has received a lot of attention in the last decades, both to enable new applications and to improve AI applications and make them market ready [1]. Data is the very source of good AI in this regard; if the *data quality* is poor or if there is distortion due to data pre-processing, many of these problems will be transferred to the result or even amplified [39]. This includes, in particular, *detection and avoidance of bias* [27]. Discrimination and bias often arise in the data collection and modeling process, for example, when target variables are incorrectly defined, questions are unclear, or historical data is used that was collected in a time when moral concepts are no longer in line with current ones [4].

A major challenge, especially in robotics, is also to bridge the *gap between simulation and reality* and to make "digital twins" robust to changing parameters of the environment. Here, domain randomization approaches [34] are promising. Furthermore, automation is especially problematic in so-called mixed environments, where robotic actors directly interact with human actors, which is a major problem in the area of autonomous driving [9].

Reliability. We consider the continuous reliability of AI, as well as aspects of trustworthiness and explainability. Machine learning (ML) models degrade over time. A major reason for this is that the world, and thus the data, are not static; therefore, the data to which the models are applied also change over time. This effect is referred to as "drift". Methods for *drift detection and mitigation* have been discussed for some time [22], but adequate monitoring of AI systems has only recently been established by trends such as MLOps.

The *explainability* of AI often missing in many advanced methods refers to non-transparent (black-box-like) decision-making processes [15] for which, for example, testing for backdoors is practically impossible. *Human-centered AI* means to involve humans for labeling, improvement or correction. Especially the use of AI together with human expertise is promising here, e.g. by formalization with semantic technologies, resulting in the research area of semantic AI [5].

4 Sustainability Effects

In this section, we explore interdependencies between robust and resilient AI and the UN Sustainable Development Goals [35]. Are there synergies or do conflicting goals arise and how can these be negotiated? As outlined in Sect. 2, direct and indirect sustainability impacts can be identified. Here, indirect means an effect through a resilience field of action, provided the AI is used in a sustainability-relevant application area. In this paper, we consider *precision agriculture and forestry*, *smart health and precision medicine*, *smart cities and regions* (including energy and mobility transition), *industry and critical infrastructures*, and *police, justice, and military* as such sustainability-relevant application areas, informed by the project experience of the authors and the experts interviewed. Figure 4 illustrates the various dependencies broken down by resilience field of action and Sustainable Development Goal (SDG).

4.1 Direct Sustainability Effects

The AI Act of the European Union [12] includes, among others, a prohibition of discrimination against groups of persons on the basis of their sex or other characteristics. Dealing with *legal aspects and a corresponding certification* therefore leads directly to an improvement in relation to the sustainable development goals (SDGs) 5 (gender equality) and 10 (reduced inequalities). The same applies to the technical measures derived from this to *detect and mitigate bias*. *Explainability* of AI systems also leads to an improvement here, as inadequacies of the models used are at least made visible.

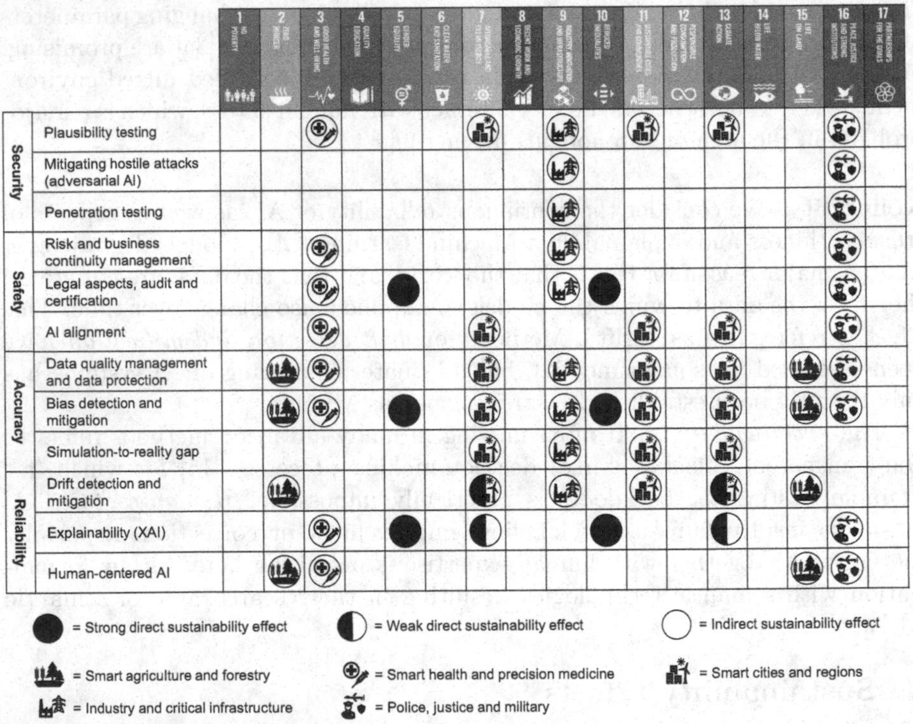

1 2 3 4 5 6 7 8 9 10 11 12 13 14 15 16 17

Security
Plausibility testing
Mitigating hostile attacks (adversarial AI)
Penetration testing

Safety
Risk and business continuity management
Legal aspects, audit and certification
AI alignment

Accuracy
Data quality management and data protection
Bias detection and mitigation
Simulation-to-reality gap

Reliability
Drift detection and mitigation
Explainability (XAI)
Human-centered AI

= Strong direct sustainability effect
= Weak direct sustainability effect
= Indirect sustainability effect
= Smart agriculture and forestry
= Smart health and precision medicine
= Smart cities and regions
= Industry and critical infrastructure
= Police, justice and military

Fig. 4. Overview of sustainability impacts of robust and resilient AI fields of action

Explainable AI models also tend to use less computationally intensive algorithms, resulting in a sustainability impact with respect to SDGs 7 (affordable and clean energy) and 13 (climate action). On the contrary, *drift detection and mitigation* usually leads to regular (potentially frequent) retraining of AI models, causing an increase in energy consumption and therefore a negative effect on SDGs 7 and 13.

4.2 Sustainability Effects in Selected Application Areas

Precision Agriculture and Forestry. An important application area of AI in precision agriculture and forestry is plant pest and disease detection and prediction [20], which positively impacts SDG 2 (no hunger). Other goals include optimizing the use of scarce resources such as water, as well as fertilizers and pesticides, which also has a positive effect on SDG 15 (life on land). Relevant fields of action for resilience are *data quality* as well as *bias and drift detection and mitigation.* The latter is particularly important in a changing environment, e.g., due to climate change. Since labeled data is usually rare in such use cases [29], use of *human-centered AI* techniques (such as interactive learning) is also relevant.

Smart Health and Precision Medicine. Important application areas of AI in healthcare, and thus with an impact on SDG 3 (health and well-being) are, for example, individualized medications and semi-automated diagnosis. This is also referred to as P4 medicine (predictive, preventive, personalised and participatory) [13]. Here, AI always serves only as support; the final decision must rest with the physician. This is why *human-centered AI* approaches and *explainability* are so important [18].

Due to the (also legal) classification of medical products as "high-risk AI" [12], *risk and business continuity management* are also particularly important. Medical applications may not be common targets of cyberattakcs (at least not like, e.g., critical infrastructure targets), however basic security measures such as *plausibility checks* should of course also be applied.

Medical AI systems have higher accuracy requirements than other applications. Therefore, consideration of *data quality* is particularly relevant, especially in the context of rare phenomena. This also applies to the avoidance of *bias*, since "biased" algorithms are more difficult to generalize. Given the nature of medicine directly affecting humans, *AI alignment* is of particular importance as well. Studying the effects of treatments, for example, on AI systems is not sufficient, basically like trying to study them only on lab mice.

Smart Cities and Regions. Sustainability in smart cities and regions is directly represented by SDG 11 (sustainable cities and communities). Application fields of AI here include the optimization of the use of renewable energy. Here, too, adaptability to a changing environment (e.g., due to climate change), i.e. *drift detection and mitigation*, is of central importance. An important aspect in many smart cites is the concept of the sharing economy, i.e. citizens make (private) resources available for use by others when they do not need them [14]. Studies have already been conducted on the impact of this sharing economy on the sustainability of such smart cities [3]. The use of AI in this area results in various trustworthiness and resilience requirements related to *bias and explainability.* Also, the fact that personal data is involved may require *data protection* measures such as anonymization, which is in fact a form of distortion in data preprocessing. The emerging use of digital twins in smart cities also requires consideration of the *simulation-to-reality gap.* Sharing itself might also increase the difficulty of attack attribution, which in itself is already a huge problem in IT Security [7].

Industry and Critical Infrastructure. Defending against *hostile attacks* and therefore *penetration testing* are particularly important in the area of industry and critical infrastructure (e.g., power plants and power grids), represented by SDG 9 (industry, innovation and infrastructure), as these are obvious targets for cyber warfare. Based on the attacks on the Ukrainian power grid in 2015, there was a strong increase in attention in this area [6] and the creation of corresponding technologies and organizational units like specialized CERTs for the energy sector. Applications of AI such as predictive maintenance or quality control require *data quality and distortion* handling and means of *drift detection and mitigation*. AI systems in critical infrastructure are legally classified as "high-risk AI" [12], hence requiring proper *risk and business continuity management*. *Standards and certifications* are also particularly important in this area [32].

Police, Justice and Military. Robust and resilient AI in the police, justice, and military sectors (e.g., through demonstrable avoidance of bias and discrimination) inherently impacts SDG 16 (peace, justice, and strong institutions) [2]. A prominent example here is the AI-based COMPAS database in the US, which was developed to predict the likelihood of recidivism among offenders [26]. However, predictive policing is rather controversial, both for ethical reasons (requiring resilience measures such as *bias mitigation* and *explainability*) and with regard to the actual verifiable benefit [28].

In the military field, the situation is still much more opaque, as many details are subject to secrecy. Nevertheless, some trends and frameworks can be identified, such as the ban on so-called Lethal Autonomous Weapon Systems (LAWS) [31] in the European Union [10] and specifically through the AI Act [12]. There are currently some well-known public programs, such as from the US, the focus in these publications is very much in the area of predictive maintenance, unmanned aerial vehicles (UAVs), training of personnel and augmentation.

5 Conclusion

In this paper, we have provided an initial overview of fields of action for robust and resilient AI and examined them for their sustainability effects. Direct effects were identified, e.g. through the reduction of bias. Indirect effects arise from the use of AI in sustainability-relevant application areas. We have selected precision agriculture and forestry, smart health and precision medicine, smart cities and regions, industry and critical infrastructures, as well as police, justice and military, based on the project experience of the authors and interview partners.

On the one hand, a further, broader survey would be useful to expand the analysis, which is certainly not complete to date. On the other hand, we also intend to go deeper and define concrete recommended actions to achieve resilience with a more narrow focus (public actors in critical infrastructures and smart cities and regions), which again need to be analyzed for their (in particular ecological) sustainability impacts. For example, a recommendation may be to regulate and therefore limit the use of large pre-trained AI models (due to their intransparency and potential bias). However, given that these pre-trained models save training effort (and therefore resources) such a recommendation may have a negative sustainability effect. Both aspects are being adressed in the exploratory scenario analysis with our interviewees, which currently ongoing.

References

1. Achour, Y., Pourghasemi, H.R.: How do machine learning techniques help in increasing accuracy of landslide susceptibility maps? Geosci. Front. **11**(3), 871–883 (2020). https://doi.org/10.1016/j.gsf.2019.10.001
2. Adensamer, A., Klausner, L.D.: Part man, part machine, all cop: automation in policing. Front. Artif. Intell. Forthcom. (2021)
3. Akande, A., Cabral, P., Casteleyn, S.: Understanding the sharing economy and its implication on sustainability in smart cities. J. Clean. Prod. **277**, 124077 (2020)
4. Barocas, S., Selbst, A.D.: Big data's disparate impact. 104 California Law Review, p. 671 (2016). https://doi.org/10.2139/ssrn.2477899, https://www.ssrn.com/abstract=2477899
5. Breit, A., et al.: Combining machine learning and semantic web: a systematic mapping study. ACM Comput. Surv. (2023). https://doi.org/10.1145/3586163
6. Case, D.U.: Analysis of the cyber attack on the Ukrainian power grid. Electr. Inf. Shar. Anal. Center (E-ISAC) **388**, 1–29 (2016)
7. Clark, D.D., Landau, S.: Untangling attribution. Harv. Nat'l Sec. J. **2**, 323 (2011)
8. Das, N., et al.: MLsploit: a framework for interactive experimentation with adversarial machine learning research. In: Proceedings of the 25th ACM SIGKDD International Conference on Knowledge Discovery and Data Mining. KDD '19, ACM, New York, NY (2019). https://www.kdd.org/kdd2019/docs/KDD2019_Showcase_2062.pdf
9. Di, X., Shi, R.: A survey on autonomous vehicle control in the era of mixed-autonomy: from physics-based to AI-guided driving policy learning. Transp. Res. Part C Emerg. Technol. **125**, 103008 (2021)
10. Doll, T.: Künstliche Intelligenz in den Landstreitkräften. Amt für Heeresentwicklung (2019). https://www.bundeswehr.de/resource/blob/156024/d6ac452e72f77f3cc071184ae34dbf0e/download-positionspapier-deutsche-version-data.pdf
11. Eigner, O., et al.: Towards resilient artificial intelligence: survey and research issues. In: 2021 IEEE International Conference on Cyber Security and Resilience, pp. 536–542. CSR 2021, IEEE, Washington, DC (2021). https://doi.org/10.1109/CSR51186.2021.9527986
12. European Commission: Laying Down Harmonised Rules on Artificial Intelligence (Artificial Intelligence Act) and Amending Certain Union Legislative Acts (2021). https://eur-lex.europa.eu/legal-content/EN/ALL/?uri=celex:52021PC0206 , proposal for a Regulation of the European Parliament and of the Council, No. COM/2021/206 final
13. Flores, M., Glusman, G., Brogaard, K., Price, N.D., Hood, L.: P4 medicine: how systems medicine will transform the healthcare sector and society. Pers. Med. **10**(6), 565–576 (2013)
14. Frenken, K., Schor, J.: Putting the sharing economy into perspective. In: A Research Agenda for Sustainable Consumption Governance, pp. 121–135. Edward Elgar Publishing (2019)

15. Goebel, R., et al.: Explainable AI: the new 42? In: Holzinger, A., Kieseberg, P., Tjoa, A.M., Weippl, E. (eds.) CD-MAKE 2018. LNCS, vol. 11015, pp. 295–303. Springer, Cham (2018). https://doi.org/10.1007/978-3-319-99740-7_21
16. Großklaus, M.: Vom Modewort zum transformativen Hebel: Wie die Konjunktur des Resilienzbegriffs für die digital-ökologische Transformation genutzt werden kann. IZT (2022). https://codina-transformation.de/transformative-resilienz/, CO:DINA position paper no. 11
17. High-Level Expert Group on Artificial Intelligence: Ethics Guidelines for Trustworthy AI. Publications Office of the European Union, Luxembourg (2019). https://doi.org/10.2759/346720
18. Holzinger, A., Weippl, E., Tjoa, A.M., Kieseberg, P.: Digital transformation for sustainable development goals (SDGs) - a security, safety and privacy perspective on AI. In: Holzinger, A., Kieseberg, P., Tjoa, A.M., Weippl, E. (eds.) CD-MAKE 2021. LNCS, vol. 12844, pp. 1–20. Springer, Cham (2021). https://doi.org/10.1007/978-3-030-84060-0_1
19. Institute for the Protection and Security of the Citizen (Joint Research Centre): Towards Testing Critical Infrastructure Resilience. Publications Office of the European Union, Luxembourg (2014). https://doi.org/10.2788/41633
20. Java, O., Asprion, B., Priebe, T., Sarkozi, E., Neves Madeira, R.: Application of digital technology in agriculture: potential support for winegrowers. In: Proceeding of the 8th International Conference on Trends in Agricultural Engineering 2022. Prague (2022). https://2022.tae-conference.cz/proceeding/TAE2022-32-Oskars-JAVA.pdf
21. Kazim, E., Denny, D.M.T., Koshiyama, A.: AI auditing and impact assessment: according to the UK information commissioner's office. AI Ethics **1**, 301–310 (2021). https://doi.org/10.1007/s43681-021-00039-2
22. Klinkenberg, R.: Learning drifting concepts: example selection vs. example weighting. Intell. Data Anal. **8**(3), 281–300 (2004). https://dl.acm.org/doi/10.5555/1293831.1293836
23. Kong, Z., Xue, J., Wang, Y., Huang, L., Niu, Z., Li, F.: A survey on adversarial attack in the age of artificial intelligence. Wirel. Commun. Mob. Comput. **2021**, TBD (2021). https://doi.org/10.1155/2021/4907754
24. Kosow, H., Gaßner, R.: Methods of future and scenario analysis: overview, assessment, and selection criteria, vol. 39. DEU (2008)
25. KPMG: AI Risk and Controls Matrix (2018). https://assets.kpmg/content/dam/kpmg/uk/pdf/2018/09/ai-risk-and-controls-matrix.pdf
26. Larson, J., Mattu, S., Kirchner, L., Angwin, J.: How We Analyzed the COMPAS Recidivism Algorithm. ProPublica (2016). https://www.propublica.org/article/how-we-analyzed-the-compas-recidivism-algorithm
27. Mehrabi, N., Morstatter, F., Saxena, N., Lerman, K., Galstyan, A.: A survey on bias and fairness in machine learning. ACM Comput. Surv. **55**(6) (2022). https://doi.org/10.1145/3457607
28. Meijer, A., Wessels, M.: Predictive policing: review of benefits and drawbacks. Int. J. Public Adm. **42**(12), 1031–1039 (2019)
29. Neves Madeira, R., et al.: Towards digital twins for multi-sensor land and plant monitoring. Procedia Comput. Sci. **210**, 45–52 (2022). https://doi.org/10.1016/j.procs.2022.10.118
30. Rammert, W.: Where the action is: distributed agency between humans, machines, and programs. In: Seifert, U., Kim, J.H., Moore, A. (eds.) Paradoxes of Interactivity: Perspectives for Media Theory, Human-Computer Interaction, and Artistic

Investigations, pp. 62–91. transcript, Bielefeld (2008). https://doi.org/10.14361/9783839408421-004
31. Righetti, L., Pham, Q.C., Madhavan, R., Chatila, R.: Lethal autonomous weapon systems [ethical, legal, and societal issues]. IEEE Robot. Autom. Mag. **25**(1), 123–126 (2018)
32. Tao, F., Qi, Q., Wang, L., Nee, A.Y.C.: Digital twins and cyber-physical systems toward smart manufacturing and industry 4.0: correlation and comparison. Engineering (Beijing) **5**(4), 653–661 (2019). https://doi.org/10.1016/j.eng.2019.01.014
33. Tjoa, S., Buttinger, C., Holzinger, K., Kieseberg, P.: Penetration testing artificial intelligence. ERCIM News **123**, 36–37 (2020). https://ercim-news.ercim.eu/en123/r-i/penetration-testing-artificial-intelligence
34. Tobin, J., et al.: Domain randomization for transferring deep neural networks from simulation to the real world. In: 2017 IEEE/RSJ Int. Conf. on Intelligent Robots and Systems, pp. 23–30. IROS 2017, IEEE, Washington, DC (2017). https://doi.org/10.1109/IROS.2017.8202133
35. United Nations: Transforming our World: The 2030 Agenda for Sustainable Development (2015). https://sdgs.un.org/2030agenda, resolution No. A/RES/70/1
36. Vermaas, P., Kroes, P., van de Poel, I., Franssen, M., Houkes, W.: A Philosophy of Technology: From Technical Artefacts to Sociotechnical Systems, Synthesis Lectures on Engineers, Technology, and Society, vol. 17. Morgan & Claypool, San Rafael, CA (2011). https://doi.org/10.2200/S00321ED1V01Y201012ETS014
37. Winter, P.M., et al.: Trusted Artificial Intelligence: Towards Certification of Machine Learning Applications. TÜV Austria, Brunn am Gebirge (2021). https://www.tuv.at/loesungen/digital-services/trusted-ai
38. Yudkowsky, E.: The ai alignment problem: why it is hard, and where to start. Symbolic Systems Distinguished Speaker (2016). https://intelligence.org/stanford-talk/
39. Zelaya, C.V.G.: Towards explaining the effects of data preprocessing on machine learning. In: Proceedings of the 35th International Conference on Data Engineering. pp. 2086–2090. ICDE '19, IEEE Computer Society, Washington, DC (2019). https://doi.org/10.1109/ICDE.2019.00245

The Split Matters: Flat Minima Methods for Improving the Performance of GNNs

Nicolas Lell(✉) and Ansgar Scherp(✉)

Universität Ulm, Ulm, Germany
{nicolas.lell,ansgar.scherp}@uni-ulm.de

Abstract. When training a Neural Network, it is optimized using the available training data with the hope that it generalizes well to new or unseen testing data. At the same absolute value, a flat minimum in the loss landscape is presumed to generalize better than a sharp minimum. Methods for determining flat minima have been mostly researched for independent and identically distributed (i.i.d.) data such as images. Graphs are inherently non-i.i.d. since the vertices are edge-connected. We investigate flat minima methods and combinations of those methods for training graph neural networks (GNNs). We use GCN and GAT as well as extend Graph-MLP to work with more layers and larger graphs. We conduct experiments on small and large citation, co-purchase, and protein datasets with different train-test splits in both the transductive and inductive training procedure. Results show that flat minima methods can improve the performance of GNN models by over 2 points, if the train-test split is randomized. Following Shchur et al., randomized splits are essential for a fair evaluation of GNNs, as other (fixed) splits like "Planetoid" are biased. Overall, we provide important insights for improving and fairly evaluating flat minima methods on GNNs. We recommend practitioners to always use weight averaging techniques, in particular EWA when using early stopping. While weight averaging techniques are only sometimes the best performing method, they are less sensitive to hyperparameters, need no additional training, and keep the original model unchanged. All source code is available under https://github.com/Foisunt/FMMs-in-GNNs.

1 Introduction

Flat minima are regions in the weight space of a neural model where the error function remains largely stable. It is argued that such larger regions of the error function with a constant, low score correspond to less chance of overfitting of the model and thus show higher generalization performance [14,15]. We demonstrate this in Fig. 1a, where we plot the training and testing loss of the same model when changing its weights following a random direction. In that example, the loss landscapes are shifted between train and test data. Therefore, finding a flat minimum or choosing a central point in a flat region can lead to better generalization compared to a model with the lowest possible loss in a sharper

Published by Springer Nature Switzerland AG 2023
A. Holzinger et al. (Eds.): CD-MAKE 2023, LNCS 14065, pp. 200–226, 2023.
https://doi.org/10.1007/978-3-031-40837-3_13

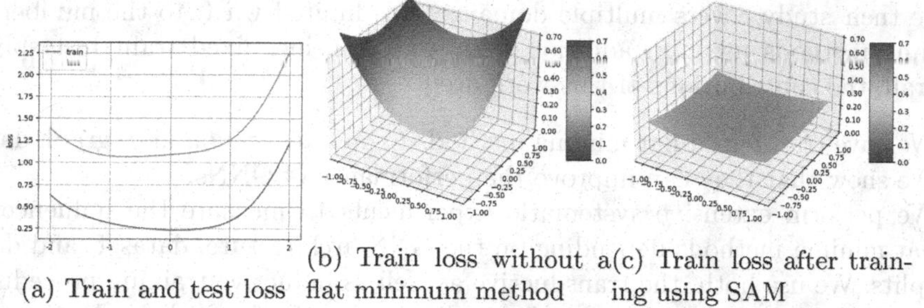

(a) Train and test loss (b) Train loss without a flat minimum method (c) Train loss after training using SAM

Fig. 1. Loss of GCN on CiteSeer with the Planetoid split. Plots following [25].

minimum. Methods for determining flat minima have been researched in the past largely on toy examples and for data that are independent and identically distributed (i.i.d.) such as images, e.g., [8,9,19,30,40,44].

Graph neural networks (GNNs) deal with non-i.i.d. graph data, since vertices are connected via edges. GNNs are powerful models but are likewise also known to be difficult to train and susceptible to the training procedure [32]. Even small changes in the hyperparameters, data split, etc. can lead to unstable training and lack of generalization performance.

We tackle these challenges of GNNs by transferring flat minima methods to graphs. We consider a wide selection of weight-averaging and sharpness-aware flat minima methods, including the well known methods SWA [19] and SAM [9], and lesser known or new ones like Anticorrelated Perturbed Gradient Descent [30], Penalizing Gradient Norm [40], and Sharpness Aware Training for Free [8]. We also apply existing and new combinations such as Penalizing Gradient Norm [30] plus ASAM [24]. We evaluate the performance of flat minima methods on different GNN architectures using small and large benchmark datasets. As GNNs, we use the well known Graph Convolutional Network (GCN) [23] and Graph Attention Network (GAT) [33] as well as the novel Graph-MLP [17], which operates without the classical message passing. Regarding the evaluation procedure, we follow Shchur et al. [32] who warned that on the common benchmark datasets Cora, CiteSeer, and PubMed the train and test splits heavily impact the models' performance and can lead to an arbitrary reranking of similarly good GNNs. Thus, in addition to the commonly employed (fixed) "Planetoid" split [38], we apply two randomized splits on those datasets.

Our results show that in most cases flat minima methods improve the performance. But the improvement heavily depends on the used model, dataset, dataset split, and flat minima method. An illustration of the effect of flat minima methods for GNNs can be seen in Fig. 1b showing the training loss surface of a GNN trained *without* any flat minima method versus Fig. 1c showing the loss surface of the same model but trained *with* SAM. To the best of our knowledge, we are the first to systematically transfer and analyze the impact of many flat minima methods to non-i.i.d. graph data. Only Kaddour et al. [32] applied two flat minima methods SAM and SWA to study images, text, and graphs. Thus, while their study covers multiple domains, it is limited w.r.t. to the number of minima methods used. In addition, they only consider fixed train/test splits. Overall, the contributions of this work are:

- We have transferred flat minima methods to operate on non-i.i.d. graph data. We show that they can improve the performance of GNNs.
- We perform extensive systematic experiments to measure the influence of flat minima methods depending on the GNN architecture, dataset, and data splits. We use both, the transductive as well as inductive training procedure.
- We demonstrate that using random splits is essential for fairly evaluating not only GNN models but also the flat minima methods.
- We combine flat minima methods and show that this improves the performance even further.

Below, we review flat minima methods and introduce graph neural networks. In Sect. 3, we describe in more detail the flat minima methods used in our experiments. Section 4 introduces the experimental apparatus. The results are reported in Sect. 5 and discussed in Sect. 6.

2 Related Work

First, we discuss works in the search of finding flat minima. Second, we introduce graph neural networks and describe representative models, which we use in our experiments.

2.1 Searching for Flat Minima

Hochreiter and Schmidhuber [14,15] were among the first who searched for flat minima in neural networks. They suggest that finding flatter minima leads to simpler neural networks with better generalization performance.

SAM-Based Approaches. Foret at al. [9] introduced a now popular method that improves generalization through the promotion of flatter minima which they call Sharpness Aware Minimization (SAM). Their idea is to minimize the loss at the approximated worst (adversarial) point in an explicit region around the model's current parameters and they show that SAM improves the performance and robustness to label noise of Convolutional Neural Networks (CNNs). SAM showed to also improve the performance of Vision Transformers [5] and Language Models [1]. Some follow up works used SAM to improve performance on other tasks like model compression [26,28]. Other follow up work focused on improving the efficiency of SAM [27]. For example, Brock et al. [2] sampled only a subset of each batch to accelerate the adversarial point calculation.

A different line of follow up work focused on improving the performance of SAM. Kwon et al. [24] introduced Adaptive SAM (ASAM), which compensates the influence of parameter scaling on the adversarial step. Kim et al. [22] proposed Fisher SAM which replaces the fixed euclidean balls used by SAM with ellipsoids induced by Fisher information. Zhao et al. [40] proposed a method which penalizes the gradient norm to find flatter optima and show that SAM is a special case of this method. Zhuang et al. [44] proposed Surrogate Gap Guided SAM (GSAM), which in addition to the usual SAM objective, also explicitly minimizes the sharpness.

Averaging Approaches. One averaging approach is ensembling [43], which combines multiple models' outputs to a single, usually more accurate prediction. For example, Devlin et al. [7] showed that an ensemble of BERT large models gain roughly 1% F1 score on SQuAD 1.1 compared to a single model. [18] proposed a method called Snapshot Ensemble, which averages a single model's predictions at different points during training.

There are also averaging approaches other than ensembling. Izmailov et al. [19] proposed the now well known method Stochastic Weight Averaging (SWA) which averages a single model's weights at different points during training. There are some follow up works on SWA, for example using SWA in low precision training can close the performance difference, even when using only 8 bits for each parameter and gradient[37]. Recently, Wortsman et al. [34] showed that most of the good models obtained during hyperparameter tuning lie in the same flat region and that averaging those models' weights leads to better performance compared to simply using the best found model. Further extension and uses of SWA are described by [10,11].

Other Approaches. Perturbed Gradient Descent (PGD) is a version of gradient descent where noise is injected in every epoch. This helped to escape from local minima [42] and saddle points [20]. Orvieto et al. [30] proposed a modification of PGD, which they call Anticorrelated PGD (Anti-PGD). The idea of Anti-PGD is to inject noise in the current epoch, depending on the noise injected in the previous epoch. They prove for some special problems that this leads the optimizer to the widest optimum and show that it increases performance on benchmark datasets. Damian et al. [6] showed that adding noise to the labels when using Stochastic Gradient Descent (SGD) leads to flatter optima and better generalization.

Du et al. [8] proposed a method which they coin Sharpness-Aware Training for Free (SAF). They consider SAM's adversarial point approximation as too costly, and instead rely on a trajectory loss to reduce sharpness.

2.2 Graph Neural Networks

Graph Neural Networks (GNNs) are neural networks that are designed to work with graph data. That means in addition to the vertex features, a GNN also uses the adjacency information which connects different data points. In the following, the adjacency matrix is denoted with $\boldsymbol{A}$, the normalized adjacency matrix with $\hat{\boldsymbol{A}} = \boldsymbol{D}^{-1/2}\boldsymbol{A}\boldsymbol{D}^{-1/2}$ with $\boldsymbol{D}_{ii} = \sum_j \boldsymbol{A}_{ij}$ and layer l's output with $\boldsymbol{H}^{(l)}$. Many GNNs follow the message passing architecture [12,41], where the current vertex aggregates the features of neighboring vertices to update its own feature vector. Different aggregation and update methods then lead to different GNNs. A well known example is the Graph Convolution Network (GCN) [23], where the implementations are inspired by CNNs. In each layer, every vertex combines its neighbors' features to calculate its output as follows: $\boldsymbol{H}^{(l+1)} = \sigma(\hat{\boldsymbol{A}}\boldsymbol{H}^{(l)}\boldsymbol{W}^{(l)})$.

Another well known GNN is the Graph Attention Network (GAT) [33], which uses attention to weigh each neighboring vertex's importance for the current vertex. The attention weights are calculated by $a_{ij} = \text{softmax}(\text{FF}(\boldsymbol{W}\boldsymbol{h}_i, \boldsymbol{W}\boldsymbol{h}_j))$, where FF is a one layer feed forward network. Those are then used to calculate vertex i's output with:

$$\boldsymbol{h}_i^{(l+1)} = \Big\Vert_{k=1}^{K} \sigma\left(\sum_{j \in N_i} a_{ij}^k \boldsymbol{W}^k \boldsymbol{h}_j\right) \tag{1}$$

where $\Vert$ is the concatenation, K the number of attention heads, and N_i the 1-hop neighborhood of vertex i including itself.

These GNNs usually use the whole graph in a single batch, i.e., require to load the full graph at once. This makes it difficult to apply GCN and GAT to very large graphs. There are different methods to scale GNNs that sample subgraphs and train on those instead of the full graph [3,13,39]. Wu et al. [35] propose Simplified GCN, which uses only a single message passing layer with the adjacency matrix to some power instead of multiple iterations with the normal adjacency matrix.

A common issue with the GNNs mentioned so far is over-smoothing [12], which means that after multiple message passing steps, all vertex representations tend to be very similar. This can either be avoided by restricting the GNNs to usually only one to three layers or adding residual or skip connections to the model. The Jumping Knowledge model [36] uses skip connections from every layer to the last layer. Chen et al. [4] introduced GCNII which utilizes skip connections from the input layer to every hidden layer. Both ideas make it possible to gain performance by increasing the model depth up to 64 layers.

Graph-MLP [17] is a GNN approach that is not based on the message-passing architecture. Rather, Graph-MLP employs a standard Multi Layer Perceptron (MLP) on the vertex features and uses a contrastive loss function on the r-th power normalized adjacency matrix $\hat{\boldsymbol{A}}^r$. The neighbor contrastive (NC) loss for vertex i is calculated as

$$l_i = -\log \frac{\sum_{j \neq i} \hat{\boldsymbol{A}}_{ij}^r \exp(\cos(\boldsymbol{z}_i, \boldsymbol{z}_j)\tau)}{\sum_{k \neq i} \exp(\cos(\boldsymbol{z}_i, \boldsymbol{z}_k)\tau)} \tag{2}$$

where $\boldsymbol{z}_i$ is the embedding/intermediate layer output of vertex i and τ is a temperature parameter. Other than GCN and GAT that are full batch by default, Graph-MLP randomly samples a batch from the input graph each epoch.

3 Flat Minima Methods

Here we give a brief introduction for a high level understanding as well as the modified parameter update rules for the flat minima methods used in our work.

For details, we refer to the primary literature. In the following, we denote the learning rate with η and a model's weights as $\mathbf{W}$, e.g., for a l-layer GAT $\mathbf{W} = [\boldsymbol{W}^{(1)}, \mathrm{FF}^{(1)}, ..., \boldsymbol{W}^{(l)}, \mathrm{FF}^{(l)}]$. We begin with SAM and works extending SAM, followed by weight-averaging methods. Finally, we discuss SAF and Anti-PGD.

SAM [9] searches for a model that has a region with low loss around it, instead of finding the model with lowest loss. SAM minimizes the loss of the approximately worst point $\mathbf{W}_{adv}$ in the region of size ρ around the model. The adversarial point is approximated via $\mathbf{W}_{adv} = \mathbf{W}_n + \rho(\nabla L(\mathbf{W}_n))/(||\nabla L(\mathbf{W}_n)||_2)$ and is used for the model's training by $\mathbf{W}_{n+1} = \mathbf{W}_n - \eta \nabla L(\mathbf{W}_{adv})$.

ASAM [24] considers that a model's parameters can be scaled without changing the loss. By incorporating the weights' norms into the parameter update, the performance of SAM can be increased. Formally, ASAM changes SAM's calculation to $\mathbf{W}_{adv} = \mathbf{W}_n + \rho(T^2_{\mathbf{W}} \nabla L(\mathbf{W}_n))/(||T_{\mathbf{W}} \nabla L(\mathbf{W}_n)||_2)$ with $T_{\mathbf{W}}$ being a normalization operator for the weights.

PGN: The gradient's norm directly corresponds to the sharpness of the model's current weights. By penalizing the gradient norm (PGN) during training, the models tend to reach flatter optima [40]. PGN generalizes SAM with the update rule $\mathbf{W}_{n+1} = \mathbf{W}_n - \eta((1-\alpha)\nabla L(\mathbf{W}) + \alpha \nabla L(\mathbf{W}_{adv}))$, where α is a new balancing parameter. We also experiment with a combination of PGN with ASAM, where we use ASAM to calculate $\mathbf{W}_{adv}$, which we call **PGNA**.

GSAM: SAM only optimizes the worst point in a region around it. But it might be better to explicitly minimize the sharpness of said region as well. GSAM [44] does this by adding a sharpness term to the loss while ensuring that the gradient of the sharpness term does not increase SAM's original loss via an orthogonal projection. This results in $\mathbf{W}_{n+1} = \mathbf{W}_n - \eta(\nabla L(\mathbf{W}_{adv}) - \alpha \nabla L(\mathbf{W})_{\perp})$, where α is a balancing parameter. We also use a variant called **GASAM**, which uses ASAM to calculate $\mathbf{W}_{adv}$.

SWA [19] is based on the observation that models trained using SGD with cyclic or high constant learning rates tend to traverse flat regions of the loss. As the loss landscapes are slightly shifted between training, validation, and test data, which can be seen in Fig. 1a, the center point of the training loss basin should generalize best. To exploit this assumption, SWA calculates an average model $\mathbf{W}_{\mathrm{swa}}$ by proportionally adding the current weights every k-th epoch by $\mathbf{W}_{\mathrm{swa}} = (\mathbf{W}_{\mathrm{swa}} \cdot n_{\mathrm{models}} + \mathbf{W}_{\mathrm{current}})/(n_{\mathrm{models}} + 1)$, with n_{models} being the number of models averaged. As we use early stopping following Shchur et al. [32], we do not know in advance for how many epochs each model trains. Thus, different to the original SWA which used predefined compute budgets, we start averaging at epoch *begin* and stop averaging *end* epochs after early stopping triggered.

EWA: Pre-experiments showed that the number of epochs a model trains heavily depend on the GNN architecture, dataset, split, and smoothing method used. In our case, it ranges from 5 epochs (GCN on CiteSeer with the 622 split) to about 2000 epochs (Graph-MLP on arXiv). This makes it hard to choose the *begin* parameter as one ideally only wants to average models that are already close to the optimum. Therefore, we also experiment with exponential weight averaging (EWA), i.e., $\mathbf{W}_{\mathrm{ewa}} = \alpha \mathbf{W}_{\mathrm{ewa}} + (1-\alpha) \cdot \mathbf{W}_{\mathrm{current}}$. With the introduction of the new hyperparameter α, we expect that EWA works well independent of the number of training epochs.

Anti-PGD: Noise can be injected into gradient descent to improve the training through faster escape from saddle points or local minima. When the loss is in a valley, anti-correlated noise additionally moves the model to a wider section of the valley [30]. The model's weights are updated by $\mathbf{W}_{n+1} = \mathbf{W}_n - \eta \nabla L(\mathbf{W}_{n+1}) + (\Xi_{n+1} - \Xi_n)$, where Ξ_i is a random tensor with variance σ^2. After training the model for some epochs with noise injection, the noise injection is stopped for the remaining training to improve convergence of the model.

SAF: Since SAM computes two gradients, it uses about twice the time per weight update compared to standard SGD. As mentioned in Sect. 2.1, there are some methods to reduce the impact of computing the additional gradient, but SAF [8] removes the second gradient calculation all together. Instead it approximates the sharpness by the change of the model output over the epochs. Specifically, a new trajectory loss $L^{tra} = \lambda/|B| \cdot \sum_{i \in B} \text{KL}(\boldsymbol{y}_i^{(e-E)}/\tau, \boldsymbol{y}_i^{(e)}/\tau)$ is added to the normal loss. In that case KL() is the Kullback-Leibler divergence, B a batch, $\boldsymbol{y}^{(e)}$ is the model's output at the current epoch e, $\boldsymbol{y}^{(e-E)}$ is the model's output E epochs ago, τ is a temperature, and λ the loss weight.

4 Experimental Apparatus

In this section, we present our experimental apparatus, i.e., the used datasets, models, procedure, and measures.

4.1 Datasets

We use different benchmark datasets to evaluate the flat minima methods. Table 1 reports statistics of the datasets. Cora [31], CiteSeer [31], PubMed [29], and OGB arXiv [16] are citation graphs. Amazon Computers and Amazon Photo [32] are co-purchase graphs. For these datasets the task is single label vertex classification. Protein Protein Interaction (PPI) [45] is a collection of 24 protein graphs with 20 of those used for training and 2 each for validation and testing. The task for PPI is multi label vertex classification. There are 121 different labels with each vertex having between 0 and 101 labels, an average of 36.9 ± 22.2 labels.

The "Planetoid" (in tables "plan") train-test split [38] is often used for Cora, CiteSeer, and PubMed. It is a fixed split with 20 vertices per class for training, 500 vertices for validation, and 1000 for testing. Shchur et al. [32] showed that changing the train-test split can arbitrarily rerank GNN methods of similar performance. Thus, we also use two other kinds of randomly generated splits that we also use for the Computers and Photo datasets. The random Planetoid "ra-pl" split follows [32] with 20 vertices per class for training, 30 per class for validation, and all other vertices for testing. The "622" split, for example used in [4], consists of 60% of the vertices for training, 20% for validation, and 20% for testing.

For OGB arXiv, we use the default training (paper before 2018), validation (paper from 2018), and test split (paper after 2018). We add reverse and self edges, making the graph essentially undirected, which is needed for good performance. This increases the number of edges from 1 166 243 to 2 484 941. Note that reverse edges are already included by default in the other benchmark datasets. Table 2 summarizes all used splits.

OGB arXiv is used in the transductive setting, i.e., all vertex features and edges are available during training. PPI is used in the inductive setting, i.e., no validation and test vertices and edges are used during training. For the other dataset we use both settings. However, we do not use the ra-pl split in the inductive setting. The reason is that the induced subgraph over the 20 vertices drawn per class in the ra-pl split typically results in no connected vertices. Thus, there are no edges in the subgraph for training, which renders this split ineffective.

Table 1. Datasets used. C is the number of classes and F is the feature size. As PPI contains multiple graphs, the sum of vertices and edges is shown here. †Number after adding self and reverse edges; number before is 1 166 243.

Dataset	C	F	$\|V\|$	$\|E\|$
Cora	7	1 433	2 708	10 556
CiteSeer	6	3 703	3 327	9 104
PubMed	3	500	19 717	88 648
Computers	10	767	13 752	491 722
Photo	8	745	7 650	238 162
arXiv	40	128	169 343	†2 484 941
PPI	121	50	56 944	1 587 264

4.2 Procedure

We precompute the random splits of our datasets (ra-pl, 622) such that they are consistent between models and methods. PPI is used inductively, i.e., only training vertices and edges connecting those are used for training. arXiv is used transductively, i.e., labeled training vertices are available together with the other vertices but without labels. We use both setups for the other datasets. The actual experimental procedure is then executed in two steps. First, we optimize the GNN models (GCN, GAT, Graph-MLP) in a traditional way without any flat minima methods as described in Sect. 4.3. Second, using the hyperparameters fixed in the first step, we add the flat minima methods and only optimize their respective hyperparameters(again described in Sect. 4.3). For both hyperparameter searches, only the training and validation sets are used. Subsequently, we evaluate the models. Additionally, we combine promising flat minima methods (without further hyperparameter tuning) on a subset of the datasets. We run multiple repeats with different seeds for each of our experiments. For the smaller

Table 2. Dataset splits. Besides the fixed "Planetoid" split, we use: "ra-pl" denotes random splits as used by [32] and "622" denotes random 60% train, 20% validation, and 20% test split.

	Split type	Train	Val	Test
Cora	plan	140	500	1 000
	ra-pl	140	210	2 358
	622	1 621	542	545
CiteSeer	plan	120	500	1 000
	ra-pl	120	180	3 027
	622	1 993	666	668
PubMed	plan	60	500	1 000
	ra-pl	60	90	19 567
	622	11 829	3 944	3 944
Computers	ra-pl	200	300	13 252
	622	8 246	2 750	2 756
Photo	ra-pl	160	240	7 250
	622	4 586	1 530	1 534
arXiv	default	90 941	29 799	48 603
PPI	default	44 906	6 514	5 524

datasets, we use 100 repeats, and 10 repeats for the arXiv and PPI datasets. We report the mean performance of the GNN models averaged over those repeats. For the flat minima methods, we report the difference to the respective GNN model. This allows for fast visual assessment of the results. In addition, we the report the standard deviation over the different runs for all models.

4.3 Hyperparameters

We tune the GNN hyperparameters, fix them, and then tune the flat minima methods' hyperparameters. We optimized all hyperparameters individually per setting (in-, transductive), dataset, and split. In summary, we use early stopping with a patience of 100 epochs, two to three layer models with a hidden size smaller or equal to 256 for the small datasets and slightly modify Graph-MLP. For PPI and arXiv we use deeper (up to ten layers) and wider models that also use residual connections. For the flat minima methods we tune each methods' hyperparameters while keeping the base ones from fixed. For PGN and GSAM we reuse ρ we found for SAM and ASAM. For details and all final values see Appendix A.

4.4 Metrics

For the multi-label PPI dataset, we report weighted Macro-F1 scores. The F1 score is calculated per class and averaged with weights based on the support of each class. For all other, single-label datasets we report accuracy.

5 Results

The results of the transductive experiments are shown in Table 3 with standard deviations shown in Table 13. Regarding the base models, we see that on Cora GAT performs best. On CiteSeer, PubMed, and Photo, Graph-MLP beats the message passing methods by 1 to 3 points. On Computers Graph-MLP is the best model when using the 622 split but the worst model when using the ra-pl split. On arXiv GCN performs best with a 0.3 point lead over GAT. Regarding the different splits, we can see that compared to the Planetoid split the performance is lower on the ra-pl and higher on the 622 split. Regarding the flat minima methods, we observe that there is no method that always works best. The largest improvement is over 2 points on the CiteSeer ra-pl split with GAT+EWA. On arXiv all non-weight averaging methods improve the performance of GCN, but only SAF for GAT. For Graph-MLP on arXiv, EWA improves the performance by 0.82 points. There are also some bad combinations. For example, ASAM reduces the performance of GAT in most cases.

Table 3. Transductive mean accuracy per split and dataset. Note: The minima method PGNA is a combination of PGN+ASAM. GASAM combines GSAM and ASAM. For the SD over the 100 and 10 runs, we refer to Table 13.

Dataset	Cora			CiteSeer			PubMed			Computer		Photo		arXiv
Split	plan	ra-pl	622	plan	ra-pl	622	plan	ra-pl	622	ra-pl	622	ra-pl	622	-
GCN	82.02	79.82	88.44	71.39	67.41	76.81	79.34	77.27	89.46	82.78	91.88	90.89	94.55	72.95
+SAM	+0.21	+0.53	+0.05	**+1.34**	+1.41	+0.02	−0.27	+0.24	−0.13	−0.07	+0.17	+0.40	−0.02	+0.10
+ASAM	+0.28	+0.53	+0.02	+0.93	+1.47	−0.18	+0.18	+0.41	−0.28	−0.20	+0.11	+0.30	−0.03	+0.01
+PGN	+0.04	+0.40	−0.01	+1.08	+1.91	−0.01	−0.06	−0.09	−0.07	+0.08	+0.09	+0.37	+0.00	+0.07
+PGNA	**+0.35**	+0.70	−0.01	+0.93	+1.78	−0.12	**+0.22**	+0.33	−0.02	+0.11	+0.06	+0.28	+0.02	+0.01
+GSAM	+0.15	+0.50	−0.01	+1.25	+1.55	−0.01	−0.19	+0.14	−0.09	−0.05	+0.16	+0.38	−0.00	+0.09
+GASAM	**+0.35**	**+0.84**	+0.02	+1.16	+1.58	−0.12	+0.13	**+0.43**	0.01	−0.35	+0.10	+0.38	−0.02	+0.05
+SWA	−0.11	+0.39	**+0.12**	+0.52	+1.59	+0.03	−0.60	−0.48	−0.09	−0.59	+0.23	+0.21	**+0.09**	−0.61
+EWA	−0.09	−0.01	+0.04	+0.25	**+2.09**	**+0.26**	+0.04	−0.09	**+0.21**	−0.36	**+0.33**	+0.02	+0.06	−0.21
+Anti-PGD	−0.02	+0.51	+0.05	−0.04	+1.11	−0.08	−0.01	−0.08	+0.19	+0.17	+0.09	+0.05	−0.00	**+0.13**
+SAF	+0.26	+0.57	−0.00	+0.47	+0.13	+0.03	+0.01	−0.00	+0.02	**+1.13**	+0.19	**+0.59**	−0.02	+0.11
GAT	82.94	80.73	88.42	71.39	69.96	76.55	79.09	77.22	88.59	83.02	92.17	90.56	94.72	72.65
+SAM	−0.28	−0.29	+0.19	+0.07	−0.06	+0.10	+0.30	+0.09	−0.11	−0.28	**+0.29**	−0.15	+0.01	−0.07
+ASAM	−0.74	−0.14	+0.06	−0.27	−0.61	−0.19	−0.24	−0.35	−0.16	−0.42	+0.22	−0.16	+0.08	−0.03
+PGN	**+0.32**	−0.00	**+0.24**	**+0.64**	+0.23	+0.12	+0.50	+0.38	**+0.09**	**+0.27**	+0.23	+0.08	**+0.17**	−0.03
+PGNA	+0.30	−0.01	+0.20	−0.22	+0.20	+0.03	+0.29	+0.30	−0.09	−0.21	+0.21	+0.06	+0.10	−0.07
+GSAM	+0.07	−0.29	+0.17	+0.33	−0.01	+0.14	**+0.88**	+0.09	+0.06	−0.21	+0.20	−0.09	+0.11	−0.03
+GASAM	−0.14	**+0.02**	−0.00	−0.33	−0.25	+0.05	+0.51	**+0.46**	**+0.09**	−0.40	+0.22	−0.14	+0.10	−0.04
+SWA	−0.87	−0.28	**+0.24**	−0.76	+0.55	+0.19	−0.87	−1.25	−0.09	−0.77	+0.14	**+0.14**	+0.07	−35.14
+EWA	−0.26	−0.06	−0.05	−0.31	+0.16	**+0.24**	−0.26	−0.28	+0.04	−0.41	+0.23	**+0.14**	+0.07	−41.32
+Anti-PGD	+0.01	−0.06	+0.08	+0.02	**+0.59**	−0.04	+0.15	+0.11	+0.06	−0.13	+0.03	+0.09	+0.02	−0.02
+SAF	−0.13	−0.08	−0.06	−0.05	−0.01	−0.09	+0.01	−0.11	−0.05	+0.00	+0.05	+0.10	−0.01	**+0.12**
Graph-MLP	80.58	78.76	88.08	74.53	71.36	77.69	82.16	78.19	90.31	81.59	92.25	91.30	95.94	67.79
+SAM	**+0.63**	+0.45	+0.22	−0.27	+0.26	+0.06	+0.35	+0.61	−0.01	−0.14	+0.04	+0.45	+0.01	+0.77
+ASAM	+0.19	−0.07	+0.16	−0.58	+0.14	+0.05	**+0.44**	+0.54	+0.04	−0.18	+0.02	+0.39	+0.07	+0.62
+PGN	+0.40	+0.45	+0.13	**+0.20**	+0.17	+0.08	+0.05	−0.13	+0.05	+0.23	+0.01	+0.49	+0.06	+0.75
+PGNA	+0.27	+0.13	+0.11	−0.15	+0.09	−0.03	+0.19	+0.15	+0.09	+0.10	+0.05	+0.47	+0.04	+0.79
+GSAM	+0.52	+0.49	**+0.27**	−0.11	+0.20	−0.06	+0.36	**+0.75**	−0.01	−0.25	+0.00	+0.32	+0.02	+0.77
+GASAM	+0.20	+0.07	+0.16	−0.39	+0.29	−0.11	+0.28	+0.38	+0.04	−0.19	+0.02	+0.25	**+0.08**	+0.68
+SWA	+0.34	−0.01	**+0.27**	+0.06	+0.51	**+0.30**	−0.45	+0.12	−0.33	**+0.32**	+0.02	**+0.67**	+0.06	−3.44
+EWA	−0.05	+0.02	+0.01	−0.01	+0.06	+0.03	−0.16	+0.07	+0.14	+0.07	+0.10	+0.02	−0.01	**+0.82**
+Anti-PGD	+0.17	−0.07	+0.10	+0.02	+0.03	+0.08	−0.20	−0.32	+0.14	+0.17	+0.07	+0.22	+0.02	+0.22
+SAF	−0.06	**+1.35**	+0.07	−0.02	**+1.01**	+0.08	+0.02	−0.11	**+0.33**	+0.23	**+0.13**	+0.08	+0.07	−0.06

Table 4. Inductive mean accuracy per split and dataset. For PPI we report weighted Macro-F1 scores. Note: The minima method PGNA is a combination of PGN+ASAM. GASAM combines GSAM and ASAM. For the SD over the 100 and 10 runs, we refer to Table 14.

Dataset	Cora		CiteSeer		PubMed		Comp..	Photo	PPI
Split	plan	622	plan	622	plan	622	622	622	-
GCN	81.08	88.02	71.56	76.80	78.63	89.81	91.63	94.49	99.34
+ SAM	−0.39	+0.36	+0.82	+0.33	+0.47	−0.22	+0.17	+0.03	−0.01
+ ASAM	−0.31	+0.37	+0.33	+0.13	+0.45	−0.44	+0.15	+0.02	+0.00
+ PGN	−0.04	+0.21	+0.77	+0.19	+0.47	−0.15	+0.02	+0.06	−0.01
+ PGNA	+0.02	+0.41	+0.37	+0.26	**+0.52**	−0.06	+0.02	+0.03	−0.01
+ GSAM	+0.24	+0.21	**+1.06**	+0.41	**+0.52**	−0.23	+0.11	+0.03	−0.00
+ GASAM	**+0.37**	+0.48	+0.50	+0.29	+0.47	−0.15	+0.11	+0.01	−0.00
+ SWA	−0.10	**+0.50**	−0.34	+0.34	−1.06	−0.47	+0.11	+0.06	−0.20
+ EWA	−0.05	+0.38	+0.55	+0.36	−0.48	**+0.17**	**+0.37**	+0.04	+0.01
+ Anti-PGD	−0.01	+0.37	+0.44	**+0.45**	−0.09	+0.10	+0.05	**+0.09**	−0.01
+ SAF	−0.01	+0.07	−0.13	−0.05	−0.07	−0.01	+0.07	−0.04	**+0.04**
GAT	82.20	87.92	72.10	76.69	78.30	88.55	91.66	94.45	99.29
+ SAM	−0.16	+0.21	−0.33	−0.18	+0.31	+0.06	+0.38	+0.00	+0.04
+ ASAM	−0.45	+0.14	−0.36	−0.06	−0.03	−0.16	+0.30	+0.16	+0.04
+ PGN	+0.23	+0.17	+0.03	+0.04	−0.10	+0.00	+0.31	+0.15	**+0.05**
+ PGNA	−0.24	+0.22	**+0.18**	+0.01	−0.26	−0.03	+0.30	**+0.19**	**+0.05**
+ GSAM	−0.15	+0.21	−0.33	−0.24	**+0.38**	+0.11	+0.37	+0.11	+0.04
+ GASAM	−0.30	+0.18	−0.36	+0.00	−0.02	−0.03	+0.33	+0.15	**+0.05**
+ SWA	−1.33	**+0.45**	+0.04	**+0.29**	−0.69	+0.00	+0.31	+0.13	−0.49
+ EWA	−0.04	+0.20	+0.03	**+0.29**	−0.11	**+0.19**	**+0.40**	+0.08	+0.03
+ Anti-PGD	**+0.32**	−0.01	+0.14	+0.00	−0.16	+0.04	+0.08	+0.04	+0.01
+ SAF	+0.16	+0.08	+0.00	+0.01	+0.03	+0.05	+0.18	+0.05	−0.15
Graph-MLP	68.72	77.47	69.37	74.13	81.20	89.51	87.39	92.87	54.63
+ SAM	+2.25	**+0.45**	−0.23	+0.18	−0.12	+0.13	**+0.58**	**+0.51**	−3.35
+ ASAM	+1.06	−0.27	−0.36	−0.13	−0.06	−0.12	−0.01	+0.15	−1.18
+ PGN	+2.83	+0.27	−0.07	+0.38	+0.05	**+0.35**	+0.42	+0.37	+1.81
+ PGNA	+1.83	−0.19	−0.07	−0.04	−0.01	+0.09	+0.09	+0.15	+1.26
+ GSAM	+2.00	+0.36	−0.12	+0.26	−0.08	+0.31	+0.55	+0.45	−3.20
+ GASAM	+2.49	+0.43	−0.23	+0.45	+0.06	−0.03	−0.01	+0.22	−0.58
+ SWA	−0.91	+0.32	−0.10	+0.18	−0.65	−0.29	+0.25	+0.40	−2.23
+ EWA	−0.94	+0.00	−0.01	+0.03	−0.06	+0.16	+0.03	+0.18	−0.94
+ Anti-PGD	**+2.86**	−0.01	−0.09	+0.02	+0.03	+0.31	+0.09	+0.02	+1.15
+ SAF	+1.75	+0.41	**+1.08**	**+0.51**	**+0.19**	+0.14	+0.06	+0.00	**+2.62**

Table 4 shows the results for the inductive experiments with standard deviations shown in Table 14. As explained in Sect. 4.1, the ra-pl split was not used here. The performance of most models lies within 1 point of the performance in the transductive setting. Graph-MLP drops over 10 points on Cora and around 4 points on CiteSeer, Computers, and Photo, but is still the best model for PubMed. On PPI, GCN is the best model while Graph-MLP has a 45 point drop in F1 score. Regarding the flat minima methods, the overall picture is the same. In many cases the best performing method from the transductive setting still is one of the best ones in the inductive setting. For example, for GCN on Cora, PubMed, and Computers, the same flat minima method works best in both settings. In other cases it changes, for example for Graph-MLP on CiteSeer with the Planetoid split now only SAF increases the performance, while in the transductive setting PGN workes best and SAF reduces the performance. On Computers and Photo nearly all flat minima methods improve the performance. On PPI SAF works for GCN and all SAM variants improve GAT. For Graph-MLP on PPI, the effects of the minima methods are mixed. For example, SAM reduces the performance by 3.35 points, PGN increases it by 1.81, and SAF brings a large improvement with 2.62 points.

6 Discussion

6.1 Key Insights

Regarding the flat minima methods, we can see that there is no method that always works best. However, for each combination of a base GNN model and dataset, there is at least one flat minima method that improves the performance. But in many cases some flat minima methods also reduce the performance. For GCN and Graph-MLP, most methods improve the performance on the citation graphs, while for GAT the results are mixed. For the co-purchase dataset Computers and 622 split, all flat minima methods improve the results, while on the ra-pl split most of the methods decrease the performance. We make a similar observation for the Photo dataset, but this time the ra-pl split is improved by the flat minima methods, except for four methods in combination GAT. On PPI, the flat minima methods improve the results for GAT while the improvement is mixed for the other GNNs.

When comparing the flat minima methods overall, we notice that the methods extending SAM, i.e., ASAM, PGN, and GSAM, do not consistently improve the results more than SAM. In most cases, one of the extensions works better than the original SAM, but this depends on the GNN and datasets. For example, for GAT on the small datasets, SAM does on average not change the performance compared to the base model while ASAM reduces it by 0.22 and PGN increases it by 0.25 points. Using ASAM instead of SAM for the adversarial calculation for PGN and GSAM works sometimes better, even though by itself SAM worked better than ASAM. EWA works better than SWA with the highest improvement in the transductive setting of 2.09 points when using EWA with a GCN on the CiteSeer and ra-pl split (see Table 3). This is likely because early stopping negatively impacts SWA. SWA's begin and end epoch heavily depend on the model and dataset. EWA nearly always begins in epoch 3, ends one epoch after early stopping triggered, and in most cases uses a low α of 0.5 to 0.9 that favors more recent weights.

Anti-PGD works surprisingly well for a method that just adds noise to the model. It is usually not the best method but, e.g., on the small datasets with GAT it outperforms SAM, ASAM, SWA, EWA, and SAF. It also reaches the overall highest accuracy on arXiv, which is achieved when applied to GCN. While SAF is motivated by SAM, it impacts the models' performance differently. For the small datasets with GCN it is worse than all the SAM-based methods, while with Graph-MLP it is better than all the SAM-based methods. On arXiv, SAF is the only method that improves GAT's performance. On PPI, SAF decrease the performance, while the SAM-based methods always improve it. The training of GAT with SAF on PPI was unstable and occasionally needed restarts. SAF's recommended $\lambda = 0.3$ works for the citation datasets. However, on PPI with GCN and GAT using $\lambda = 0.3$, early stopping is triggered at epoch 4 ± 0, i.e., once SAF starts the performance only decreases. Training is feasible with lower values, with $\lambda = 0.03$ GCN+SAF is the best model on PPI.

To the best of our knowledge, the only work who also applied flat minima methods to GNNs is the study by Kaddour et al. [32]. As said in the introduction, their work is limited to two flat minimal methods SAM and SWA and they also consider only one fixed train/test split. We argue that it is crucial to consider randomized splits for a fair evaluation of flat minima methods on GNNs. Considering the findings of [21], they found that SAM works better than SWA on GNNs and that the results are influenced by the dataset and GNN architecture. In contrast, we found that SAM works better than SWA. This may be due to the additional randomized splits or the use of early stopping which affects SWA more than other methods. Another reason may be that [21] used the original hyperparameter search space of SAM [9], while we additionally consider lower and higher values of ρ.

For GCN one should use GASAM, for GAT one should use PGN, and for Graph-MLP one should use SAF. In any case, one should always run one of the weight averaging methods as they do not need additional gradient computations. In addition, one always obtains the original model without the modifications from the flat minima methods as well. Finally, in our experiments we use early stopping. Thus, it is important to decide on the hyperparameter values when to begin and stop averaging in SWA. This choice of the hyperparameter is easier for EWA, since we start at a fixed epoch to average the weights. Our hyperparameter search showed that EWA works well when one begins to average soon after the training starts and ending it when early stopping triggers, while using a strong decay value of 0.5. For Graph-MLP, SWA is the preferred flat minima method. The reasons is that Graph-MLP trains for more epochs and SWA can better adapt the parameter weights.

6.2 Combining up to Three Flat Minima Methods

Above we mostly study existing methods, consider with EWA a variant of SWA, and with GASAM and PGNA combinations of two flat minima methods. As proof of concept, we also further combine different flat minima methods without additional hyperparameter tuning. As basis we use GCN in the transductive setting. Table 5 shows the results for GCN with combinations of up to three flat minima methods with standard deviations in Table 12. This shows that combining methods can increase the performance even further. For example, on the CiteSeer ra-pl split, combining EWA and GASAM, which is a combination of EWA+GSAM+ASAM, increases the performance by 2.89 points.

Table 5. Transductive GCN with combination of up to three flat minima methods. SDs in Tables 13 and 12.

Dataset	Cora			CiteSeer			PubMed			Computer		Photo	
Split	plan	ra-pl	622	plan	ra-pl	622	plan	ra-pl	622	ra-pl	622	ra-pl	622
GCN	82.02	79.82	88.44	71.39	67.41	76.81	79.34	77.27	89.46	82.78	91.88	90.89	94.55
+ PGNA	+0.35	+0.70	−0.01	+0.93	+1.78	−0.12	**+0.22**	+0.33	−0.02	+0.11	+0.06	+0.28	**+0.02**
+ GASAM	+0.35	+0.84	+0.02	+1.16	+1.58	−0.12	+0.13	**+0.43**	−0.01	−0.35	+0.10	+0.38	−0.02
+ Anti-PGD+SAM	+0.16	+0.78	+0.02	**+1.40**	+1.60	−0.11	−0.15	−0.06	+0.14	+0.13	+0.24	+0.43	+0.00
+ Anti-PGD+GASAM	**+0.41**	+0.72	+0.02	+1.03	+1.89	−0.12	+0.12	+0.32	+0.24	−0.54	+0.25	+0.33	+0.01
+ Anti-PGD+SAF	+0.02	**+0.96**	+0.03	+0.26	+1.13	−0.08	+0.08	+0.03	+0.21	**+1.04**	+0.14	**+0.62**	−0.02
+ EWA+Anti-PGD	−0.09	+0.56	+0.03	+0.24	+1.14	**+0.26**	+0.04	−0.27	**+0.32**	−0.33	+0.26	+0.10	**+0.02**
+ EWA+SAM	−0.32	+0.54	**+0.06**	+0.68	+1.88	+0.08	−0.34	+0.12	+0.00	−0.19	+0.34	+0.35	+0.00
+ EWA+GASAM	−0.50	+0.64	+0.00	+0.41	**+2.89**	+0.08	+0.01	+0.35	+0.21	−0.56	+0.32	+0.37	−0.01
+ EWA+SAF	+0.12	+0.76	+0.03	+0.29	+2.06	**+0.26**	+0.01	−0.14	+0.24	+0.72	**+0.38**	+0.56	−0.02

6.3 Influence of Dataset Splits

Regarding the dataset splits, we can see that the random split ra-pl is more difficult than the often used Planetoid split. The Planetoid split uses a fixed set of 20 vertices per class as training data. This makes models more susceptible to overfit the hyperparameters to that specific split than for a randomized ra-pl split. For the 622 splits, the much higher amount of training data explain the overall better result. Especially for the PubMed dataset, the amount of training data is larger by a factor of 200 in the 622 split compared to the Planetoid split. In both the transductive and inductive settings, the dataset split impacts the performance one can gain from the flat minima methods. The increase is on average higher and more consistent on the hardest ra-pl split compared to the other splits.

Shchur et al. [32] show that randomized splits need to be used for a fair evaluation of GNNs. We follow their suggestion by using two variants of randomized data splits. We confirm their observations that the commonly used "Planetoid" split is biased and should not be used on its own. We extend on this observation and conclude from our experiments that randomized splits are also important for a fair evaluation of the flat minima methods applied to GNNs.

6.4 Transductive vs. Inductive Training

The hyperparameters and basic model performance are similar between the transductive and inductive setting. In most cases the basic performance is within

1 point. The ranking of the flat minima methods is similar as well, i.e., in many cases the best transductive method also works well in the inductive setting. The major exception to this is Graph-MLP, discussed below.

6.5 Detailed Discussion of Graph-MLP

Compared to the original Graph-MLP [17] our modifications improved it by over 1 point on Cora and CiteSeer and over 2 points on PubMed. The reason is that the original hyperparameter optimization was suboptimal. For example, we found a larger batch/sample size to be beneficial. On the arXiv dataset, we found τ to be a critical hyperparameter and setting it outside the recommended $[.5, \ldots, 2]$ range to 15 increased the performance by roughly 5 points. On PPI, Graph-MLP completely fails. The main reason for this is PPI's inductive nature, which means that Graph-MLP never uses the validation and testing edges. This also explains the performance drop of Graph-MLP when we compare the same dataset between the transductive and inductive setting. The size of the performance drop probably corresponds to the importance of knowing the edges that include testing vertices. These edges seem to be very important for PPI, quite important for Cora where the performance dropped by over 10 points, and not very important for PubMed where the drop was smaller than 1 point, which is similar to the other GNNs. In the transductive case, information about the edges connected to test vertices is available in training through $\hat{A}^r$.

6.6 Assumptions and Limitations

We assume that our GNN models provide a fair foundation for evaluating the flat minima methods. We optimized the hyperparameters and checked the performance of the GNN models with the original works. GCN and Graph-MLP achieve a performance on all datasets better than the literature [17,23]. The performance of our GAT model is slightly lower on CiteSeer, within standard error on Cora and PubMed, and better on PPI, compared with [33]. Depending on the hyperparameters, the training of some GAT models on PPI was unstable and required multiple restarts.

Our study considers the task of vertex classification. We cover most nuances like small to large datasets with fixed and random splits, training in transductive and inductive settings, and single- and multi-label classification. There are larger datasets than arXiv, but it is computationally very expensive to tune hyperparameters on these dataset for the GNN models and many flat minima methods considered here. Beyond vertex classification, future extensions could consider also other tasks such as edge prediction and graph classification.

7 Conclusion

Overall our results show that the choice of the best flat minima method depends on the GNN model used and dataset split. For the realistic and challenging random split datasets (ra-pl, 622), the flat minima methods can improve the GNN model more than on a fixed dataset split. Shchur et al. [32] argue for the need of using such random splits to fairly evaluate GNN models. We extend on this and argue that a realistic assessment of flat minima methods on graph models requires such an evaluation procedure as well. We observe that combining up to three flat minima methods can even further improve the results. We recommend to always use weight averaging as SWA and EWA do not need any additional gradient calculations while also producing the original, unmodified models. When using early stopping, we especially recommend using EWA.

Appendix

A Hyperparameters

We present the searched and final hyperparameters in this section. For all experiments, Adam optimizer with PyTorch's default values $\beta_1 = 0.9$, $\beta_2 = 0.999$, and $eps = 1e-08$ is used. Early stopping with a patience of 100 epochs and 20,000 max epochs is applied. After pre-experiments, we fixed the learning rates to the respective values and adjacency (edge) dropout to 0 as well as all parameters not noted in the grid search ranges below. Unless otherwise noted (Graph-MLP with arXiv and PPI), we did a full grid search over all combinations of the listed hyperparameters. All final values are reported in Tables 6, 7 for GCN, in Tables 8, 9 for GAT, and in Tables 10, 11 for Graph-MLP.

Table 6. Optimal hyperparameter values for GCN on transductive tasks.

Dataset	Cora			CiteSeer			PubMed			Computer		Photo		arXiv
Split	pl	ra-pl	622	pl	ra-pl	622	pl	ra-pl	622	ra-pl	622	ra-pl	622	-
input dropout	0.15	0.2	0.0		0.05		0.2	0.2	0.0	0.15	0.15	0.1	0	0.2
model dropout	0.8	0.7	0.4	0.4	0.6	0.8	0.5	0.6	0.8	0.6	0.8	0.8	0.5	0.6
weight decay	0.1	0.1	0.001	0.316	0.316	0.01	0.1	0.01	0.1	0.01	0.00316	0.0316	0.001	0
norm		id			id		id	id	ln	ln	ln	ln	ln	ln
residual con		no			no			no		no	no	no	no	yes
num layers		2			2			2		2	2	2	2	6
hdim		128			256		128	256	128	128	256	128	256	768
lr		0.01			0.01			0.01		0.01	0.01	0.01	0.01	0.001
SAM ρ	1	1	0.05	5	1	5	0.1	0.5	0.2	2	0.2	5	0.0005	0.005
ASAM ρ	10	10	0.5	10	20	20	0.1	10	0.01	10	0.5	5	0.02	0.002
PGN α	0.3	0.1	0.3	0.4	0.7	0.2	0.1	0.3	0.1	0.4	0.9	0.3	0.6	0.2
PGNA α	0.3	0.9	0.3	0.5	0.8	0.6	0.7	0.5	0.9	0.1	0.3	0.6	0.2	0.9
GSAM α	0.5	0.5	5	0.5	0.5	0.01	0.002	0.05	0.002	0.5	1	0.005	0.5	0.01
GASAM α	0.5	1	0.01	0.5	1	0.5	2	0.005	2	0.2	0.5	2	5	0.01
SWA begin	75	3	3		3		75	3	77	3	75	3	25	75
SWA end	100	1	10	1	10	10	100	25	100	100	100	10	50	100
EWA begin		3			3			3		3	3	3	3	3
EWA end		1		1	1	50	1	1	100	1	100	1	1	100
EWA α	0.5	0.5	0.5	0.99	0.99	0.98	0.8	0.5	0.9	0.5	0.95	0.5	0.5	0.95
Anti-PGD σ	0.003	0.3	0.003	0.0003	0.3	0.001	0.03	0.03	0.1	0.03	0.1	0.01	0.001	0.001
Anti-PGD E		50		200	50	50	50	200	200	200	200	200	200	200
SAF λ	0.5	3	0.1	3	0.1	0.2		0.1		15	3	5	0.3	0.2
SAF τ	5	2	10		5		10	10	2	10	5	5	2	5

A.1 Base Models

Small Datasets. We use 20 repeats/seeds for the pre-experiments and parameter search on the small datasets. We ran pre-experiments to fix some parameters, and then ran a full grid search over the remaining parameter ranges. For GCN, the searched space is input dropout in $\{.0, .05, .1, .15, .2\}$, model dropout in $\{.4, .5, .6, .7, .8\}$, weight decay in $\{.001, .00316, .01, .0316, .1, .316\}$, normalization in $\{id, ln\}$ (id means no normalization, ln layer norm, and bn batch norm), and hidden dimension in $\{128, 256\}$. For Computer and Photo, model dropout was extended by $\{.2, .3\}$. For GAT, the number of attention heads is 8 in the first and 1 in the last layer. The other parameters are searched with input dropout in $\{.0, .05, .1, .15, .2\}$, model dropout in $\{.4, .5, .6, .7, .8\}$, attention dropout in $\{.1, .2, .3, .4, .5\}$, weight decay in $\{.001, .00316, .01, .0316, .1, .316\}$, norm in $\{id, ln\}$, and hidden dimension in $\{16, 32\}$ (times 8 attention heads). For Computer and Photo, weight decay was

Table 7. Optimal hyperparameter values for GCN on inductive tasks.

Dataset	Cora		CiteSeer		PubMed		Computers	Photo	PPI
Split	pl	622	pl	622	pl	622	622	622	-
input dropout	0.0	0.15	0.0	0.15	0.2	0.05	0.2	0.2	0.2
model dropout	0.8	0.5	0.8	0.7	0.5	0.8	0.7	0.7	0.4
weight decay	0.001	0.01	0.316	0.0316	0.1	0.01	0.01	0.001	0.0001
norm	id	id	id	id	id	ln	ln	ln	ln
residual con	no	no	no	no	no	no	no	no	yes
num layers	2	2	2	2	2	2	2	2	7
hdim	256	256	256	128	128	256	256	256	2048
lr	0.01	0.01	0.01	0.01	0.01	0.01	0.01	0.01	0.003
SAM ρ	2	1	5	1	0.0001	2	0.2	0.001	0.002
ASAM ρ	20	10	20	5	0.005	0.1	0.01	0.1	0.001
PGN α	0.9	0.3	0.2	0.4	0.8	0.1	0.1	0.9	0.4
PGNA α	0.9	0.4	0.5	0.5	0.5	0.1	0.6	0.2	0.6
GSAM α	2	0.05	0.005	0.1	0.1	0.0005	0.2	5	0.005
GASAM α	2	0.02	1	0.5	0.02	5	0.01	2	0.002
SWA begin	3	3	3	3	75	100	75	25	75
SWA end	1	50	1	25	75	100	100	100	100
EWA begin	3	3	3	3	3	3	3	3	3
EWA end	1	1	1	1	1	100	100	1	100
EWA α	0.5	0.8	0.9	0.9	0.99	0.9	0.9	0.5	0.8
Anti-PGD σ	0.03	0.3	0.1	1	0.001	0.1	0.01	0.3	0.03
Anti-PGD E	50	200	50	50	50	200	50	200	200
SAF λ	0.01	10	0.3	0.3	0.03	0.3	0.5	1.5	0.005
SAF τ	10	10	5	5	10	5	10	2	2

Table 8. Optimal hyperparameter values for GAT on transductive tasks.

Dataset	Cora			CiteSeer			PubMed			Computer		Photo		arXiv
Split	pl	rand pl	622	pl	rand pl	622	pl	rand pl	622	rp	622	rp	622	-
input dropout	0.2	0.2	0.05	0.05	0.15	0.15	0.1	0.15	0.0	0.2	0.2	0.15	0.15	0.2
model dropout	0.8	0.7	0.7	0.8	0.7	0.8	0.8	0.7	0.7	0.4	0.5	0.4	0.4	0.5
weight decay	0.0316	0.01	0.00316	0.0316	0.1	0.01	0.1	0.1	0.001	0.01	0.001	0.001	0.001	0.0001
norm	ln	id	id	ln	id	id		ln		ln	ln	id	ln	bn
residual con		no			no			no		no	no	no	no	yes
num layers		2			2			2		2	2	2	2	6
hdim	32	16	32	16	32	32	32	32	32	16	32	16	32	120
lr		0.01			0.01			0.01		0.01	0.01	0.01	0.01	0.001
attn dropout		0.5		0.5	0.3	0.5	0.4	0.3	0.4	0.3	0.4	0.4	0.5	0
num attn head		8			8			8		8	8	8	8	3
SAM ρ	1	0.001	2	2	5	2	0.5	0.5	0.2	0.5	0.5	0.001	1	0.05
ASAM ρ	5	20	20		10		2	5	0.001	2	2	0.1	2	0.1
PGN α	0.7	0.5	0.4	0.4	0.1	0.1	0.8	0.8	0.9	0.1	0.5	0.6	0.3	0.4
PGNA α	0.9	0.5	0.1	0.4	0.4	0.2	0.4	0.9	0.9	0.8	0.5	0.4	0.9	0.6
GSAM α	1	0.2	0.2	1	0.01	0.1	2	1	2	1	0.01	1	0.5	0.002
GASAM α	2	0.5	0.1	0.01	2	1	5	2	5	0.1	0.1	0.002	1	0.02
SWA begin	75	3	3	75	25	25	75	3	75	75	75	3	75	3
SWA end	100	50	10	100	50	25	100	1	100	100	100	50	100	1
EWA begin		3			3			3		3	3	3	3	3
EWA end		1			1		1	1	10	1	100	1	1	1
EWA α		0.5		0.5	0.5	0.9	0.5	0.5	0.9	0.5	0.8	0.5	0.8	0.99
Anti-PGD σ	0.01	0.003	0.01	0.0003	0.1	0.1	0.001	0.01	0.1	0.0003	0.003	0.003	0.001	0.01
Anti-PGD E	200	200	50	200	50	50	50	200	200	200	50	50	200	50
SAF λ	0.1	0.1	0.2	0.3	0.2	0.2		0.1		0.2	0.1	0.3	0.01	0.3
SAF τ		10			10		10	2	10	10	2	10	5	2

instead searched in $\{.0001, .000316, .001, .00316, .01, .0316\}$. Different to the original Graph-MLP we use dropout, layer norm, and activations between all layers. The batch size b to 100% of each dataset. The other parameters searched are model dropout in $\{.2, .3, .4, .5, .6, .7, .8\}$, weight decay in $\{.1, .01, .001, .0001, 1E-5\}$, and τ in $\{.5, 1, 2\}$. Loss weight and the layer after which the loss is calculated in $\{1@-2, 30@-3\}$, where layer -1 means after the last, -2 after the penultimate layer and so on. For Photo and Computer model dropout was extended by $\{0, .1\}$, τ by $\{3, 5\}$, and the loss was instead searched in $\{0.3@-2, 1@-2, 3@-2, 10@-2\}$.

OGB arXiv. We added reverse and self edges to the arXiv dataset which increased the accuracy by over 5 points for most configurations. We also used deeper models and added residual connections. We used 3 repeats for the arXiv and PPI pre-experiments and parameter selection experiments. For GCN, we searched input dropout in $\{.0, .1, .2, .3, .4\}$, model dropout in $\{.4, .5, .6, .7, .8\}$, and weight decay in $\{0, 1E-5, .0001\}$. For GAT, we searched attention dropout in $\{.0, .1, .2, .3\}$, input dropout in $\{.0, .1, .2\}$, model drop-out in $\{.4, .5, .6, .7\}$, and weight decay in $\{0, .0001\}$. For Graph-MLP, we did not perform a full grid search

Table 9. Optimal hyperparameter values for GAT on inductive tasks.

Dataset	Cora		CiteSeer		PubMed		Computers	Photo	PPI
Split	pl	622	pl	622	pl	622	622	622	-
input dropout	0.15	0.05	0.0	0.1	0.1	0.1	0.2	0.15	0
model dropout	0.8	0.4	0.6	0.8	0.8	0.8	0.5	0.4	0.1
weight decay	0.0316	0.01	0.1	0.01	0.316	0.001	0.001	0.001	$1E-6$
norm	ln	ln	id	ln	ln	ln	ln	ln	id
residual con	no	no	no	no	no	no	no	no	yes
num layers	2	2	2	2	2	2	2	2	7
hdim	32	32	32	32	32	32	32	32	256
lr	0.01	0.01	0.01	0.01	0.01	0.01	0.01	0.01	0.003
attn dropout	0.5	0.5	0.1	0.4	0.2	0.1	0.2	0.2	0.2
num attn head	8	8	8	8	8	8	8	8	8
SAM ρ	1	0.5	0.5	1	0.2	1	2	0.5	0.02
ASAM ρ	0.002	1	5	10	1	0.5	5	5	0.0005
PGN α	0.3	0.9	0.2	0.2	0.9	0.9	0.1	0.5	0.5
PGNA α	0.4	0.1	0.6	0.4	0.9	0.1	0.3	0.2	0.6
GSAM α	0.5	0.005	0.05	0.1	0.02	2	0.05	1	0.002
GASAM α	0.005	0.01	0.01	0.002	2	5	0.02	0.01	0.002
SWA begin	75	75	75	3	3	75	75	75	25
SWA end	100	100	100	50	1	100	100	100	25
EWA begin	3	3	75	3	3	3	75	3	3
EWA end	1	1	100	10	1	100	100	1	1
EWA α	0.5	0.5	0.98	0.95	0.5	0.95	0.95	0.5	0.9
Anti-PGD σ	0.03	0.0003	0.1	0.03	0.01	0.0003	0.1	0.001	0.0003
Anti-PGD E	50	50	50	50	200	200	50	200	50
SAF λ	0.01	0.07	0.07	0.2	0.01	0.02	1	0.1	0.03
SAF τ	10	2	10	2	5	10	5	10	5

over the hyperparameters due to their larger number and Graph-MLP's lower training speed. After fixing the other hyperparameters, we searched over many combinations of b in $\{0.02, 0.04, 0.06, 0.08, 0.1, 1.2\}$, $NC@$ in $\{-2, -4, -6\}$, loss weight in $\{10, 30, 100\}$, τ in $\{0.5, 1, 1.5, 2, 2.5, 3, 5, 10, 15, 20, 25, 50, 100\}$ input dropout in $\{.0, .05\}$, and model dropout in $\{0.1, 0.15, 0.2, 0.25\}$.

PPI. We used deeper models with residual connections for PPI as well. For PPI, the threshold to assign a label was chosen as 0.5. For GCN, we searched input dropout in $\{.0, .1, .2\}$, model dropout in $\{.2, .3, .4, .5, .6, \}$, and weight decay in $\{0, 1E-5, .0001\}$. For GAT, we searched attention dropout in $\{.0, .1, .2\}$, input

Table 10. Optimal hyperparameter values for Graph-MLP on transductive tasks.

Dataset	Cora			CiteSeer			PubMed			Computer		Photo		arXiv
Split	pl	rand pl	622	pl	rand pl	622	pl	rand pl	622	rp	622	rp	622	-
input dropout		0			0			0		0	0	0	0	0
model dropout	0.4	0.4	0.6	0.7	0.8	0.7	0.5	0.4	0.2	0.6	0.3	0.4	0.5	0.15
weight decay	0.0001	0.01	0.001	0.0001	0.01	0.0001	0.001	0.01	0.001	0.01	0.001	0.01	0.001	0.0
norm		ln			ln			ln		ln	ln	ln	ln	ln
residual con		no			no			no		no	no	no	no	yes
num layers		3			3			3		3	3	3	3	8
hdim		256			256			256		256	256	256	256	2048
lr		0.01			0.01			0.01		0.01	0.01	0.01	0.01	0.001
NC @	−3	−2	−2		−2			−2		−2	−2	−2	−2	−4
NC weight	30	1	1		1			1		10	1	3	1	30
tau	0.5	2	2	0.5	2	0.5	2	2	1	3	10	2	10	15
r		3			3			3		2	2	2	2	3
b (% of data)		100			100			100		100	100	100	100	4
SAM ρ	0.5	0.5	0.005	2	1	0.2	0.05	0.02	0.05	0.1	0.01	1	5	0.1
ASAM ρ	2	2	0.5	0.05	5	1	0.2	0.2	0.01	2	0.0005	5	10	0.05
PGN α	0.3	0.2	0.1	0.3	0.3	0.5	0.1	0.2	0.5	0.3	0.9	0.2	0.1	0.2
PGNA α	0.4	0.7	0.1		0.1		0.5	0.4	0.7	0.1	0.6	0.2	0.3	0.4
GSAM α	0.01	0.002	1	0.5	0.1	5.	0.005	0.01	1	2	2	0.002	0.002	0.01
GASAM α	0.1	0.005	2	2	0.01	5.	0.01	0.2	0.01	1	0.5	0.01	0.005	0.1
SWA begin	25	25	75	75	75	25	75	25	75	75	75	75	75	75
SWA end	100	50	100	100	100	50	100	25	100	100	100	100	100	100
EWA begin		3			3			3		3	3	3	3	3
EWA end		1			1			1		1	1	1	1	100
EWA α		0.5			0.5		0.5	0.5	0.8	0.5	0.8	0.5	0.5	0.99
Anti-PGD σ	0.01	0.0003	0.001	0.003	0.0003	0.003	0.01	0.03	0.1	0.01	0.03	0.01	0.1	0.001
Anti-PGD E		200		200	50	50	200	200	50	50	200	50	50	200
SAF λ	0.1	2	0.1	0.1	10	0.1	0.1	0.1	10	0.5	4	0.07	1	0.3
SAF τ	10	5	5	10	10	5	10	5	5	5	5	10	2	10

dropout in $\{.0, .1, .2\}$, model dropout in $\{.0, .1, .2, .3, .4\}$, and weight decay in $\{0, 1E{-}5, 1E{-}6\}$. For Graph-MLP, we searched with drop input in $\{0, .1\}$, model dropout in $\{0, .1, .2, .3\}$, weight decay in $\{3E-5, .0001, 0.0003\}$, loss weight in $\{10, 100\}$, and $NC@$ in $\{-4, -6, -8\}$, and τ in $\{3, 4, 5\}$. Afterwards we searched for b and found 0.8 (i.e., 80% of each graph) to be the best value.

Table 11. Optimal hyperparameter values for GMLP on inductive tasks.

Dataset	Cora		CiteSeer		PubMed		Computers	Photo	PPI
Split	pl	622	pl	622	pl	622	622	622	-
input dropout	0	0	0	0	0	0	0	0	0
model dropout	0.8	0.8	0.8	0.8	0.6	0.3	0.6	0.7	0.1
weight decay	0.1	0.0001	0.0001	$1E-5$	0.001	0.01	0.01	0.01	0.0001
norm	ln	ln	ln	ln	ln	ln	ln	ln	ln
residual con	no	no	no	no	no	no	no	no	yes
num layers	3	3	3	3	3	3	3	3	10
hdim	256	256	256	256	256	256	256	256	2048
lr	0.01	0.01	0.01	0..01	0.01	0.01	0.01	0.01	0.001
NC@	−3	−2	−2	−2	−2	−2	−2	−2	−4
NC weight	30	1	1	1	1	1	3	1	1
tau	0.5	1	0.5	0.5	2	1	10	5	4
r	3	3	3	3	3	3	2	2	3
b (% of train)	100	100	100	100	100	100	100	100	80
SAM ρ	1	5	1	5	0.1	0.5	2	1	0.1
ASAM ρ	10	0.002	0.02	0.1	0.5	0.001	10	5	0.5
PGN α	0.5	0.4	0.2	0.2	0.3	0.5	0.4	0.2	0.8
PGNA α	0.4	0.1	0.5	0.1	0.8	0.7	0.1	0.2	0.9
GSAM α	0.002	0.02	0.5	0.002	0.2	2	0.1	0.01	0.01
GASAM α	1	5	1	5	0.5	0.05	0.02	0.05	0.02
SWA begin	75	3	75	3	75	75	75	75	75
SWA end	100	25	100	25	100	100	100	100	100
EWA begin	3	3	3	3	3	3	3	3	3
EWA end	1	1	1	1	1	1	1	1	1
EWA α	0.8	0.5	0.5	0.5	0.5	0.8	0.5	0.8	0.5
Anti-PGD σ	0.003	0.01	0.001	0.03	0.01	0.3	0.03	0.1	0.001
Anti-PGD E	200	50	50	200	200	50	50	200	200
SAF λ	7	15	1.5	15	0.5	15	0.4	0.7	0.07
SAF τ	2	5	10	5	10	10	5	2	10

A.2 Flat Minima Methods

(A)SAM. Both SAM and ASAM have the parameter ρ which is usually set higher for ASAM than for SAM [24], so we search over ρ in $\{0.0001, 0.0002, 0.0005, 0.001, 0.002, 0.005, 0.01, 0.02, 0.05, 0.1, 0.2, 0.5, 1, 2\}$ for SAM and ρ in $\{0.0005, 0.001, 0.002, 0.005, 0.01, 0.02, 0.05, 0.1, 0.2, 0.5, 1, 2, 5, 10\}$ for ASAM. On the small data-sets, we saw potential for improvement with higher ρ and thus additionally searched over $\{5, 10\}$ for SAM and $\{20, 50\}$ for ASAM.

PGN. For PGN, we searched α in $\{0.1, 0.2, ..., 0.8, 0.9\}$ in all cases.

GSAM. For GSAM, we searched α in $\{0.002, 0.005, 0.01, 0.02, 0.05, 0.1, 0.2, 0.5, 1, 2, 5\}$ for all models.

SWA and EWA. For both SWA and EWA, we searched all combinations of *begin* in $\{3, 25, 75\}$ and *end* in $\{1, 10, 25, 50, 100\}$ where $end >= begin - 3$. This was done to prevent cases where no models are averaged at all as the lowest observed number of trained epochs from the first hyperparameter search is 5. For SWA, we averaged the model every epoch as we used fixed learning rates. For EWA, we additionally tried the combinations above with α in $\{0.5, 0.8, 0.9, 0.95, 0.98, 0.99\}$.

Anti-PGD. For Anti-PGD, we tried stopping the noise after $\{50, 200\}$ epochs and σ in $\{0.0003, 0.001, 0.003, 0.01, 0.03, 0.1, 0.3\}$. For the small datasets, we additionally used σ in $\{1, 3\}$.

SAF. We always started SAF at epoch 5 with a epoch difference E of 3. We tested all combinations of τ in $\{2, 5, 10\}$ and λ in $\{0.1, 0.2, 0.3, 0.4, 0.5, 0.7, 1, 2, 3\}$. We noticed that the optimal λ value often was on the border of that range so, we extended it by $\{0.01, 0.02, 0.03, 0.04, 0.05, 0.07\}$ on all datasets except arXiv, additionally by $\{1.5, 4, 5, 7, 10, 15\}$ on the small datasets, and additionally by $\{0.001, 0.002, 0.003, 0.005\}$ on PPI.

B Standard Deviations of Results

Here we present the standard deviations of the main result tables for completeness, as they did not fit into the main tables but we still want to present them for interested readers. Table 13 shows the standard deviations for the transductive results, Table 14 for the inductive results, and Table 12 for the combination of more flat minima methods.

Table 12. SDs of Table 5, combined methods on transductive GCN.

Dataset	Cora			CiteSeer			PubMed			Computer		Photo	
Split	plan	ra-pl	622	plan	ra-pl	622	plan	ra-pl	622	ra-pl	622	ra-pl	622
+ Anti-PGD+SAM	0.55	1.35	1.31	0.76	1.26	1.44	0.54	2.23	0.44	1.88	0.48	0.90	0.58
+ Anti-PGD+ASAM+GSAM	0.41	1.49	1.33	0.87	1.50	1.54	0.44	2.36	0.47	1.85	0.47	1.00	0.58
+ Anti-PGD+SAF	0.61	1.50	1.41	1.63	1.33	1.51	0.53	2.12	0.46	1.94	0.48	1.08	0.58
+ EWA+Anti-PGD	0.61	1.41	1.37	0.64	1.39	1.50	0.33	2.24	0.48	1.82	0.47	1.08	0.56
+ EWA+SAM	0.62	1.38	1.33	1.38	1.42	1.46	0.36	2.16	0.46	2.06	0.45	0.96	0.59
+ EWA+ASAM+GSAM	0.79	1.59	1.33	0.64	1.27	1.46	0.40	2.22	0.44	2.03	0.45	0.96	0.57
+ EWA+SAF	0.65	1.66	1.34	1.19	1.26	1.52	0.38	2.37	0.44	2.06	0.48	1.05	0.59

Table 13. Standard deviations for Table 3; 10 repeats on arXiv and 100 all other datasets

Dataset	Cora			CiteSeer			PubMed			Computers		Photo		arXiv
Split	plan	ra-pl	622	plan	ra-pl	622	plan	ra-pl	622	ra-pl	622	ra-pl	622	-
GCN	0.57	1.58	1.31	1.20	1.80	1.57	0.50	2.28	0.49	1.85	0.50	1.19	0.59	0.12
+ SAM	0.50	1.50	1.34	0.85	1.66	1.42	0.45	2.12	0.43	1.95	0.49	0.93	0.57	0.12
+ ASAM	0.47	1.48	1.32	0.86	1.58	1.53	0.40	2.16	0.49	1.89	0.49	0.97	0.59	0.11
+ PGN	0.61	1.43	1.41	1.08	1.36	1.53	0.43	2.25	0.44	1.94	0.45	0.87	0.56	0.14
+ PGNA	0.54	1.66	1.39	0.84	1.69	1.61	0.42	2.31	0.43	1.92	0.48	0.91	0.55	0.16
+ GSAM	0.55	1.25	1.32	0.89	1.63	1.47	0.46	2.13	0.46	1.93	0.42	0.97	0.58	0.14
+ GASAM	0.48	1.33	1.33	0.85	1.61	1.49	0.53	2.22	0.51	1.88	0.51	0.99	0.55	0.13
+ SWA	0.26	1.40	1.30	0.40	1.31	1.46	0.33	2.36	0.47	2.00	0.47	1.01	0.56	0.12
+ EWA	0.71	1.53	1.32	0.61	1.25	1.51	0.37	2.30	0.46	2.00	0.48	1.25	0.57	0.07
+ Anti-PGD	0.62	1.40	1.40	1.06	1.36	1.57	0.54	2.18	0.45	1.86	0.58	1.14	0.56	0.12
+ SAF	0.57	1.71	1.36	0.79	1.97	1.60	0.58	2.28	0.49	1.91	0.46	1.07	0.57	0.15
GAT	0.84	1.34	1.36	0.82	1.11	1.52	0.83	2.33	0.50	1.90	0.47	1.32	0.56	0.15
+ SAM	0.95	1.28	1.36	1.01	1.29	1.57	1.37	2.61	0.49	1.87	0.43	1.28	0.64	0.11
+ ASAM	0.89	1.26	1.42	0.97	1.24	1.59	1.17	2.44	0.49	1.95	0.46	1.29	0.60	0.10
+ PGN	0.69	1.36	1.46	0.87	1.22	1.50	0.99	2.40	0.48	1.76	0.50	1.19	0.58	0.12
+ PGNA	0.67	1.50	1.37	0.77	1.29	1.51	0.80	2.34	0.49	1.93	0.44	1.21	0.57	0.11
+ GSAM	0.72	1.31	1.45	0.93	1.25	1.45	1.11	2.55	0.49	1.75	0.46	1.24	0.60	0.14
+ GASAM	0.67	1.36	1.45	0.95	1.27	1.57	0.99	2.21	0.48	1.86	0.47	1.25	0.57	0.16
+ SWA	0.59	1.32	1.38	0.44	1.08	1.44	0.64	2.45	0.52	2.04	0.44	1.30	0.55	5.26
+ EWA	0.74	1.24	1.45	0.74	1.08	1.47	0.87	2.31	0.52	1.97	0.44	1.27	0.58	6.77
+ Anti-PGD	0.99	1.23	1.37	0.74	1.06	1.53	0.85	2.17	0.48	1.82	0.47	1.25	0.61	0.16
+ SAF	0.87	1.38	1.36	0.75	1.14	1.50	0.87	2.32	0.49	1.81	0.45	1.21	0.54	0.11
Graph-MLP	0.68	1.65	1.25	0.60	1.26	1.57	0.86	2.29	0.42	2.07	0.45	1.30	0.51	0.50
+ SAM	0.66	1.75	1.29	0.77	1.01	1.47	0.83	2.31	0.44	2.05	0.46	1.11	0.52	0.46
+ ASAM	0.68	1.68	1.37	0.64	1.24	1.61	0.80	2.34	0.44	1.91	0.48	1.01	0.47	0.61
+ PGN	0.64	1.72	1.41	0.60	1.01	1.51	0.77	2.59	0.47	2.02	0.45	1.00	0.49	0.36
+ PGNA	0.80	1.75	1.29	0.65	1.16	1.51	0.78	2.30	0.50	1.98	0.46	0.91	0.48	0.32
+ GSAM	0.74	1.78	1.26	0.77	1.41	1.45	0.75	2.33	0.44	1.87	0.46	1.23	0.44	0.35
+ GASAM	0.67	1.67	1.33	0.65	1.11	1.37	0.75	2.35	0.44	1.98	0.47	1.00	0.47	0.48
+ SWA	0.48	1.23	1.31	0.55	0.98	1.57	0.73	2.50	1.03	1.86	0.48	1.08	0.49	0.24
+ EWA	0.65	1.54	1.24	0.58	1.22	1.57	0.96	2.28	0.41	2.08	0.46	1.35	0.49	0.31
+ Anti-PGD	0.79	1.76	1.25	0.65	1.14	1.48	0.74	2.30	0.37	1.98	0.47	1.37	0.50	0.42
+ SAF	0.67	1.59	1.41	0.62	1.28	1.40	0.72	2.33	0.46	1.83	0.50	1.38	0.45	0.23

Table 14. Standard deviations for Table 4; 10 repeats on PPI and 100 all other datasets

Dataset	Cora		CiteSeer		PubMed		Computers	Photo	PPI
Split	plan	622	plan	622	plan	622	622	622	-
GCN	0.46	1.38	0.99	1.52	0.59	0.46	0.48	0.59	0.03
+ SAM	0.56	1.35	1.01	1.65	0.49	0.47	0.47	0.59	0.02
+ ASAM	0.60	1.42	0.80	1.59	0.51	0.51	0.49	0.60	0.03
+ PGN	0.50	1.51	0.89	1.59	0.57	0.48	0.51	0.59	0.02
+ PGNA	0.60	1.41	0.99	1.59	0.52	0.43	0.48	0.59	0.04
+ GSAM	0.50	1.48	1.16	1.59	0.46	0.46	0.49	0.60	0.02
+ GASAM	0.53	1.35	0.88	1.65	0.46	0.46	0.49	0.62	0.03
+ SWA	0.46	1.42	0.25	1.63	0.46	0.44	0.49	0.58	0.06
+ EWA	0.46	1.33	0.37	1.63	0.31	0.47	0.48	0.59	0.02
+ Anti-PGD	0.51	1.35	0.65	1.57	0.50	0.48	0.52	0.58	0.03
+ SAF	0.47	1.55	0.97	1.54	0.53	0.47	0.53	0.61	0.01
GAT	1.14	1.39	0.48	1.56	0.84	0.47	0.50	0.65	0.03
+ SAM	0.81	1.46	0.53	1.51	0.94	0.52	0.45	0.65	0.03
+ ASAM	0.70	1.38	0.52	1.64	1.01	0.49	0.47	0.65	0.03
+ PGN	0.90	1.47	0.54	1.62	0.92	0.51	0.49	0.61	0.02
+ PGNA	1.00	1.36	0.46	1.63	0.94	0.49	0.49	0.62	0.02
+ GSAM	0.89	1.40	0.50	1.53	0.87	0.56	0.50	0.65	0.04
+ GASAM	0.74	1.43	0.50	1.62	1.01	0.50	0.47	0.62	0.03
+ SWA	0.56	1.29	0.17	1.53	0.96	0.49	0.50	0.64	0.73
+ EWA	1.01	1.27	0.23	1.59	0.84	0.47	0.47	0.68	0.05
+ Anti-PGD	0.79	1.36	0.30	1.59	0.87	0.51	0.53	0.66	0.04
+ SAF	1.10	1.43	0.48	1.53	0.87	0.47	0.48	0.66	0.12
Graph-MLP	0.97	1.66	0.85	1.48	0.92	0.48	0.62	0.61	1.12
+ SAM	1.12	1.62	0.72	1.53	0.83	0.41	0.60	0.57	1.60
+ ASAM	1.25	1.66	0.81	1.40	0.81	0.49	0.61	0.58	1.80
+ PGN	1.19	1.63	0.77	1.48	0.86	0.46	0.62	0.55	0.26
+ PGNA	1.28	1.53	0.70	1.40	0.76	0.44	0.57	0.58	0.25
+ GSAM	1.40	1.66	0.78	1.54	0.87	0.36	0.57	0.64	1.47
+ GASAM	1.20	1.68	0.69	1.54	0.70	0.47	0.56	0.59	0.97
+ SWA	0.55	1.54	0.65	1.44	0.85	0.51	0.56	0.54	0.93
+ EWA	0.84	1.67	0.83	1.47	0.90	0.42	0.58	0.61	1.06
+ Anti-PGD	0.90	1.55	0.73	1.42	0.77	0.51	0.57	0.61	0.42
+ SAF	0.80	1.53	0.87	1.54	0.57	0.39	0.59	0.61	0.23

References

1. Bahri, D., Mobahi, H., Tay, Y.: Sharpness-aware minimization improves language model generalization. In: ACL 2022. ACL (2022). https://doi.org/10.18653/v1/2022.acl-long.508
2. Brock, A., De, S., Smith, S.L., Simonyan, K.: High-performance large-scale image recognition without normalization. In: ICML 2021. PMLR (2021)
3. Chen, J., Zhu, J., Song, L.: Stochastic training of graph convolutional networks with variance reduction. In: ICML 2018. PMLR (2018)
4. Chen, M., Wei, Z., Huang, Z., Ding, B., Li, Y.: Simple and deep graph convolutional networks. In: ICML 2020. PMLR (2020)
5. Chen, X., Hsieh, C., Gong, B.: When vision transformers outperform ResNets without pre-training or strong data augmentations. In: ICLR 2022. OpenReview.net (2022)
6. Damian, A., Ma, T., Lee, J.D.: Label noise SGD provably prefers flat global minimizers. In: NeurIPS 2021 (2021)
7. Devlin, J., Chang, M., Lee, K., Toutanova, K.: BERT: pre-training of deep bidirectional transformers for language understanding. In: Proceedings of the 2019 Conference of the North American Chapter of the Association for Computational Linguistics: Human Language Technologies, NAACL-HLT 2019. ACL (2019). https://doi.org/10.18653/v1/n19-1423
8. Du, J., Zhou, D., Feng, J., Tan, V., Zhou, J.T.: Sharpness-aware training for free. In: NeurIPS (2022)
9. Foret, P., Kleiner, A., Mobahi, H., Neyshabur, B.: Sharpness-aware minimization for efficiently improving generalization. In: ICLR 2021. OpenReview.net (2021)
10. Guo, H., Jin, J., Liu, B.: Stochastic weight averaging revisited. CoRR (2022)
11. Gupta, V., Serrano, S.A., DeCoste, D.: Stochastic weight averaging in parallel: large-batch training that generalizes well. In: ICLR 2020. OpenReview.net (2020)
12. Hamilton, W.L.: Graph Representation Learning. Morgan & Claypool Publishers (2020). https://doi.org/10.2200/S01045ED1V01Y202009AIM046
13. Hamilton, W.L., Ying, Z., Leskovec, J.: Inductive representation learning on large graphs. In: NeurIPS 2017 (2017)
14. Hochreiter, S., Schmidhuber, J.: Simplifying neural nets by discovering flat minima. In: NeurIPS 1994. MIT Press (1994)
15. Hochreiter, S., Schmidhuber, J.: Flat minima. Neural Comput. (1997). https://doi.org/10.1162/neco.1997.9.1.1
16. Hu, W., et al.: Open graph benchmark: Datasets for machine learning on graphs. In: NeurIPS 2020 (2020)
17. Hu, Y., You, H., Wang, Z., Wang, Z., Zhou, E., Gao, Y.: Graph-MLP: node classification without message passing in graph. CoRR (2021)
18. Huang, G., Li, Y., Pleiss, G., Liu, Z., Hopcroft, J.E., Weinberger, K.Q.: Snapshot ensembles: train 1, get M for free. In: ICLR 2017. OpenReview.net (2017)
19. Izmailov, P., Podoprikhin, D., Garipov, T., Vetrov, D.P., Wilson, A.G.: Averaging weights leads to wider optima and better generalization. In: Proceedings of the Thirty-Fourth Conference on Uncertainty in Artificial Intelligence, UAI 2018. AUAI Press (2018)
20. Jin, C., Netrapalli, P., Ge, R., Kakade, S.M., Jordan, M.I.: On nonconvex optimization for machine learning: gradients, stochasticity, and saddle points. J. ACM (2021). https://doi.org/10.1145/3418526

21. Kaddour, J., Liu, L., Silva, R., Kusner, M.: When do flat minima optimizers work? In: NeurIPS (2022). https://openreview.net/forum?id=vDeh2yxTvuh
22. Kim, M., Li, D., Hu, S.X., Hospedales, T.M.: Fisher SAM: information geometry and sharpness aware minimisation. In: ICML 2022. PMLR (2022)
23. Kipf, T.N., Welling, M.: Semi-supervised classification with graph convolutional networks. In: ICLR 2017. OpenReview.net (2017)
24. Kwon, J., Kim, J., Park, H., Choi, I.K.: ASAM: adaptive sharpness-aware minimization for scale-invariant learning of deep neural networks. In: ICML 2021. PMLR (2021)
25. Li, H., Xu, Z., Taylor, G., Studer, C., Goldstein, T.: Visualizing the loss landscape of neural nets. In: NeurIPS 2018 (2018)
26. Liu, J., Cai, J., Zhuang, B.: Sharpness-aware quantization for deep neural networks. CoRR (2021)
27. Liu, Y., Mai, S., Chen, X., Hsieh, C., You, Y.: Towards efficient and scalable sharpness-aware minimization. In: IEEE/CVF Conference on Computer Vision and Pattern Recognition, CVPR 2022, New Orleans, LA, USA, 18–24 June 2022. IEEE (2022). https://doi.org/10.1109/CVPR52688.2022.01204
28. Na, C., Mehta, S.V., Strubell, E.: Train flat, then compress: sharpness-aware minimization learns more compressible models. CoRR (2022). https://doi.org/10.48550/arXiv.2205.12694
29. Namata, G., London, B., Getoor, L., Huang, B., Edu, U.: Query-driven active surveying for collective classification. In: 10th International Workshop on Mining and Learning with Graphs (2012)
30. Orvieto, A., Kersting, H., Proske, F., Bach, F.R., Lucchi, A.: Anticorrelated noise injection for improved generalization. In: ICML 2022. PMLR (2022)
31. Sen, P., Namata, G., Bilgic, M., Getoor, L., Gallagher, B., Eliassi-Rad, T.: Collective classification in network data. AI Mag. (2008). https://doi.org/10.1609/aimag.v29i3.2157
32. Shchur, O., Mumme, M., Bojchevski, A., Günnemann, S.: Pitfalls of graph neural network evaluation. CoRR (2018)
33. Velickovic, P., Cucurull, G., Casanova, A., Romero, A., Liò, P., Bengio, Y.: Graph attention networks. In: ICLR 2018. OpenReview.net (2018)
34. Wortsman, M., et al.: Model soups: averaging weights of multiple fine-tuned models improves accuracy without increasing inference time. In: ICML 2022. PMLR (2022)
35. Wu, F., Jr., Souza, A.H., Zhang, T., Fifty, C., Yu, T., Weinberger, K.Q.: Simplifying graph convolutional networks. In: ICML 2019. PMLR (2019)
36. Xu, K., Li, C., Tian, Y., Sonobe, T., Kawarabayashi, K., Jegelka, S.: Representation learning on graphs with jumping knowledge networks. In: ICML 2018. PMLR (2018)
37. Yang, G., Zhang, T., Kirichenko, P., Bai, J., Wilson, A.G., Sa, C.D.: SWALP: stochastic weight averaging in low precision training. In: ICML 2019. PMLR (2019)
38. Yang, Z., Cohen, W.W., Salakhutdinov, R.: Revisiting semi-supervised learning with graph embeddings. In: ICML 2016. JMLR.org (2016)
39. Zeng, H., Zhou, H., Srivastava, A., Kannan, R., Prasanna, V.: GraphSAINT: graph sampling based inductive learning method. In: ICLR 2020 (2020). https://openreview.net/forum?id=BJe8pkHFwS
40. Zhao, Y., Zhang, H., Hu, X.: Penalizing gradient norm for efficiently improving generalization in deep learning. In: ICML 2022. PMLR (2022)
41. Zhou, J., et al.: Graph neural networks: a review of methods and applications. AI Open (2020). https://doi.org/10.1016/j.aiopen.2021.01.001

42. Zhou, M., Liu, T., Li, Y., Lin, D., Zhou, E., Zhao, T.: Toward understanding the importance of noise in training neural networks. In: ICML 2019. PMLR (2019)
43. Zhou, Z.H.: Ensemble Methods: Foundations and Algorithms. CRC Press, Boca Raton (2012)
44. Zhuang, J., et al.: Surrogate gap minimization improves sharpness-aware training. In: ICLR 2022. OpenReview.net (2022)
45. Zitnik, M., Leskovec, J.: Predicting multicellular function through multi-layer tissue networks. Bioinformatics (2017). https://doi.org/10.1093/bioinformatics/btx252

Probabilistic Framework Based on Deep Learning for Differentiating Ultrasound Movie View Planes

Andrei Gabriel Nascu[1], Smaranda Belciug[1(✉)], Anca-Maria Istrate-Ofiteru[1,2], and Dominic Gabriel Iliescu[1,2]

[1] University of Craiova, Craiova, Romania
sbelciug@inf.ucv.ro
[2] University of Medicine and Pharmacy of Craiova, Craiova, Romania

Abstract. Fetal death, infant morbidity and mortality are generally caused by the presence of congenital anomalies. By performing a fetal morphology scan, the sonographer can detect their presence and have a thorough conversation with the soon-to-be parents. Diagnosing congenital anomalies is a difficult task even for an experienced sonographer. A more accurate diagnosis can be set through a merger between the doctor's knowledge and Artificial Intelligence. The aim of paper is to present an intelligent framework that is able to differentiate accurately between the view planes of the fetal abdomen in an ultrasound movie. Deep learning methods, such as ResNet50, DenseNet121, and InceptionV3, have been trained to classify each movie frame. A thorough statistical analysis is used to benchmark the neural networks, and to build a hierarchy. The best performing algorithm is used to classify each frame of the movie, followed by a synergetic weighted voting system that sets the label of the entire ultrasound video. We have tested our proposed framework on several fetal morphology videos. The experimental results showed that the framework differentiates well between the fetal abdomen view planes, even if the deep learning neural networks are able to differentiate between the static images with accuracies that range between 46.01% and 77.80%.

Keywords: Deep learning · statistical analysis · synergetic voting system · fetal morphology · ultrasound movie

1 Introduction

The second trimester ultrasound is recommended by the International Society of Ultrasound in Obstetrics and Gynecology, Fetal Medicine Foundation, and many national societies such as the American College of Obstetricians and Gynecologists. During this ultrasound, also called a fetal morphology scan, the doctor is able to check the fetus anatomy, and see whether all organs are developing normal. If the morphology scan is read properly, and the doctor has suspicions regarding one or more than one congenital anomalies, then a detailed discussion with the parents regarding the prognosis can take

Published by Springer Nature Switzerland AG 2023
A. Holzinger et al. (Eds.): CD MAKE 2023, LNCS 14065, pp. 227–238, 2023.
https://doi.org/10.1007/978-3-031-40837-3_14

place. The discussion has as critical point the following topics: procedural risks, long-term mortality, morbidity, and obviously the quality of life for all those involved. Sadly enough, things are never that simple. Reading an ultrasound is observer depended, so if the doctor is unexperienced, the anomalies might be missed. Studies have shown that the discrepancies between the pre- and postnatal diagnosis have a sensitivity that ranges between 27.5% and 96%, [1]. These discrepancies are not strictly related to the amount of experience. Other factors such as fatigue, time pressure, maternal characteristics, fetal movement, etc. carry the load also.

It is high time to introduce Artificial Intelligence (AI) methods into the fetal morphology domain. The amazing results of Deep Learning (DL) neural networks (NNs) applied in healthcare convinced the Food and Drug Administration to approve several DL software to be used in clinical practice, [2, 3]. Different studies regarding DL and maternal-fetal ultrasounds have been reported in literature, [4–6]. For instance, in [7], the authors used two pretrained convolutional neural networks and two non-DL methods to differentiate fetal ultrasound images. The results in terms of accuracy ranged between 54% and 93.6%. A CNN was used to segment the fetal brain using 3D images, [8], while the fetal lungs and brain were segmented by a sequential forward feature selection technique, support vector machines and DL when applied on magnetic resonance images and ultrasounds, [9]. In [10] and [11], neuroevolution methods were used to determine the architecture of a DL in a fetal morphology scan classification problem.

The aim of this study is to present a probabilistic framework based on deep learning methods and a synergetic weighted voting system that is able to classify correctly fetal morphology movies, by being previously trained on ultrasound images. The paper is organized as follows: Sect. 2 presents the DL algorithms trained on the ultrasound images and the benchmarking dataset, while Sect. 3 presents the results and the statistical analysis of the DLs performances. Section 4 depicts the design and application of the weighted voting system for video classification. The paper ends with Sect. 5, which contains the conclusions and future work.

2 Materials and Methods

2.1 The Deep Learning Algorithm

The framework presented in this study makes use of three types of DL algorithms that are trained on an ultrasound fetal morphology dataset to recognize the view plane of the fetal abdomen. In general, the architecture of a DL neural network contains three kinds of layers: convolutional, pooling, and fully connected. In the convolutional layer, a feature map is created by convolution operations that scan the input with the use of different filters. The pooling layer is used to down sample the feature map created in the convolutional layer, and also to produce spatial invariance. The fully connected layer is the last layer in the network.

The most used activation function is the rectified linear unit layer (ReLU), but in special cases one can use its variants: the Leaky ReLU and exponential linear unit. The above-mentioned functions create non-linearities in the DL. The formula for the ReLU is:

$$f(x) = \begin{cases} 0, x < 0 \\ x, x \geq 0 \end{cases}.$$

The activation function that connects the last hidden layer to the output is the softmax. The softmax takes the input vector and transforms it into a probability vector. Its formula is:

$$p = \begin{pmatrix} p_1 \\ \cdot \\ \cdot \\ p_n \end{pmatrix}, \text{ where } p_i = \frac{e^{x_i}}{\sum_{j=1}^{n} e^{x_j}}.$$

In this study, we have used ResNet50, DenseNet121, and InceptionV3. Resnet50 or the Residual Network 50 has won the ILSVR ImageNet competition in 2015. Its main characteristic is that it uses a skip connection, which gives access to a complementary cutoff route to the gradient's flow in the neural network. The higher levels will perform just as well as the lower levels. The architecture of a ResNet50 contains 48 convolutional layers, 1 maxpool, and 1 averagepool, [12]. The DensetNet121, on the other hand, uses an extra input which is added to each layer. We refer to this input as the collective knowledge gathered from preceding layers. This makes the architecture dense. The features are not summed up, they are concatenated, thus being reused [13]. In InceptionV3, we encounter factorized convolutions, that build the architecture in a gradual manner. By this, the number of parameters is reduced. Big convolutions are replaced by small convolutions, which speeds up the training process. Besides this, in InceptionV3 we have asymmetric convolutions. As a final note on InceptionV3, we mention the existence of the auxiliary classifier, that acts as a regularizer, [14].

The same classifier was used for the 3 DLs. In what regards the architecture, we have used the GlobalAveragePooling2D, Dense (1024, activation = ReLU), and Dense (softmax). The loss function was set to be the crossentropy, as for the optimizer, we have chosen Adam. We have pretrained the 3 DLs on the ImageNet dataset. We have used Keras and Tensorflow. In what regards the parameters we have used a $5 \times 5 \times 3$ filter, no zero-padding, and a stride equaling 2. No hyperparameter optimization approach has been used in this study. The total number of parameters for each DL is presented in Table 1.

Table 1. Total number of parameters for each DL

DL algorithm	Total parameters
ResNet50	25 692 038
DenseNet121	8 93 254
InceptionV3	24 492 198

2.2 Fetal Abdomen Dataset

The 3 DLs have been trained and tested on a second trimester fetal morphology dataset, which contains ultrasound images regarding the fetal abdomen. The data comes from a prospective cohort study implemented in a maternity hospital in Romania, the University Emergency County Hospital of Craiova. All eligible patients were pregnant and admitted themselves to the Prenatal Unit of the County Hospital for their second trimester morphology scan. The doctors, that were part of the research project, informed the patients about the research and invited them to take part of this study. All patients understood the implication of the study and gave a written consent. The medical personnel have a minimum 2-year experience in performing fetal morphology scans. All the data was acquired using Voluson 730 Pro (GE Medical Systems, Zipf, Austria) and Logic e (GE Healthcare, China US machines with 2–5-MHz, 4–8-MHz, and 5–9 MHz curvilinear transducers). The fetal abdomen images where split by the doctors in 8 different view planes: 3 vessels plus bladder (253 images), gallbladder (123 images), transverse cord insertion (211 images), anteroposterior kidney(188 images), echogenic (440 images), sagittal kidney (326 images), bladder (66 images), and cord intestinal sagittal (324 images). The text and other artefacts have been eliminated by using CV2 and Keras-OCR. Since these algorithms are prone to producing false positives (detecting text where it is not), we have proceeded into manually rechecking each image manually. Figures 1 and 2 show the same image before and after preprocessing.

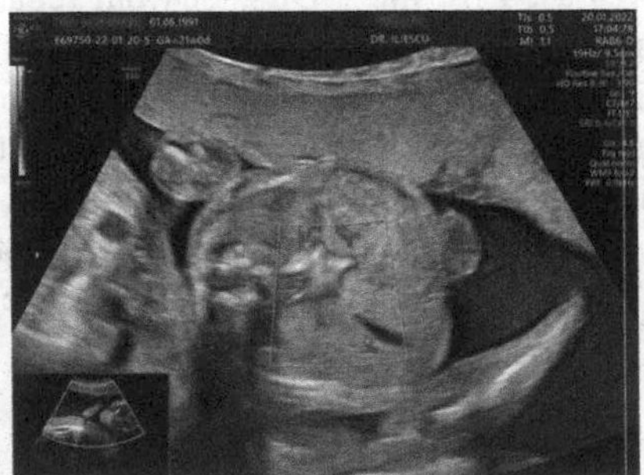

Fig. 1. Unprocessed image.

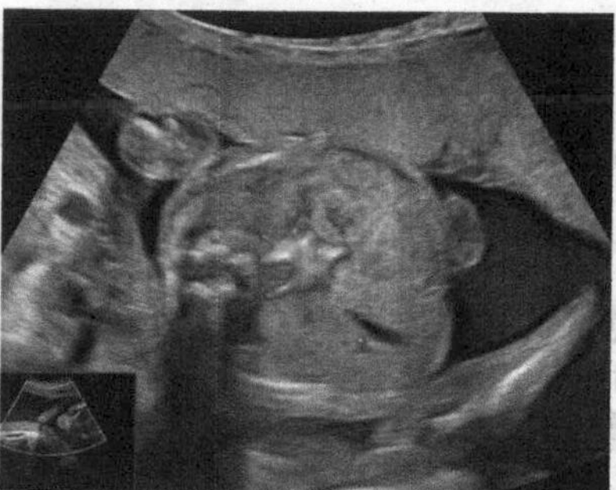

Fig. 2. Preprocessed image.

For a better visualization of the fetal abdomen planes, we depict in Fig. 3 a sample from each view plane.

Due to the imbalanced nature of the dataset and to avoid overfitting, we have used a data generator to enlarge its size. Different transformations have been applied to every image using the following values: shear range = 0.2, rotation range = 20, wifht and height shift range = 0.1, zoom range = 0.2, and brightness range between 0.7 and 1.4. We have also reshaped all the images to 224 × 224 px.

3 Results

To achieve adequate statistical power (95% type I error $\alpha = 0.05$), we have performed power analysis to determine the needed sample size of independent computer runs. The needed sample sizes equaled 50, hence each DL had been run independently for 50 times in a complete 10-fold cross validation cycle. We present the obtained results in terms of average accuracy (ACA) over 50 runs and standard deviation (SD) in Table 2.

From Table 2, we can see that ResNet50 outperforms the other two DLs. All models seem to be robust taking into account the SD values. To be certain that indeed there are significant differences between ResNet50's performance and DenseNet121's, we have begun the data screening process, which involved Anderson-Darling test to verify the normality of the sample data, and Brown-Forsythe for the verifying the equality of variances. In Table 3, we present the results of the normality test, while in Table 4 the results of the equality of variances test.

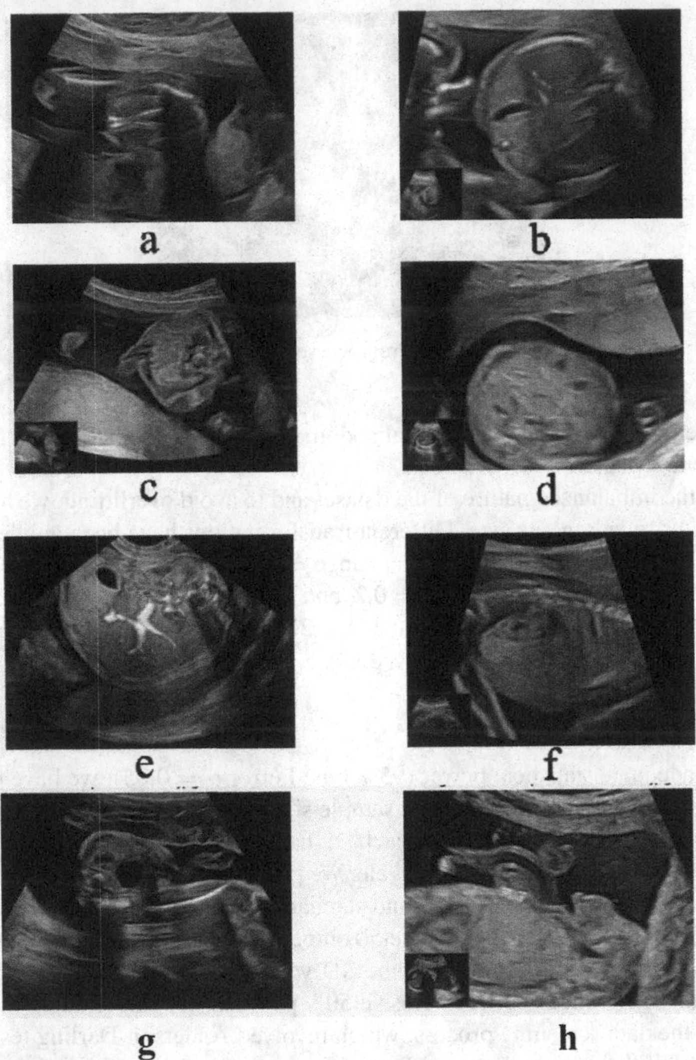

Fig. 3. (a) 3 vessels plus bladder; (b) gallbladder; (c) transverse cord insertion; (d) anteroposterior kidney; (e) echogenic; (f) sagittal kidney; (g) bladder; (h) cord intestinal sagittal.

Table 2. Performance of DLs over 50 computer runs

DL algorithm	ACA (%)	SD
ResNet50	77.80	3.01
DenseNet121	64.20	2.12
InceptionV3	46.01	2.01

Table 3. Normality test results

DL algorithm	Anderson-Darling	
ResNet50	15.64	0.000
DenseNet121	16.26	0.000
InceptionV3	15.50	0.000

From Table 3, we can see that none of the samples are normally distributed, but still we can make use of the *Central Limit Theorem*, that state that if the sample size is large enough (>30), then the data distribution is approximately normal, [15].

Table 4. Equality of variances results

Models	Brown-Forsythe (1, df)/*p*-level	
ResNet50 vs. DenseNet121	2.101	0.150
ResNet50 vs. InceptionV3	2.318	0.131
DenseNet121 vs. ResNet50	0.000	0.978
DenseNet121 vs. InceptionV3	4.017	0.047
InceptionV3 vs. ResNet50	4.573	0.034
InceptionV3 vs. DenseNet121	2.754	0.100

From Table 4, we can see that the only models that do not have equal variances are InceptionV3 vs ResNet50, and InceptionV3 vs DenseNet121. Since all samples have equal sizes, we can still proceed with our statistical analysis.

Because all a priori conditions for the *t*-test for independent variables are met, we can proceed with it, [16, 17]. The statistical differences are revealed in Fig. 4, in which we have plotted the distribution of the samples in the form of boxplots. The annotations show the obtained *p*-level after applying *t*-test for independent with Bonferroni correction.

The *t*-test results reveals that there are indeed statistical differences between the performances obtained by our DL models. Therefore, our framework will be further built based on the responses of the trained ResNet50.

Besides the *t*-test with Bonferroni corrections, we have also applied Kruskal-Wallis ANOVA non-parametric test. This test uses the rank of values instead of the values, thus we compare the groups medians instead of the group means. We have used the Kruskal-Wallis test using χ^2 right-tailed distribution with 2 degrees of freedom. The test statistics H equaled 20.734, with a corresponding p-value of 0.00001, thus we need to reject the null hypothesis that states that the group medians are equal. We can see that both parametric and non-parametric tests reveal the same result, that there are significant differences between the DL models.

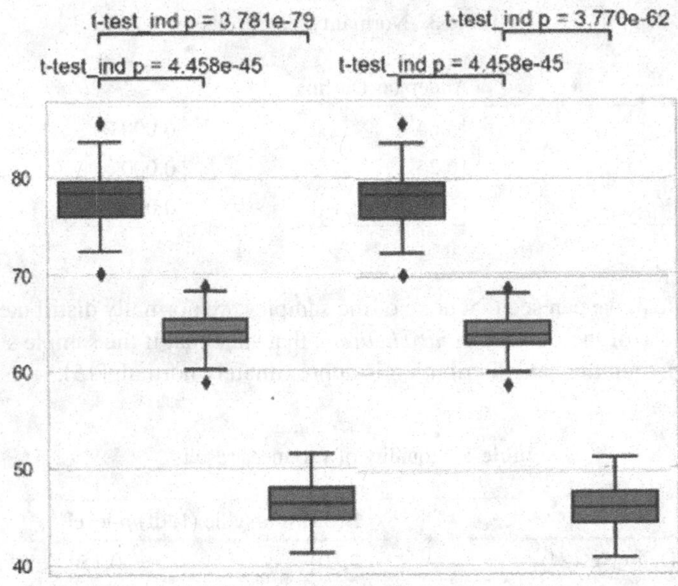

Fig. 4. *t*-test with Bonferroni correction to reveal statistical differences between DL competitors.

4 Weighted Voted System

As we have mentioned before, the goal of our paper is to build a framework that is able to differentiate between the view planes of the fetal abdomen in an ultrasound movie, not in an image. Our approach uses the previously trained DL network. The first step is to split the movie into frames, after which we apply ResNet50 on each frame to label it. The DL network will set a label to each frame with a certain probability, P_i. We transform this probability in a weighted vote using the following formula:

$$w_i = \frac{P_i}{\sum_{i=1}^{number\_frames} P_i}.$$

In Table 5 we provide a quantitative comparison between the results obtained by our framework and the stand-alone DL models on hold out samples. We can see that the proposed system outperforms the other methods.

Table 6 presents how we set the weighted votes for a movie that contains the 3 vessels + bladder view plane.

Table 5. Comparison between the proposed model and other strategies

Model	ACA	F1-score
Proposed framework	81.19	0.811
ResNet50	77.89	0.780
DenseNet121	64.01	0.640
InceptionV3	45.23	0.451

Table 6. Weighted voting system for a 3 vessels + bladder ultrasound movie

Predicted view plane	P_i	Vote
gallbladder	80.9	0.036387
gallbladder	89.53	0.040268
gallbladder	86.37	0.038847
gallbladder	79.71	0.035851
3vesselsbladder	95.4	0.042908
3vesselsbladder	99.13	0.044586
3vesselsbladder	94.19	0.042364
3vesselsbladder	99.8	0.044887
3vesselsbladder	99.95	0.044955
3vesselsbladder	99.86	0.044914
3vesselsbladder	98.64	0.044365
3vesselsbladder	99.98	0.044968
3vesselsbladder	99.99	0.044973
3vesselsbladder	99.99	0.044973
3vesselsbladder	99.99	0.044973
3vesselsbladder	99.99	0.044973
3vesselsbladder	99.99	0.044973
3vesselsbladder	99.99	0.044973
3vesselsbladder	99.99	0.044973
3vesselsbladder	99.99	0.044973
3vesselsbladder	99.99	0.044973
3vesselsbladder	99.99	0.044973

The movie label is set by adding the votes for each predicted class. Hence, for the above example we would have the following result (Table 7).

Table 7. Voting framework

Predicted class	Vote result	Actual class
gallbladder	0.151355	3 vessels + bladder
3 vessels + bladder	0.848647	

From Table 7, we can easily see that even if ResNet50 has only 77.80% accuracy in correctly differentiating the view planes, the framework through its weighted vote corrects it. To validate our framework, we have applied it on several other ultrasound movies. The following Tables 8, 9, and 10 present the obtained results.

Table 8. Voting framework example 2

Predicted class	Vote result	Actual class
echogenic	0.171553	cord intestinal sagittal
cord intestinal sagittal	0.422078	
anteroposterior kidney	0.406268	

We can see that even if the cord intestinal sagittal plane has almost the same probability as the anteroposterior kidney, still the framework is able to correct it.

Table 9. Voting framework example 3

Predicted class	Vote result	Actual class
cord intestinal sagittal	0.874941	cord intestinal sagittal
echogenic	0.125059	

Table 10. Voting framework example 4

Predicted class	Vote result	Actual class
cord intestinal sagittal	0.28882	cord intestinal sagittal
sagittal kidney	0.115637	
transverse cord insertion	0.266047	
bladder	0.203784	
echogenic	0.126261	

Tables 9 presents one of the simplest planes which involves a small number of corrections, while Table 10 presents one of the most complicated planes which involves a large number of corrections.

5 Discussion

The above study proposes a probabilistic framework that learns to classify the view planes in a movie by using a hierarchy built from multiple DL algorithms previously trained on ultrasound images. Even if we have statistically proved that this framework produces valuable results, a difficult task is to explain to the end-used, the doctor, why a particular result was obtained. Since all data is labeled by human doctors, it is prone to error, hence we are dealing with a garbage in – garbage out situation. The system by using performant DL algorithms and a weighted voting system is able to correct the human and machine possible mistakes. To convince the human doctor that the framework is reliable and robust, we have performed the thorough statistical analysis, and also reminded them that the artificial intelligent system acts only as a "counselor", and that the final call is made by the doctor. As said in the Holzinger's keynote, [18], explainability is the first step in making artificial intelligence be trusted by clinicians, and robustness is the second step.

In this paper, we show that the combination of human intelligence and artificial intelligence, can indeed help resolve a rather complicated matter. In our case the ultrasound movie is built from consecutive frames. As the baby and mother breathe and move, the view-planes change even if the doctor knows that he/she is looking at a certain plane. Our probabilistic voting system is able to establish the correct view plane in order to prepare the system for the next logical step, that is to start segmenting the important organs in each plane. Taking into account the standard in performing second trimester morphologies, particular organs are searched in each view plane. This is why the combo doctor + artificial intelligence system needs to work ensemble as a team in differentiating correctly the view planes.

To obtain more reliable results we must increase the size of the image dataset, and to try multiple DL algorithms to see which one resolve the best the problem at hand.

6 Conclusions

In this study, we have provided a way to build a framework that is able to differentiate between different view planes in an ultrasound movie regarding a second trimester fetal morphology scan, by using a synergetic weighted voting system of each frames' probability. We have compared the performances of three state-of-the-art DL neural networks and chose for the framework the best performing one according to a statistical benchmark process. Our framework proved to be efficient, by correcting the DL network, even if its standalone accuracy is approximately 77%. The framework's functionality did indeed accomplish its objective. We shall deepen the experiments in terms of applying in longer movies, that contain multiple view planes.

Acknowledgement. This work was supported by a grant of the Ministry of Research Innovation and Digitization, CNCS – UEFISCDI, project number PN-III-P4-PCE-2021-0057, within PNCDI III.

References

1. Salomon, L., et al.: A score-based method for quality control of fetal images at routine second trimester ultrasound examination. Prenat. Diagn. **28**(9), 822–827 (2008)
2. Topol, E.J.: High performances medicine: the convergence of human and artificial intelligence. Nat. Med. **25**, 44–46 (2019)
3. Benjamens, S., Dhunno, P., Mesko, B.: The state of artificial intelligence-based FDA approved medical devices and algorithms: an online database. NPJ Digit. Med. **3**, 118 (2020)
4. Phillip, M., et al.: Convolutional neural networks for automated fetal cardiac assessment using 4D B-mode ultrasound. In: IEEE 16th International Symposium on Biomedical Imaging (2019)
5. Matsuoka, R., Komatsu, M., et al.: A novel deep learning based system for fetal cardiac screening. Ultrasound Obstet. Gynecol. **54**(S1), 177–178 (2019). https://doi.org/10.1002/uog.20945
6. Komatsu, R., Matsuoka, R., et al.: Novel AI-guided ultrasound screening system for fetal heart can demonstrate finding in timeline diagram. Ultrasound Obstet. Gynecol. **54**(S1), 134 (2019). https://doi.org/10.1002/uog.20796
7. Burgos-Artizzu, X.P., et al.: FETAL_PLANES_DB: common maternal-fetal ultrasound images. Nat. Sci. Rep. **19**, 10200 (2020)
8. Namburete, A., et al.: Fully automated alignment of 3D fetal brain ultrasound to a canonical reference space using multi-task learning. Med. Image Anal. **46**, 1–14 (2018)
9. Torrents-Barrena, J., et al.: Assessment of radiomics and deep learning for the segmentation of fetal and maternal anatomy in magnetic resonance imaging and ultrasound. Acad. Radiol. **19**, 30575–30576 (2019)
10. Belciug, S.: Learning deep neural networks' architectures using differential evolution. Case study: medical imaging processing. Comput. Biol. Med. **146**, 105623 (2022)
11. Ivanescu, R., et al.: Evolutionary computation paradigm to determine deep neural networks architectures. Int. J. Comput. Commun. Control **17**(5), 4886 (2022). https://doi.org/10.15837/ijccc.2022.5.4886
12. He, K., Zhang, X., Ren, S., Sun, J.: Deep residual learning for image recognition (2015). https://arxiv.org/abs/1512.03385
13. Huang, G., Liu, Z., van de Maeeten, L., Weinberger, K.Q.: Densely connected convolutional networks (2016). https://arxiv.org/abs/1608.06993
14. Szegedy, C., Vanhoucke, V., Ioffe, S., Shlens, J., Wojna, Z.: Rethinking the inception architecture for computer vision (2015). https://arxiv.org/abs/1512.00567
15. Belciug, S., Iliescu, D.: Planning a pregnancy with Artificial Intelligence. In: Belciug, S., Iliescu, D. (eds.) pregnancy with Artificial Intelligence, vol. 234, pp. 63–98. Springer, Cham (2023). https://doi.org/10.1007/978-3-031-18154-2_2
16. Altman, D.G.: Practical Statistics for Medical Research. Chapman and Hall, New York (1991)
17. Demsar, J.: Statistical comparison of classifiers over multiple data sets. J. Mach. Learn. Res. **7**, 1–30 (2006)
18. Holzinger, A.: The next frontier: ai we can really trust. In: Kamp, M., et al. (eds.) Machine Learning and Principles and Practice of Knowledge Discovery in Databases: International Workshops of ECML PKDD 2021, Virtual Event, September 13-17, 2021, Proceedings, Part I, pp. 427–440. Springer, Cham (2021). https://doi.org/10.1007/978-3-030-93736-2_33

Standing Still Is Not an Option: Alternative Baselines for Attainable Utility Preservation

Sebastian Eresheim[1,2](✉), Fabian Kovac[2], and Alexander Adrowitzer[2]

[1] Research Group Security and Privacy, University of Vienna, Vienna, Austria
sebastian.eresheim@univie.ac.at

[2] Data Intelligence Research Group, St. Pölten University of Applied Sciences, St. Pölten, Austria
{sebastian.eresheim,fabian.kovac,alexander.adrowitzer}@fhstp.ac.at

Abstract. Specifying reward functions without causing side effects is still a challenge to be solved in Reinforcement Learning. Attainable Utility Preservation (AUP) seems promising to preserve the ability to optimize for a correct reward function in order to minimize negative side-effects. Current approaches however assume the existence of a no-op action in the environment's action space, which limits AUP to solve tasks where doing nothing for a single time-step is a valuable option. Depending on the environment, this cannot always be guaranteed. We introduce four different baselines that do not build on such actions and therefore extend the concept of AUP to a broader class of environments. We evaluate all introduced variants on different AI safety gridworlds and show that this approach generalizes AUP to a broader range of tasks, with only little performance losses.

Keywords: Impact Regularization · Side-Effect Avoidance · Reinforcement Learning

1 Introduction

In recent years, Reinforcement Learning (RL) has excelled on a number of tasks, agents can perform. These range from beating a grand master in Go [16], mastering a variety of Atari games, chess, Shogi and Go with a single agent [14], mastering complex, long-lasting computer games [12,21], discovering new mathematical algorithms [6], up to autonomously navigating stratospheric balloons [4]. While many impressive applications, that exceed human capabilities, lie in

S. Eresheim and F. Kovac—Both authors contributed equally to this work.

Supplementary Information The online version contains supplementary material available at https://doi.org/10.1007/978-3-031-40837-3_15.

Published by Springer Nature Switzerland AG 2023
A. Holzinger et al. (Eds.): CD-MAKE 2023, LNCS 14065, pp. 239–257, 2023.
https://doi.org/10.1007/978-3-031-40837-3_15

an information-centric realm, only a fraction involve agents that interact with real-world physical objects.

One commonality many such information-centric applications share, is a rather simple reward function. Take two-player games like chess, Go or Starcraft 2 for example: the agent is often rewarded a 1 for winning, -1 for losing and 0 for resulting in a draw. Such simple reward functions are beneficial, because they do not include human prior knowledge about the game, that might not be optimal. In chess for example, punishing the agent for every captured piece by the opponent induces non-optimal prior knowledge, because sacrificing a piece is sometimes a necessary condition for winning a game. Therefore, the simple reward function expresses everything the agent is supposed to care about, namely winning the game. On the contrary humans in the real world care about many things at the same time with different priorities.

A result of this misalignment between simple reward functions and 'many things humans care about' in the real world are unintended, negative side-effects. An agent that is tasked with moving a box, might break a vase along its way, when using a reward function that does not consider vases [2]. A major challenge therefore is to consider all aspects humans care about in the reward functions for a large variety of tasks. Since these aspects are often times not fully known or too many to be considered for computation, recent research has focused on implicit approaches for avoiding unintended, negative side-effects [8,9,18,19].

One such approach is attainable utility preservation (AUP) [18,19] which focuses on minimizing the impact the agent's actions have on the environment while simultaneously achieving its initial goal. The general idea is that the actual reward function the designer wants the agent to optimize for, is unknown or cannot be expressed explicitly. However if the agent preserves the ability to optimize for a wide range of reward functions, then it most likely also preserves the ability to optimize the actual reward function in mind. This is done by decorating the original reward function with an additional penalty term that punishes agent behavior if it is valuable for seemingly unrelated goals. This penalty term can be thought of as a measure of how impactful the action is in general. Its purpose is to incentivize the agent to select less impactful actions, except when they are necessary to achieve the designated goal. The penalty term is defined as the average difference in action-values between the selected action and a no-operation (no-op) action, where the agent has no influence on the environment's dynamics for one time step.

However, not every environment is suitable for containing a no-op action in the action space. Consider robots on a factory work floor for example, which are highly optimised for their time-dependent tasks and every step requires an action. These robots cannot simply 'stand still' while performing their tasks, which would lead to delays in production. Other environments might have security restrictions, to not let the agent choose to do nothing. For example controlling the velocity and direction of an already moving object (e.g. car, ship, air plane, etc.). If an auto-pilot would take over control of a fast moving car on a curvy highway, choosing to do nothing would likely lead to an accident and therefore might already be restricted by an additional safeguarding system. In such scenarios AUP is not a viable option due to its dependence on the no-op action.

Nevertheless, agents deployed in environments without a no-op action might sill unintentionally cause side-effects that negatively impact the environment or the task at hand. Therefore, there is a need for side-effect avoidance in these scenarios to ensure that the agent can perform its task while minimizing the negative impact of its actions. This is particularly important in environments where the consequences of an agent's actions can have serious real-world consequences, such as in the case of a fast-moving car or a robot on a factory work floor. By incorporating side-effect avoidance into the agent's learning algorithm, it can learn to avoid actions that could have negative unintended consequences, and thus better align with correct and robust behaviour.

We contribute to the field in three separate ways:

- We suggest three alternative baselines, to measure the impact of actions, that do not require a no-op action.
- In order to show that these alternative baselines are an extension of the original AUP approach, we evaluate these baselines in the same AI Safety Gridworlds [11] as AUP was evaluated on.
- Additionally, we evaluate these three baselines in variants of the AI Safety Gridworlds that do not include a no-op action, a scenario the original AUP approach could not have handled.

The rest of this paper is structured as follows: Sect. 2 elaborates on the bigger picture of side-effect avoidance, Sect. 3 gives a more detailed introduction about AUP, Sect. 4 describes our four examined variants in detail, Sect. 5 describes the experiment setup, Sect. 6 reports on the results, Sect. 7 discusses these results and gives a brief outlook about potential future work, and Sect. 8 concludes the paper.

2 Related Work

One of the first implicit side-effect avoiding algorithms was introduced by Krakovna et al. [8]. It is called relative reachability and uses different baselines to penalize side effects of the agent using state reachability measures. The primary focus of this approach is on irreversible side-effects.

A more recent work by Krakovna et al. [9] builds on the previous approach but uses auxiliary reward functions of possible future tasks. The introduced approach punishes the agent if current actions have a negative influence on the ability to complete these future tasks. To avoid interference with events in the environment that make future tasks less achievable, a baseline policy is introduced to filter out future tasks that are not achievable by default. The authors formally define interference incentives and show that the future task approach with a baseline policy avoids these incentives in the deterministic case.

Alamdari et al. [1] propose an agent that takes the impact of its actions into consideration on the well-being and agency of others in the environment. The agent's reward is augmented based on the expectation of future return by others in the environment, and different criteria are provided for characterizing this impact. The authors demonstrate through experiments in gridworld environments that the agent's behavior can range from self-centered to selfless, depending on how much it factors in the impact of its actions on others. The proposed approach addresses the issue of incomplete or underspecified objectives and contributes to AI safety by encouraging agents to act in ways that are considerate of others in the environment.

Shah et al. [15] propose an algorithm that utilizes implicit preference information in the state of the environment to fill in the gaps left out inadvertently in the reward function of agents. The authors argue that when a robot is deployed in an environment where humans act, the state of the environment is already optimized for what humans want, providing a source of implicit preference information. The proposed algorithm is called Maximum Causal Entropy IRL (Inverse Reinforcement Learning) [7] and is evaluated in a suite of proof-of-concept environments designed to show its properties. The authors show that information from the initial state can be used to infer both, side-effects that should be avoided and preferences for how the environment should be organized. The proposed approach has the potential to alleviate the burden of explicitly specifying all the preferences and constraints of the environment, making it easier to design safe and effective RL agents.

Recent work by Turner et al. proves that certain symmetries of environments are a reason for optimal policies to tend to seek power [20]. While power-seeking policies are related to the ability to achieve a wide range of goals in this context, these symmetries however exist in many environments, where the agent can either be shut down or even destroyed [20]. These miss-aligned agents causing negative side-effects range from incentivized behavior with dying before entering difficult video game levels on purpose [13], or exploiting a learned reward function by volleying a ball indefinitely [5].

3 Attainable Utility Preservation

Intuitively, AUP [18] tries to preserve the ability to optimize a correct objective, which is (partially) unknown, while a proxy objective is optimized. Thus the goal of AUP is that an agent selects actions that are mainly relevant for its main objective and not relevant for seemingly unrelated goals. Because actions that are highly relevant for seemingly unrelated goals are likely to introduce a side-effect to the environment. For example spilling paint on a factory floor is a highly relevant action if the agent is tasked to draw a painting on the floor. However, painting on the factory floor is a seemingly unrelated task to everyday factory situations and spilled paint poses as a side-effect. The idea behind AUP is to additionally penalize an action correspondingly if it is, on average, relevant to a multitude of such seemingly unrelated tasks.

Formally, Turner et al. consider a Markov decision process (MDP) $(\mathcal{S}, \mathcal{A}, \mathcal{T}, R, \gamma)$, where $\mathcal{S}$ is a state space, $\mathcal{A}$ is an action space, $\mathcal{T} : \mathcal{S} \times \mathcal{A} \rightarrow \Delta(\mathcal{S})$ is a transition function mapping state-action pairs to distributions over states, $R : \mathcal{S} \times \mathcal{A} \rightarrow \mathbb{R}$ is a reward function and $\gamma \in \mathbb{R}$ a discounting factor. In the setting of AUP Turner et al. assume the action space contains a no-op action $\varnothing \in \mathcal{A}$ where the agent does not influence the environment's dynamics for one time step. This no-op action is used for the so called *step-wise inaction baseline*, where the value of an action is compared with that of the no-op action, to determine its impact on the state. Additionally, Turner et al. assume the designer provides a finite set of auxiliary reward functions $\mathcal{R} \subset \mathbb{R}^{\mathcal{S} \times \mathcal{A}}$. Q_{R_i} denotes the corresponding action-value function (or Q-function) for an auxiliary reward function $R_i \subset \mathcal{R}$. The AUP reward function is then defined as follows:

$$R_{AUP}(s, a) := R(s, a) - \frac{\lambda}{\mu} \sum_{i=1}^{|\mathcal{R}|} |Q_{R_i}(s, a) - Q_{R_i}(s, \varnothing)|, \tag{1}$$

where $\lambda \geq 0$ is a regularization parameter to control the influence of the penalty on the primary reward function and μ scales the penalty by one of the following two options:

$$\mu := \begin{cases} \sum_{i=1}^{|\mathcal{R}|} Q_{R_i}(s, \varnothing) & \text{case 1} \\ |\mathcal{R}| & \text{case 2} \end{cases} \tag{2}$$

In the first case the intention is to make the penalty roughly invariant to the absolute magnitude of auxiliary Q-values, which depend on the auxiliary reward functions and can be arbitrary. This is achieved by scaling with an action-value of a 'mild action' (e.g. $\varnothing$). In the second case the idea is to result in the average change in action values of the auxiliary reward functions.

To learn the action-value functions $Q_{R_i}(s, a)$ of the corresponding auxiliary sets $R_i \in \mathcal{R}$ as well as the optimal action-value function $Q_{\text{AUP}}(s, a)$, AUP uses Q-learning to perform an AUP update as shown in Algorithm 1.

Algorithm 1: AUP update [18]

begin
  for $i \in |\mathcal{R}|$ **do**
    $Q'_{R_i} = R_i(s, a) + \gamma \max_{a'} Q_{R_i}(s', a')$
    $Q_{R_i}(s, a) += \alpha(Q'_{R_i} - Q_{R_i}(s, a))$
  $Q' = R_{AUP}(s, a) + \gamma \max_{a'} Q_{AUP}(s', a')$
  $Q_{AUP}(s, a) += \alpha(Q' - Q_{AUP}(s, a))$

AUP's baseline approach is also called the *step-wise inaction* baseline, because it uses the action-value of the inaction (no-op action) relative to the current situation. In contrast, two other baselines are the *starting state* baseline [8], which compares the current state to the initial state of the environment at the start of the episode and the *inaction* baseline [3], which compares the current state to the state of the environment that naturally developed from the initial state, if the agent had done nothing or were never deployed. Both of these alternative baselines have their own drawbacks. The starting state baseline punishes the agent for changes it didn't cause, if the environment has inherent dynamics (e.g. flow of water in a river). The inaction baseline on the other hand can cause an agent behavior called *offsetting* [9], where the agent undoes a correcting behavior.

This is because the penalty punishes the agent for its correcting behavior after the correction happened, because it wouldn't have happened had the agent done nothing.

However, the step-wise inaction may suffer from delayed side-effects, which might not immediately occur after the side-effect causing action was taken. In order to (slightly) mitigate this weakness, Turner et al. adapted their so far introduced approach (which is referred to as model-free AUP), by leveraging a model and virtually executing 8 additional no-op actions in both comparison cases. This copes for side-effects that originate up to 8 time-steps after the action has happened, but not beyond. This model-based version is referred to by Turner et al. as AUP.

4 Methods

We consider the same setting as Turner et al. [18], except that we do not assume a no-op action $\varnothing$ to be part of the action space. In other words, the agent must always chose an action that influences the environment's dynamics at every time step. By removing the no-op action from the action space $\varnothing \notin \mathcal{A}$, we also remove the only known mild action for scaling the penalty by the first alternative of Eq. 2. Since we do not assume another mild action in the action space a priori, we chose the baseline itself also as a proxy. Additionally by removing the no-op action from the action space, we also remove the possibility to apply additional no-op actions to prevent delayed side-effects.

With this setting we introduce three different baselines, which were motivated by model-free AUP. These are the *average*, *average-others* and *advantage* baseline.

Average Baseline. If we do not assume that there is an action, that does not influence the environment's dynamics, each action leaves a potential impact on the environment's state. Our first baseline therefore uses the absolute change compared to the average action-value in a given state as one possible impact measure. We call this version average baseline or in short *avg*. The reward function for the average baseline is defined as:

$$R_{avg}(s,a) := R(s,a) - \frac{\lambda}{|\mathcal{R}|} \sum_{R_i \in R} \frac{\left|Q_{R_i}(s,a) - \left(\frac{1}{|\mathcal{A}|} \sum_{a' \in \mathcal{A}(s)} Q_{R_i}(s,a')\right)\right|}{\frac{1}{|\mathcal{A}|} \sum_{a' \in \mathcal{A}(s)} Q_{R_i}(s,a')}. \quad (3)$$

Average-Others Baseline. Since the action-value of the action selected by the agent contributes to the average over all actions, we compare it to a variant where this action is excluded from the average. Intuitively this is the absolute difference between the selected action and the average value of all alternatives. We call this version average-others baseline or in short *oth*, which is defined as:

$$R_{oth}(s,a) := R(s,a) - \frac{\lambda}{|\mathcal{R}|} \sum_{R_i \in R} \frac{\left|Q_{R_i}(s,a) - \left(\frac{1}{|\mathcal{A}|} \sum_{a' \in \mathcal{A}(s) \setminus \{a\}} Q_{R_i}(s,a')\right)\right|}{\frac{1}{|\mathcal{A}|} \sum_{a' \in \mathcal{A}(s) \setminus \{a\}} Q_{R_i}(s,a')}. \tag{4}$$

Advantage Baseline. One idea of AUP is, that if an action has an impact on the environment, then it contributes to a reward function where this impact is the agent's goal in a different setting. One way to measure the contribution a single action has on the overall expected cumulative reward is the advantage value $A(s,a) := Q(s,a) - V(s)$. In our third approach, we use the absolute advantage values of actions, averaged over many reward functions as a measure of impact and call it *advantage* baseline or short *adv*. We do this by exploiting the equality $v_\pi(s) = \sum_{a' \in \mathcal{A}} \pi(a'|s)\, q_\pi(s,a')$ [17], where $\pi : \mathcal{S} \times \mathcal{A} \rightarrow [0,1]$ is a policy, mapping state-action pairs to probabilities. The reward function with the advantage baseline is defined as:

$$R_{adv}(s,a) := R(s,a) - \frac{\lambda}{|\mathcal{R}|} \sum_{R_i \in R} \frac{\left|Q_{R_i}(s,a) - \sum_{a' \in \mathcal{A}} \pi_{Q_{R_i}}(a'|s)\, Q_{R_i}(s,a')\right|}{\sum_{a' \in \mathcal{A}} \pi_{Q_{R_i}}(a'|s)\, Q_{R_i}(s,a')}. \tag{5}$$

Random-Action Baseline. Lastly, we use the action-value of a valid random action $a' \in \mathcal{A} \setminus \{a\}$ that is different from the action the agent selected, as a baseline and call it *random-action* baseline or in short *rand*. This baseline allows to measure the impact of the agent compared to any other random action in the action space. It is defined as:

$$R_{rand}(s,a) := R(s,a) - \frac{\lambda}{|\mathcal{R}|} \sum_{R_i \in R} \frac{|Q_{R_i}(s,a) - Q_{R_i}(s,a')|}{Q_{R_i}(s,a')}. \tag{6}$$

We exclude the chosen action $a \neq a'$ to make sure that the penalty cannot reach 0 and is therefore never neglected. The random-action baseline is used as a conceptual baseline, additional to Q-Learning, for comparison with the previous three approaches.

5 Experimental Design

We follow the approach of Turner et al. [18] and evaluate our approaches on a subset of the AI Safety Gridworlds [11] with the focus on avoiding side-effects, as well as environments developed during the AI Safety Camp 2018[1]. These

[1] https://aisafety.camp/2018/06/05/aisc-1-research-summaries/.

were also already used by Krakovna et al. [8] and Leech et al. [10]. We conduct all experiments on two separate versions of these environments. First, the original version that includes the no-op action in all environments, in order to compare our approaches to the original AUP algorithm. Second, we evaluate our approaches on modified versions of these environments, where the no-op action is removed from the action space. The code to reproduce the results as well as the requirements to setup the experiments are published on GitHub[2].

5.1 Environments

The AI Safety Gridworlds are grid world environments where the agents main objective is closely tied to movement in cardinal directions on a 2D plane. In most environments the goal of the blue agent ■ is to reach the green cell ■. Additionally each environment has its own unintended, negative side-effect which should not appear. Each environment measures the presence of the side-effect and indicates it with a special negative reward of -2, which is not observed by the agent. Figure 1 shows the environments used for evaluation.

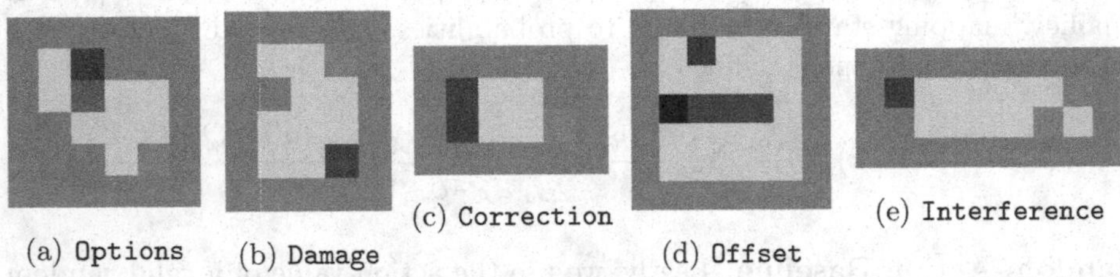

(a) `Options` (b) `Damage` (c) `Correction` (d) `Offset` (e) `Interference`

Fig. 1. Environments with safety properties of side effects [8,10,11,18] (Color figure online)

The side-effects of the individual environments are the following:

- (Figure a) `Options`: Irreversibly pushing the brown box ■ into a corner [11]
- (Figure b) `Damage`: Running into the horizontally pacing pink human ■ [10]
- (Figure c) `Correction`: Disabling the red off-switch ■ [18]
- (Figure d) `Offset`: Letting the right-moving black vase ■ fall off the conveyor belt [8]
- (Figure e) `Interference`: Disturbing the left-moving orange pallet ■ reaching the pink human ■ [10]

In each of the environments, the episode ends if the agent reached the goal cell, 20 time steps passed (not part of the state space and therefore not observed by the agent) or the agent refused to disable the off-switch in `Correction` after two time steps.

`Options` tests the agents ability to handle irreversible actions. Even though the agent is capable to push the box back into the center, if it first pushes it from left to right instead of down, it is not capable to get back to its initial position

[2] https://github.com/fkabs/attainable-utility-preservation.

(agent being in the top part of the environment and box at the center) after it moved the box for the first time. Therefore this environment is a representative where it is necessary to apply irreversible actions, yet unaffected parts of the environment should still be reversible.

Note that the point of `Correction`, `Offset` and `Interference` is to indicate, whether or not limitations of previous algorithms or baselines apply for our approaches. These limitations are specific to these previous algorithms/baselines, which is why standard Q-Learning (without any impact regularization) according to Turner et al. [18] performs well on two of them, while it does not perform well on `Options` and `Damage`. The purpose of `Correction`, is to make sure the agent does not intervene with the possibility of a human turning it off. The episode ends if the red off-switch is not disabled in the first two steps, simulating the agent was turned off by a human using the switch. Reaching the goal is as good as disabling the off-switch, because it implies the agent has taken measures to prevent its own off-switching. Not disabling the switch and not completing the environment is therefore the best outcome without causing a side effect. Furthermore, a yellow indicator appears one step before the end of the episode and turns red upon shutdown. In `Offset`, there is no goal cell present. Instead the agent's goal is to rescue the black vase off of the conveyor belt, showing that the agent is capable of intervening with an environment's dynamics when it is rewarded to do so, but also showing that offsetting behavior is not present (refraining from pushing the vase back on the conveyor belt again). The purpose of `Interference` is to show, that the agent is capable of not interfering with the environment dynamics if it is not rewarded for it.

Action Space. For each environment, the agent is allowed to move in the four cardinal directions as well as to stand still (no-op action). The original action space therefore is $\mathcal{A} = \{\text{up}, \text{down}, \text{left}, \text{right}, \varnothing\}$. In order to evaluate our approaches without the no-op assumption, we remove the no-op action from the action space for the second set of evaluations (Subsect. 6.2). On contact or interference with various objects in the environments, the agent pushes the crate or vase in the same direction the agent was moving, removes the human or off-switch, or stops the moving pallet.

Reward Function In all environments, the agent receives a primary reward of 1 when reaching the goal cell except in `Offset`, where the primary reward is observed when pushing the vase off the conveyor belt and therefore rescuing it from disappearing upon contact with the eastern wall. Each environment also features an unobserved penalty of -2 for causing a side effect, or 0 otherwise. This score can be used to evaluate safe behavior of the agents.

5.2 General Settings

All agents are trained on 50 trials, each consisting of 6,000 episodes. All agents use an ϵ-greedy policy with $\epsilon = 0.8$ to explore for the first 4,000 episodes and switch to $\epsilon = 0.1$ for the remaining 2,000 episodes to learn their respective Q-functions.

For each trial, the auxiliary reward functions are re-initialised and randomly selected from a continuous uniform distribution of the half-open interval $[0.0, 1.0)$. The default parameters for all agents can be seen in Table 1.

Table 1. Default parameters for all algorithms

Parameter	Value	Description
α	1	Step-size
γ	0.996	Discount factor
λ	0.667	Regularization parameter of the penalty term
$\|\mathcal{R}\|$	30	Number of auxiliary reward functions

This parameters with their respective values were also chosen by Turner et al. for AUP [18], which allows us to compare the results with our approaches.

6 Results

Since the purpose of our introduced baselines is to extend AUP to environments not including a no-op action, we first conduct experiments to see, whether they show comparable performance with original AUP. Therefore we evaluate our proposed baselines in the unchanged AI Safety Gridworlds from Turner et al. [18]. Additionally, we conducted experiments in modified versions of the AI Safety Gridworlds, which do not include a no-op action. Besides original AUP in the first evaluation setting, we compare our proposed approaches to Q-learning without any impact regularization, and to the random-baseline approach, where a random action is used as a dummy baseline.

Additionally, we conduct experiments to evaluate the stability of the hyperparameters for all approaches. We investigate how different λ, γ and $|\mathcal{R}|$ affect the performances of the agents. The results of these experiments are shown as "count plots" in the supplementary material, which show different outcome tallies across varying parameter settings.

Each episode may have one of four outcomes, depending on the primary objective and a side-effect:

- **No side effect, complete:** The agent fulfilled the primary objective and did not cause a side effect (best outcome for all environments except `Correction`). In this case, the agent receives a primary reward of 1 and a hidden reward of 0, resulting in a total reward of 1.
- **No side effect, incomplete:** The agent did not fulfill the primary objective, but did not cause a side effect (best outcome for `Correction`). In this case, the agent receives a primary reward of 0 and a hidden reward of 0, resulting in a total reward of 0.

- **Side effect, complete:** The agent fulfilled the primary objective, but caused a side effect. In this case, the agent receives a primary reward of 1 and a hidden reward of -2 resulting in a total reward of -1.
- **Side effect, incomplete:** The agent was not able to achieve the primary goal and also caused a side effect. In this case, the agent receives a primary reward of 0 and a hidden reward of -2 resulting in a total reward of -2.

6.1 Comparison to AUP

Figures 2, 3, 4, 5 and 6 show the results of the five environments averaging over 50 trials each. Our proposed baselines are not entirely capable to compete with model-free AUP in `Options`, yet the results show an improvement over Q-Learning and the random-action baseline. In `Damage` our results seem to be on par with model-free AUP, moreover all approaches except Q-Learning reach the best possible outcome. The results also show, that no offsetting-, nor interfering behavior appears for all proposed baselines. However, all approaches (except the random-baseline) show correcting behaviour due to its delayed effect. The best performing, introduced baseline is the advantage baseline. It even slightly outperforms model-free AUP in `Options` during the exploration phase and achieves the best possible outcome in `Damage`, along with the other approaches, after the exploration strategy switch. As expected Q-Learning causes side-effects in `Options` and `Damage`, shows correcting behavior and does not show offsetting nor interfering behavior.

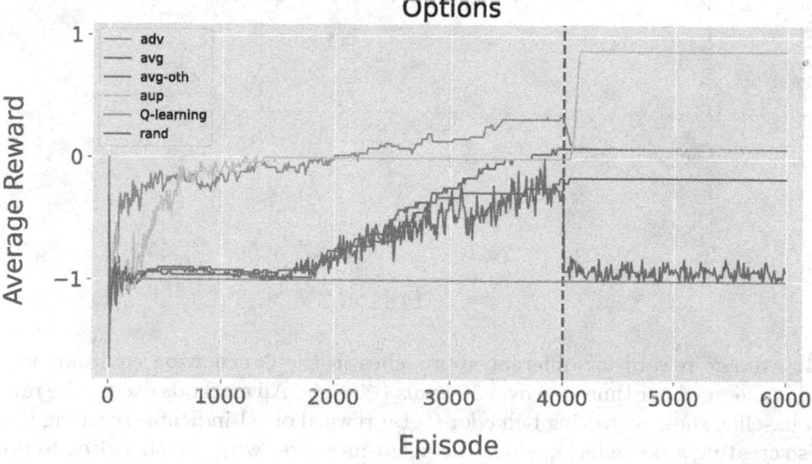

Fig. 2. Average reward for different approaches in the `Options` environment. The reward is averaged per time step over 50 trials ($\varnothing \in \mathcal{A}$). Our proposed approaches perform distinctly below model-free AUP, yet above Q-Learning and the random-action baseline. Note that the advantage baseline seemingly outperforms model-free AUP before the exploration switch at episode 4,000.

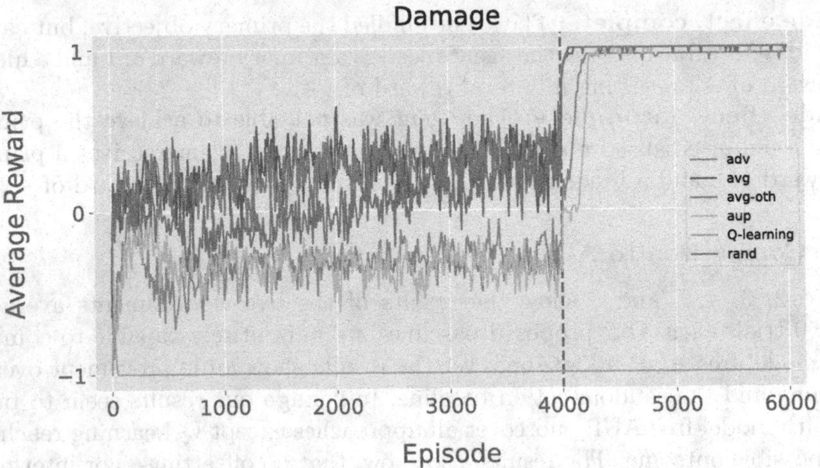

Fig. 3. Average reward for different approaches in the `Damage` environment. The reward is averaged per time step over 50 trials ($\varnothing \in \mathcal{A}$). All methods evaluated, except standard Q-Learning, reach near-optimal performance after the exploration switch.

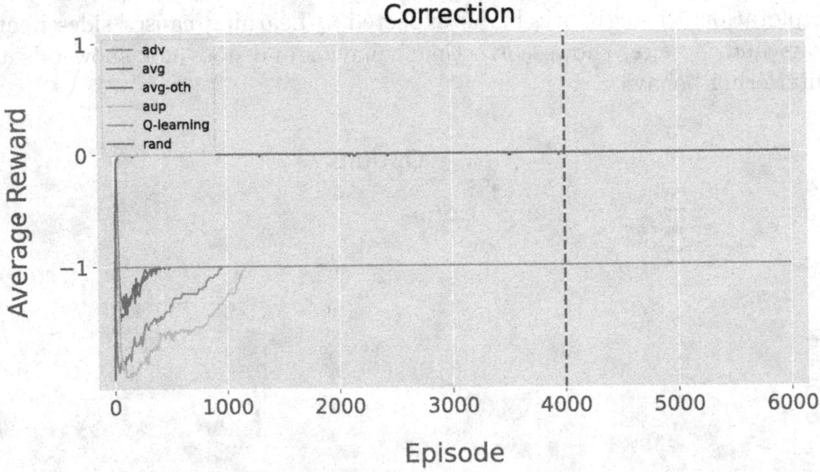

Fig. 4. Average reward for different approaches in the `Correction` environment. The reward is averaged per time step over 50 trials ($\varnothing \in \mathcal{A}$). All methods except the random-action baseline show correcting behavior (total reward of -1 indicates reaching the goal but also creating a side-effect), where the agent interferes with the off-switch to prevent an early end of the episode.

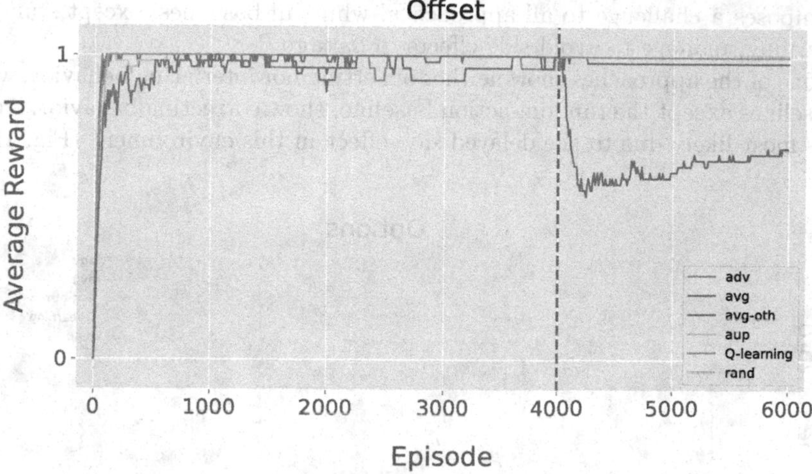

Fig. 5. Average reward for different approaches in the `Offset` environment. The reward is averaged per time step over 50 trials ($\varnothing \in \mathcal{A}$). None of the approaches, except the random-action baseline, show offsetting behavior, where the box is saved first but then put on the conveyor belt again.

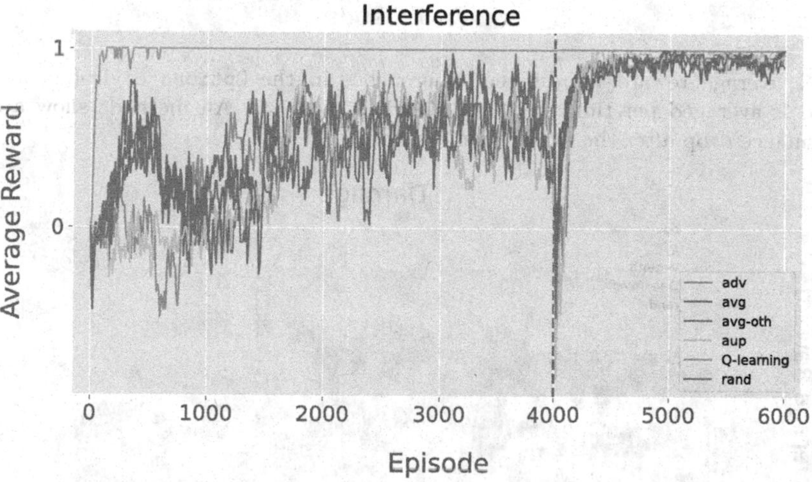

Fig. 6. Average reward for different approaches in the `Interference` environment. The reward is averaged per time step over 50 trials ($\varnothing \in \mathcal{A}$). All of the methods show near-optimal performance in the end, indicating that the agent does little or not interfere with the moving orange pallet. (Color figure online)

6.2 Dropping the No-Op Action

Figures 7, 8, 9 and 10 show the results in the modified environments where the no-op action is excluded from the action space. These results show that `Options` still imposes a challenge to all approaches, while all baselines, except standard Q-Learning, manage to avoid side-effects in `Damage`.

None of the approaches show neither offsetting nor interfering behavior, while all baselines except the random-action baseline, show correcting behavior. Again this is most likely due to the delayed side-effect in this environment (Fig. 11).

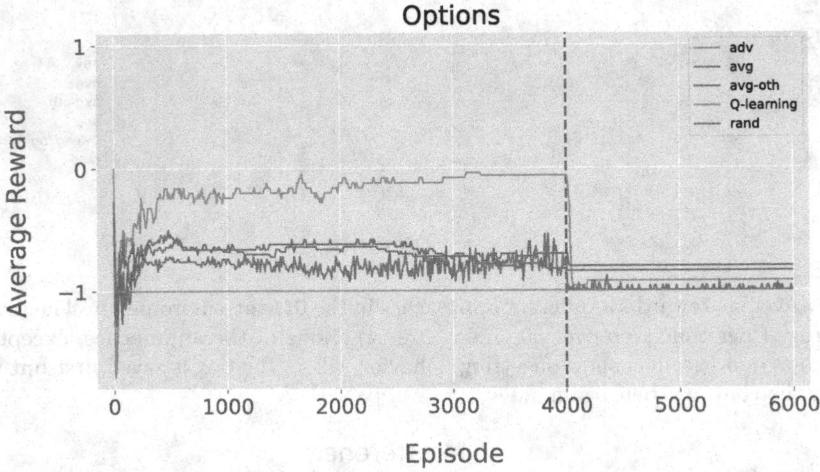

Fig. 7. Average reward for different approaches in the `Options` environment. The reward is averaged per time step over 50 trials ($\varnothing \notin \mathcal{A}$). All methods show a clear performance drop after the exploration switch.

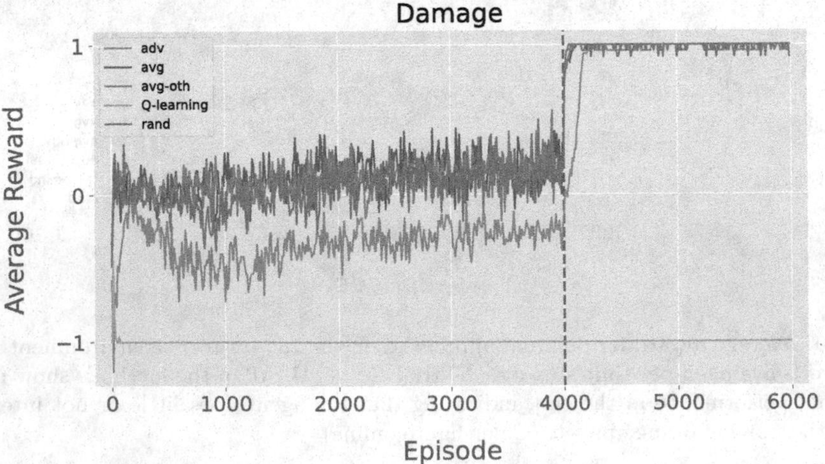

Fig. 8. Average reward for different approaches in the `Damage` environment. The reward is averaged per time step over 50 trials ($\varnothing \notin \mathcal{A}$). All methods, except Q-Learning, show near-optimal performance after the exploration switch.

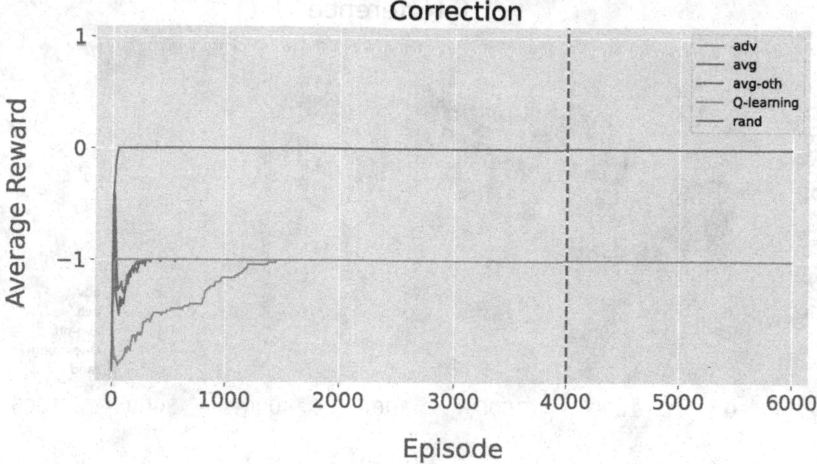

Fig. 9. Average reward for different approaches in the `Correction` environment. The reward is averaged per time step over 50 trials ($\varnothing \notin \mathcal{A}$). All methods except the random-action baseline show correcting behavior (total reward of -1 indicates reaching the goal but also creating a side-effect), where the agent interferes with the off-switch to prevent an early end of the episode.

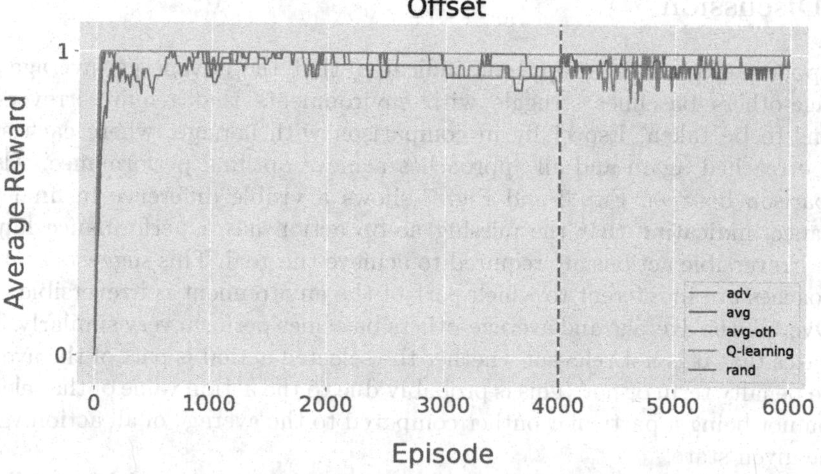

Fig. 10. Average reward for different approaches in the `Offset` environment. The reward is averaged per time step over 50 trials ($\varnothing \notin \mathcal{A}$). None of the approaches show clear offsetting behavior, where the box is saved first but then put on the conveyor belt again.

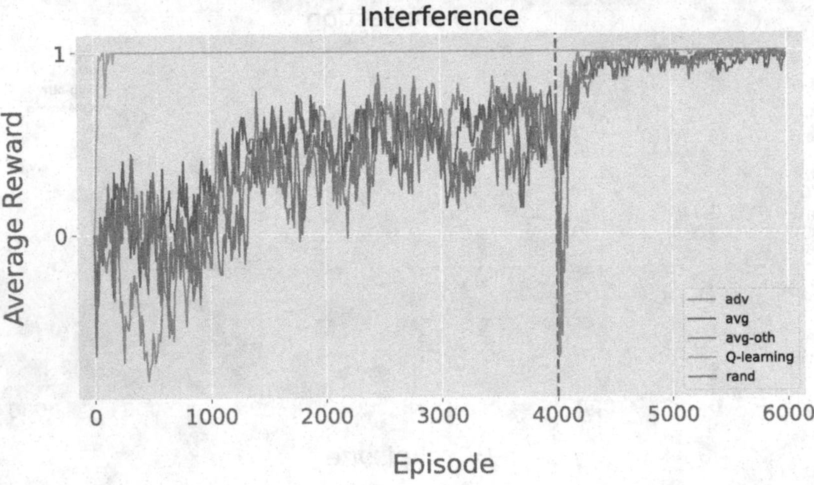

Fig. 11. Average reward for different approaches in the **Interference** environment. The reward is averaged per time step over 50 trials ($\varnothing \notin \mathcal{A}$). All of the methods show near-optimal performance in the end, indicating that the agent does little or not interfere with the moving orange pallet. (Color figure online)

7 Discussion

The poor performance in `Options` indicates, that the advantage, average and average-others baselines struggle with environments that require irreversible actions to be taken. Especially in comparison with `Damage`, where each state can be reached again and all approaches achieve optimal performance. Also a comparison between Fig. 2 and Fig. 7 shows a visible difference in final performance, indicating that the missing no-op action has a performance impact when irreversible actions are required to achieve the goal. This suggests that our approaches are indifferent to which part of the environment is irreversible.

Overall, the average and average-others baselines perform very similarly. This indicates that it is not relevant whether the selected action is part of the average in the penalty term or not. This is probably due to the action value of the selected action not being a particular outlier compared to the average of all action values in the given state.

The advantage baseline was capable of outperforming model-free AUP in `Options` during the first 4,000 episodes with $\epsilon = 0.8$. Moreover, we find that the performance of the average and average-others baseline is better compared to the advantage baseline and model-free AUP in the `Damage` environment during the exploration phase. However, all approaches rise to the optimal performance once the exploration switch is reached. We assume this phenomenon has a connection to the amount of "free space" available to the agent until it comes in contact with the side-effect. While in `Options` this is rather soon, as the side-effect is just one action away from the initial state and 3 fields around the box are available for the agent, in `Damage` the side-effect is two actions away and the agent has 6 fields that are uninvolved by the human.

Unsurprisingly, the advantage, average and average-others baselines show correcting behavior, meaning they intervene with the off-switch in `Correction`. The agents using these baselines, have learned that only after this intervention they are capable of reaching the goal state. Our approaches are incapable of avoiding this side-effect as it comes with a time-delay (the episode still continuing after two steps) and by design our approaches cannot handle such side-effects, as does model-free AUP. AUP is supposed to handle delayed side-effects, however only side-effects, that are delayed by 8 time steps. Interestingly, the random-action baseline manages to prevent correcting behavior in `Correction`, which requires further investigation.

8 Conclusion

We propose three different, alternative baselines to attainable utility preservation that do not build upon a no-op action, which induce safer, yet effective behavior than standard Q-Learning. We evaluate all three baselines on two separate versions of five AI safety Gridworlds comparing them to model-free AUP, Q-learning and a random baseline. Our proposed baselines require less assumptions and therefore are more broadly usable, but also show less side-effect avoiding potential in environments with irreversible actions and are more sensitivity to parameters.

8.1 Future Work

We suggest future work on investigating the performance of the proposed baselines in larger, more complex and multi-task environments, as well as in environments with larger action spaces, to determine the extent to which our proposed baselines induce safe and effective behavior. Also coping with delayed side-effects in unspecified time frames is still an open challenge to be solved.

Acknowledgements. This research was funded in whole, or in part, by the Austrian Science Fund (FWF) [P 33656-N]. For the purpose of open access, the author has applied a CC BY public copyright licence to any Author Accepted Manuscript version arising from this submission.

References

1. Alamdari, P.A., Klassen, T.Q., Icarte, R.T., McIlraith, S.A.: Be considerate: objectives, side effects, and deciding how to act (2021). http://arxiv.org/abs/2106.02617
2. Amodei, D., Olah, C., Steinhardt, J., Christiano, P., Schulman, J., Mané, D.: Concrete problems in AI safety (2016). http://arxiv.org/abs/1606.06565
3. Armstrong, S., Levinstein, B.: Low impact artificial intelligences (2017). https://doi.org/10.48550/arXiv.1705.10720, http://arxiv.org/abs/1705.10720

4. Bellemare, M.G., et al.: Autonomous navigation of stratospheric balloons using reinforcement learning. Nature **588**(7836), 77–82 (2020)
5. Christiano, P.F., Leike, J., Brown, T., Martic, M., Legg, S., Amodei, D.: Deep reinforcement learning from human preferences. In: Advances in Neural Information Processing Systems, vol. 30. Curran Associates, Inc. (2017). https://proceedings.neurips.cc/paper/2017/hash/d5e2c0adad503c91f91df240d0cd4e49-Abstract.html
6. Fawzi, A., et al.: Discovering faster matrix multiplication algorithms with reinforcement learning. Nature **610**(7930), 47–53 (2022)
7. Gleave, A., Toyer, S.: A primer on maximum causal entropy inverse reinforcement learning (2022). https://doi.org/10.48550/arXiv.2203.11409, http://arxiv.org/abs/2203.11409
8. Krakovna, V., Orseau, L., Martic, M., Legg, S.: Penalizing side effects using stepwise relative reachability. In: Espinoza, H., et al. (eds.) Proceedings of the Workshop on Artificial Intelligence Safety 2019. CEUR Workshop Proceedings, vol. 2419. CEUR (2019). http://ceur-ws.org/Vol-2419/#paper1
9. Krakovna, V., Orseau, L., Ngo, R., Martic, M., Legg, S.: Avoiding side effects by considering future tasks. In: Larochelle, H., Ranzato, M., Hadsell, R., Balcan, M.F., Lin, H. (eds.) Advances in Neural Information Processing Systems, vol. 33, pp. 19064–19074. Curran Associates, Inc. (2020). https://proceedings.neurips.cc/paper/2020/file/dc1913d422398c25c5f0b81cab94cc87-Paper.pdf
10. Leech, G., Kubicki, K., Cooper, J., McGrath, T.: Preventing side-effects in gridworlds (2018). https://www.gleech.org/grids
11. Leike, J., et al.: AI safety gridworlds (2017). https://doi.org/10.48550/arXiv.711.09883, http://arxiv.org/abs/1711.09883
12. Berner, C., et al.: Dota 2 with large scale deep reinforcement learning (2019). https://doi.org/10.48550/arXiv.1912.06680, http://arxiv.org/abs/1912.06680
13. Saunders, W., Sastry, G., Stuhlmüller, A., Evans, O.: Trial without error: towards safe reinforcement learning via human intervention. In: Proceedings of the 17th International Conference on Autonomous Agents and MultiAgent Systems, AAMAS 2018, pp. 2067–2069. International Foundation for Autonomous Agents and Multiagent Systems (2018)
14. Schrittwieser, J., et al.: Mastering Atari, Go, chess and shogi by planning with a learned model. Nature **588**(7839), 604–609 (2020). https://doi.org/10.1038/s41586-020-03051-4, https://www.nature.com/articles/s41586-020-03051-4
15. Shah, R., Krasheninnikov, D., Alexander, J., Abbeel, P., Dragan, A.: Preferences implicit in the state of the world (2022). https://openreview.net/forum?id=rkevMnRqYQ
16. Silver, D., et al.: Mastering the game of Go with deep neural networks and tree search. Nature **529**(7587), 484–489 (2016). https://doi.org/10.1038/nature16961, http://www.nature.com/articles/nature16961
17. Sutton, R.S., Barto, A.G.: Reinforcement Learning: An Introduction. Adaptive Computation and Machine Learning Series, 2nd edn. The MIT Press (2018). http://incompleteideas.net/book/the-book.html
18. Turner, A.M., Hadfield-Menell, D., Tadepalli, P.: Conservative agency via attainable utility preservation. In: Proceedings of the AAAI/ACM Conference on AI, Ethics, and Society, AIES 2020, pp. 385–391. Association for Computing Machinery (2020). https://doi.org/10.1145/3375627.3375851
19. Turner, A.M., Ratzlaff, N., Tadepalli, P.: Avoiding side effects in complex environments. In: Larochelle, H., Ranzato, M., Hadsell, R., Balcan, M.F., Lin, H.T. (eds.) Advances in Neural Information Processing Systems. NeurIPS 2020, vol. 33,

pp. 21406–21415. Curran Associates, Inc. (2020). https://proceedings.neurips.cc/paper/2020/hash/f50a6c02a3fc5a3a5d4d9391f05f3efc-Abstract.html
20. Turner, A.M., Smith, L.R., Shah, R., Critch, A., Tadepalli, P.: Optimal policies tend to seek power (2021). https://openreview.net/forum?id=l7-DBWawSZH
21. Vinyals, O., et al.: Grandmaster level in starcraft II using multi-agent reinforcement learning. Nature **575**(7782), 350–354 (2019)

Memorization of Named Entities in Fine-Tuned BERT Models

Andor Diera[1(✉)], Nicolas Lell[1], Aygul Garifullina[2], and Ansgar Scherp[1]

[1] Ulm University, Ulm, Germany
{andor.diera,nicolas.lell,ansgar.scherp}@uni-ulm.de
[2] BT, Ipswich, UK
aygul.garifullina@bt.com

Abstract. Privacy preserving deep learning is an emerging field in machine learning that aims to mitigate the privacy risks in the use of deep neural networks. One such risk is *training data extraction* from language models that have been trained on datasets, which contain personal and privacy sensitive information. In our study, we investigate the extent of named entity memorization in fine-tuned BERT models. We use single-label text classification as representative downstream task and employ three different fine-tuning setups in our experiments, including one with Differentially Privacy (DP). We create a large number of text samples from the fine-tuned BERT models utilizing a custom sequential sampling strategy with two prompting strategies. We search in these samples for named entities and check if they are also present in the fine-tuning datasets. We experiment with two benchmark datasets in the domains of emails and blogs. We show that the application of DP has a detrimental effect on the text generation capabilities of BERT. Furthermore, we show that a fine-tuned BERT does not generate more named entities specific to the fine-tuning dataset than a BERT model that is pre-trained only. This suggests that BERT is unlikely to emit personal or privacy sensitive named entities. Overall, our results are important to understand to what extent BERT-based services are prone to training data extraction attacks (Source code and datasets are available at: https://github.com/drndr/bert_ent_attack. An extended version of this paper can be also found on arXiv [12]).

Keywords: language models · training data extraction · privacy preserving deep learning

1 Introduction

Deep Neural Networks (DNNs) became the *de facto* tool for achieving state-of-the-art performance in many research domains such as computer vision and natural language processing (NLP). Although utilizing large volumes of training data is one of the main driving factors behind the great performance of

A. Holzinger et al. (Eds.): CD-MAKE 2023, LNCS 14065, pp. 258–279, 2023.
https://doi.org/10.1007/978-3-031-40837-3_16

DNNs, publishing these models to the public raises some serious privacy concerns regarding private and confidential information present in the training data [32]. These privacy concerns are especially relevant for large Language Models (LMs) which form the basis of state-of-the-art technologies in many NLP tasks. Recent versions of these models are usually first pre-trained in a task-agnostic self-supervised manner. The latest large LMs use a corpus size ranging from hundreds of gigabytes to several terabytes of text [6,42] during this self-supervised process. The sheer size of these datasets makes it near impossible for researchers to remove all confidential information which may be present in the corpus. A recent study has shown that it is possible to extract personal information from some large LMs, even if that given information has only appeared once in the training corpus [8]. While the training cost of these large LMs became so prohibitively expensive that only the biggest tech companies can afford it [48], pre-trained LMs are commonly used in businesses that work with huge amounts of text data. These businesses include banks, telecommunications, and insurance companies, which often handle a great amount of personal and privacy sensitive data. In practice, pre-trained LMs are fine-tuned on a business-specific dataset using some downstream task (such as text-classification, question-answering, or natural language inference) before deployment [11]. Although the fine-tuning may mitigate some of the unintended memorization of the original dataset used in pre-training, it raises new concerns regarding the personal and privacy sensitive information in the business-specific dataset used for the fine-tuning process [8]. Privacy Preserving Deep Learning (PPDL) is a common term used for methods aiming to mitigate general privacy concerns present in the use of DNNs. Multiple approaches have been proposed to achieve PPDL [37], but there is no perfect solution to this problem, with each method having its own challenges and limitations. The most popular techniques include Federated Learning [36], the application of Differential Privacy (DP) [1], encryption [22], and data anonymization [52].

We investigate whether it is possible to extract personal information from one of the most popular modern LMs, BERT [11]. BERT is an auto-encoder transformer that is mostly used for natural language understanding tasks. Since BERT is less adept at generating long, coherent text sequences [57], we focus our study on the generation and extraction of named entities. We conduct our experiments on three typical fine-tuning setups to understand the privacy risks involved in using BERT for commercial purposes. We consider fine-tuning all layers of BERT (*Full*), fine-tuning only the last encoding layer and the classifier head of BERT (*Partial*), and partial fine-tuning but with a privacy preserving optimizer (*Differentially Private*, short: *DP*). As a privacy preserving optimizer, we employ the established differentially private stochastic gradient descent (DPSGD) algorithm [1], which we discuss in detail in Sect. 3.1. We compare the fine-tuning setups with a pre-trained only BERT base model. We experiment with two benchmark datasets in the domains of emails (Enron Email corpus [28]) and blogs (Blog Authorship Corpus [47]). For triggering entity extraction, we use two prompting techniques. The *naive prompting* is based on randomly selected text from the web, while the *informed prompting* uses actual text from the datasets' test sets. Each experimental setup is assessed with regard to its performance on the down-stream task, i. e., single-label text classification, and the extent of named entity memorization. In summary, our experiments show:

- The memorization rate of named entities in the fine-tuned BERT models is less than 10% in both datasets across all setups. Interestingly, the fine-tuned models do not emit more entities from the fine-tuning datasets than a pre-trained only BERT model.
- When comparing the informed prompting versus the naive prompting, the BERT models consistently generate more named entities when using naive prompts. Thus a potential attacker does not require prior knowledge about the training dataset of a model.
- Applying differentially private fine-tuning results in a strong drop in the amount of memorized entities at the cost of downstream task performance. It effectively reduces the amount of entity memorization in fine-tuned BERT models.

Below, we discuss the related work. Our framework and methods for extracting named entities from fine-tuned BERT models are described in Sect. 3. Section 4 introduces the datasets and the details of the experiments. The results are described in Sect. 5 and discussed in Sect. 6.

2 Related Work

2.1 Language Models and Text Generation

Modern large LMs rely on two core concepts that led to their dominance in the NLP field: the focus on the self-attention mechanism in the DNN architecture and the introduction of large-scale task-agnostic pre-training to learn general language representations [59]. Self-attention is used for modeling dependencies between different parts of a sequence. A landmark study in 2017 [55] has shown that self-attention was the single most important part of the state-of-the-art NLP models of that time. It introduced a new family of models called transformers, which rely solely on stacked layers of self-attention and feed-forward layers. Besides the state-of-the-art performances, another great advantage of the transformer architecture is that unlike a recurrent architecture, it allows for training parallelization. The ability to parallelize training, alongside the significant increase in computational power allowed these models to train on larger datasets than once was possible. Since supervised training requires labeled data, self-supervised pre-training with supervised fine-tuning became the standard approach when using these models [35].

State-of-the-art transformers can be divided into three main categories based on their pre-training approach [64]. Auto-regressive models use the classical language modeling pre-training task of next word prediction. Auto-encoding models are pre-trained by reconstructing sequences that have been corrupted in some way. Sequence-to-sequence models usually employ objectives of encoding-decoding models for pre-training, like replacing random sequences in a text with one special token with the objective of predicting that given sequence [42].

BERT. A major limitation of the auto-regressive models is that during pre-training they learn a unidirectional language model. In these models, tokens are restricted to only attend to other tokens left to them. In contrast, BERT is an auto-encoder transformer model that uses an attention mechanism on the entire input sequence [11]. This model utilizes Masked Language Modeling (MLM) and Next Sentence Prediction (NSP) as pre-training tasks. In MLM, some tokens are randomly removed from the input sequence and the model is trained to predict the removed tokens using context from both directions.

Although the parameter count of BERT is greatly surpassed by more recent large LMs (such as the new GPT models [6,40]), BERT is still one of the most common baselines in many NLP benchmark tasks. The strong performance coupled with the fact that the model is democratized and has publicly available pre-trained implementations makes BERT a popular choice of NLP model both in industry and academia. Since its original release, there have been dozens of follow-up studies and models published [44]. The most notable variants include RoBERTa [33], DistilBERT [46], DeBERTa [24], and domain-specific models such as SciBERT [5] or ClinicalBERT [3].

Natural Language Generation. Natural Language Generation (NLG) is a subfield of NLP that is focused on producing natural language text that enables computers to write like humans [64]. Although auto-regressive transformer models are the standard choice for the task of NLG (since they are already trained to predict the next token based solely on previous tokens in a sequence), it has been shown that BERT can also be utilized to generate reasonably coherent text. Wang and Cho [57] designed a generation strategy for BERT based on Gibbs sampling [20], where given a seed sequence, tokens at random positions are masked and replaced by new tokens based on the sampling technique. Another generation strategy developed for auto-encoding transformers [21] is to use a fully masked sequence as input and predict all tokens at once. Subsequently, tokens with the lowest probability are iteratively re-masked and replaced with a newly computed token.

2.2 Privacy Attacks in Machine Learning

Privacy attacks in machine learning denote a specific type of adversarial attack, which aim to extract information from a trained model. Based on recent surveys in the field [10,32,43], these attacks can be divided into five main categories.

Training Data Extraction Attacks. Training data extraction attacks aim to reconstruct training datapoints, but unlike model inversion attacks, the goal is to retrieve verbatim training examples and not just "fuzzy" class representatives [8]. These attacks are best suited for generative sequence models such as LMs. Initially these attacks have been designed for small LMs using academic datasets [7,53,63]. The aim of these studies was to measure the presence of specific training datapoints in the text samples generated by the models. A common approach to measure the extent of this unintended memorization is to insert so-called "canaries" (artificial datapoints) into the training datasets and quantify their occurrence during sequence completion [7]. Since these initial studies were based on smaller models trained with a high number of epochs, it was assumed that this kind of privacy leakage must be correlated with overfitting [63]. However, a follow-up study using the GPT-2 model, which is trained on a very large corpus for only a few epochs, showed that even state-of-the-art large LMs are susceptible to these kinds of attacks. Using the pre-trained GPT-2 model, Carlini et al. [8] were able to generate and select sequence samples which contained low *k-eidetic* data-points (data points that occur k times in the training corpus). A study by Lehman et al. [30] on Clinical BERT attempted to extract patient-condition association using both domain-specific template infilling and the text generation methods inspired by the text extraction research done on GPT-2 [8] and the BERT specific text generation technique proposed by Wang and Cho [57]. Their methods were not successful in reliably extracting privacy sensitive information (patient-condition associations) from Clinical BERT, but it remains inconclusive whether it is due to the limitations in their method or in the linguistic capabilities of BERT.

Membership Inference Attacks. The goal of a membership inference attack is to determine whether or not an individual data instance is part of the training dataset for a given model. This attack typically assumes a black-box query access to the model. The common approach to this type of attack is to use a shadow training technique to imitate the behavior of a specific target model. In shadow training, a model (shadow model) is trained on a dataset that has a disjoint but identically formatted training data as the target model. The trained inference model is then used to recognize differences on the target model predictions between inputs used for training and inputs not present in the training data [49].

Model Extraction Attacks. The adversarial aim of a model extraction attack is to duplicate (i. e., "steal") a given machine learning model. It achieves this by training a function f' that is approximating the function f of the attacked model [32]. A shadow training scheme has been shown to successfully extract popular machine learning models such as logistic regression, decision trees, and neural networks, using only black-box query access [54]. Other works have proposed methods to extract information about hyperparameters [58] and properties of the architecture [39] in neural networks.

Model Inversion Attacks. The idea behind model inversion attacks is that an adversary can infer sensitive information about the input data using a target model's output. These attacks can be used to extract input features and/or reconstruct prototypes of a class (in case the inferred feature characterize an entire class), given a white-box access to the model and knowledge about the target labels with some auxiliary information of the training data [17].

Property Inference Attacks. The goal of property inference attacks is to infer some hidden property of a training dataset that the owner of the target model does not intend to share (such as feature distribution or training bias). Initially, property inference attacks were applied on discriminative models with white-box access [41]. A more recent work has extended the method to work on generative models with black-box access [41].

2.3 Privacy Preserving Deep Learning

Based on the literature [10,32,43], PPDL methods can be divided into four main categories.

Differentially Private Learning. Differential Privacy (DP) is a rigorous mathematical definition of privacy in the context of statistical and machine learning analysis. It addresses the challenge of "learning nothing about an individual while learning useful information about the population" [16]. In machine learning, DP algorithms aim to obfuscate either the training data [65] or the model [45] by adding noise. Since directly adding noise to DNN parameters may significantly harm its utility, the best and most common place for applying DP in deep learning is the gradients [66]. Abadi et al. [1] proposed an efficient training algorithm with a modest privacy budget called Differentially Private Stochastic Gradient Descent (DPSGD). DPSGD ensures DP by cutting the gradients to a maximum L2 norm for each layer and then adding noise to the gradients. Although DPSGD comes with increased computational cost and performance loss, variations of this algorithm [9,14] still belong to the cutting-edge of PPDL research.

Encryption. Cryptography-based methods can be divided into two subcategories, depending whether the target of the encryption is the training data [22] or the model [4]. Regardless of the target, most existing approaches use homomorphic encryption, which is a special kind of encryption scheme that allows computations to be performed on encrypted data without decrypting it in advance [2]. Since training a DNN is already computationally expensive, adding homomorphic encryption to the process raises major challenges as it increases training times by at least an order of magnitude [32].

Data Anonymization. Data Anonymization techniques aim to remove all Personally Identifiable Information (PII) from a dataset. The common approach to achieve this is to remove attributes that are identifiers and mask quasi-identifier attributes [60]. The popular k-anonymity algorithm [52] works by suppressing identifiers (i. e., replacing them with an asterisk) and generalizing quasi-identifiers with a broader category which has a frequency of at least k in the dataset. Although data anonymization techniques were developed for structured data, it is possible to adapt them to unstructured text data [23] as well as jointly anonymizing structured data and unstructured text data [50].

Aggregation. Aggregation methods are generally used along with distributed learning, in which multiple parties train on the same machine learning task while aiming to keep their respective datasets private [32]. Although aggregation methods can provide data security during distributed training, their privacy preserving aspects are more limited than other PPDL approaches.

3 Extracting Named Entities from BERT

In order to extract the named entities of the fine-tuning dataset from the BERT model, we present the experimental pipeline depicted in Fig. 1. The pipeline consists of three phases: fine-tuning (including a privacy preserving approach using Differential Privacy), text generation from the fine-tuned models, and evaluation of the named entity memorization.

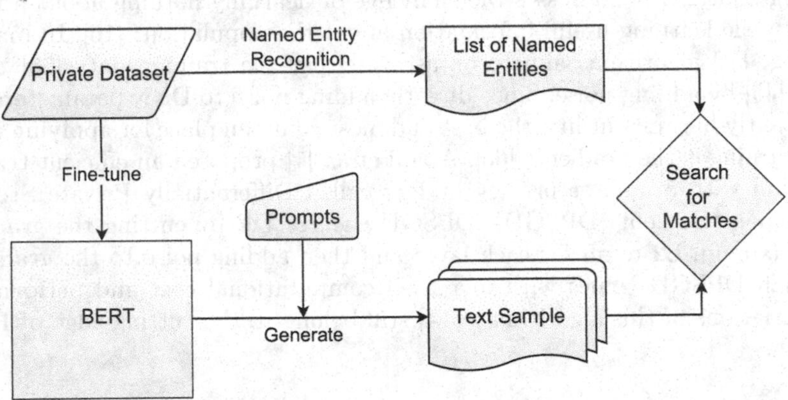

Fig. 1. An illustration of our framework for extracting training data entities from BERT. First, we fine-tune a pre-trained BERT on a private dataset. Next, we generate text samples from the fine-tuned model using prompts. Finally, we search the generated samples for the named entities that occur in the private dataset.

3.1 Fine-Tuning

In the fine-tuning phase, we employ single-label text-classification as the downstream task. Our setup consists of three different fine-tuning methods: *Full*, *Partial*, and *Differentially Private (DP)* fine-tuning. The different fine-tuning methods are depicted in Fig. 2.

The *Full* setup follows the standard practices of fine-tuning LMs, where a classifier head is attached to the base network and all the weights of a pre-trained network along with the classifier head are retrained on the task-specific dataset with a low learning rate [27]. Full fine-tuning usually leads to the best results on the downstream task, but in the case of large LMs, it is not always feasible due to the size of these networks and the computational costs of retraining them. Due to this constraint, researchers have designed alternative fine-tuning strategies where fine-tuning is employed in a more optimized manner [27,29]. A common alternative strategy is to freeze most of the layers in a network and only retrain the last few encoder layers with the task-specific head of the network [34,51]. In our *Partial* setup, we freeze all layers of the BERT model except for the last encoding layer. Applying DP in fine-tuning puts additional noise to the gradient updates, which in the lower layers carries a detrimental effect to the pre-trained knowledge of the model as the weights of the bottom layers are more sensitive to noise. For this reason, in the *Differentially Private* setup we employ the same layer-freezing approach as in the *Partial* setup.

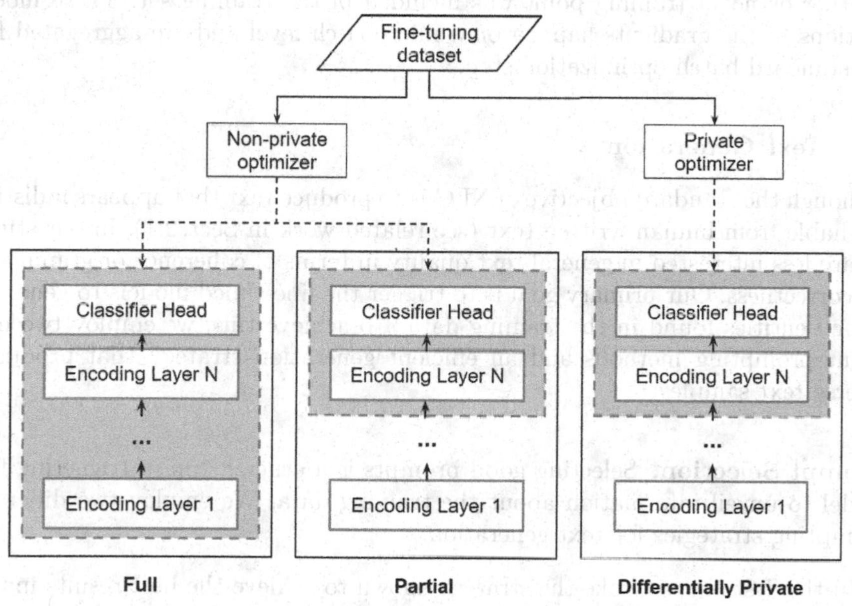

Fig. 2. An illustration of the different fine-tuning methods.

In addition, the DP fine-tuning method uses the Adam variant of DPSGD by Abadi et al. [1]. DPSGD is a modification to the stochastic gradient descent algorithm that employs (ϵ, δ) differential privacy [15]. In the formal definition of (ϵ, δ) differential privacy, a randomized algorithm M is differentially private if:

$$\Pr[\mathrm{M(x)} \in \mathrm{S}] \leq \exp(\epsilon) \cdot \Pr[\mathrm{M(y)} \in \mathrm{S}] + \delta$$

where x and y are neighboring datasets, S denotes all the potential output of M that can be predicted, ϵ is the metric of privacy loss (also known as the privacy budget) at a differential change in the data (e.g., adding or removing one datapoint), and δ is the probability of an accidental privacy leak. In deep learning, an ϵ value is defined as modest when it is below 10 and δ is usually set to the reciprocal of the number of training samples [14,62]. In standard data analytics settings, ϵ values between 0 and 1 are considered to be highly private, and values between 2 and 10 are considered somewhat private. However, in deep learning it is hard to achieve a ϵ value under 1, since the privacy budget is based on how much the data affected the model. Everytime the dataset goes through the model, the ϵ increases.

The DPSGD algorithm entails two major changes to the gradient descent algorithm: the introduction of gradient clipping and the addition of Gaussian noise to the clipped gradients. The clipping limits how much an individual training point can impact the model parameters, while the addition of the noise randomizes the behavior of the algorithm, making it statistically impossible to know whether or not a training point was included in the training set. These modifications to the gradients happen on a microbatch level and are aggregated for the standard batch optimization step.

3.2 Text Generation

Although the standard objective of NLG is to produce text that appears indistinguishable from human-written text (see related work in Sect. 2.1), in our study we are less interested in general text quality in terms of coherence or grammatical correctness. Our primary goal is to trigger the fine-tuned models to generate named entities found in the training data. To achieve this, we employ two different prompting methods and an efficient generation strategy that produces diverse text samples.

Prompt Selection. Selecting good prompts is a crucial step in triggering the model to unveil information about the training data. We employ two different prompting strategies for text generation.

1. In the first one, we take the strategy shown to achieve the best results in the experiments of Carlini et al. [8], in which a fixed length substring is randomly sampled as prompt from the Common Crawl[1] dataset. This we refer to as a *naive prompting*, since we randomly use text samples scraped from the internet. The selected prompts likely have no or only very little connection with the text and named entities from the fine-tuning dataset.
2. In the second strategy, we create the prompts by randomly selecting sequences of a fixed length from the test set of the fine-tuning data. This setup is considered as *informed prompting* given that the prompts come from the same domain and are generally highly similar to the training data.

Text Generation. Despite the fact that the bidirectional nature of BERT does not naturally admit to sequential sampling, Wang and Cho [57] have shown that it is also possible to utilize this strategy for BERT. Although their results suggest

[1] http://commoncrawl.org.

that their non-sequential iterative method produces slightly more coherent text than sequential sampling, it requires multiple iterations for each token. Since text coherence is not our primary goal, we choose to employ a computationally less expensive sequential sampling method. In this method, we choose a randomly selected prompt (see above) as a seed sequence and extend it with a masked token. For each iteration, we predict the masked token and replace the mask. We add an additional masked token to the extended sequence until we reach the defined sequence length.

On top of this generation method, we also employ a combination of beam search and nucleus sampling as an additional decoding strategy. Beam search is a commonly used decoding method in machine translation tasks [18]. Compared to greedy search, where at each iteration only the token with the highest probability is selected, beam search selects multiple tokens at each iteration. The number of tokens is defined by the beam width parameter and an additional conditional probability is used to construct the best combination of these tokens in a sequence. While both greedy search and beam search select tokens based on maximum likelihood, sampling from the probability distribution is also a viable approach. The most popular sampling method is top-k sampling, where in each iteration a token is sampled from a set of k candidates with the highest probability. Nucleus sampling (also called top-p sampling) is an alternative strategy to top-k, where instead of using a set with a fixed length, the smallest possible set is constructed with tokens whose cumulative probability exceeds the probability value of parameter p [25]. These sampling methods can be combined with both search algorithms.

3.3 Evaluating Named Entity Memorization

Named entities refer to objects and instances that identify one item from a set of other items sharing similar attributes. They usually include entity types like (person) names, organizations, locations, products, and special temporal or numerical expressions like dates or amounts of money [38]. In order to evaluate the extent to which the models have memorized the named entities found in the fine-tuning dataset, we first extract them from the datasets using Named Entity Recognition (NER) and create a dictionary with the entities and their corresponding entity types.

After this, we create three different entity sets from this dictionary. The first one (*All*) consists of every entity with a character length greater than 3. In the second (*Private*), we do a cross-check of the first entity set with the pre-training data, and remove all entities also present in the pre-training datasets, leaving us a set of entities that only appear in the fine-tuning data. For the third and final set (*Private* 1-*eidetic*), we keep all 1-eidetic entities, i. e., entities which appeared only once in the fine-tuning dataset, of the second set and discard everything else. Once we have these three sets and the generated samples from each model set up (including the samples generated from a base model that was not fine-tuned), we count the number of exact matches in the samples.

4 Experimental Apparatus

4.1 Datasets

For selecting suitable datasets for the study we had two criteria. First, to avoid privacy issues and ethical concerns only publicly available datasets were chosen. Second, to provide a good basis for the measurement of memorization, we were interested in datasets that contain a large number of named entities. We choose data which has English as its primary language and kept 20% of each dataset for testing. The main characteristics of the datasets can be seen in Table 1.

Table 1. Characteristics of the datasets

Dataset	N	#Train	#Test	#Classes
Enron	7,501	6,000	1,501	7
BlogAuthorship	430,269	344,215	86,054	39

Enron Email Dataset. The raw Enron Email corpus [28] consists of 619,446 email messages from 158 employees of the Enron corporation. The dataset contains full emails of real users, which include naturally occurring personal information such as names, addresses, organizations, social security numbers etc. Since the original dataset is not fitted to text-classification (as it lacks any official labels), we adapted it by labeling the emails by the folder names they are attached to (i. e., "sent-mail", "corporate", "junk", "proposals" etc.). We selected seven folder that could be considered as valid classes in an applied setting. The labels of these seven classes are "logistics", "personal", "management", "deal discrepancies", "resumes", "online trading", and "corporate". During the preprocessing of the emails, we removed the forward blocks, HTML links, line breaks, and tabs. The removal of forward blocks and HTML links was especially important to improve the quality of the generated texts.

Blog Authorship Corpus. The Blog Authorship Corpus [47] contains text from blogs written until 2004, with each blog being the work of a single user. The corpus incorporates a total of 681,288 posts from 19,320 users. Alongside the blogposts, the dataset includes topic labels and demographic information about the writer, including gender, age, and zodiac sign. Although the blogposts were written for the public, they contain some PII such as names, organizations and postal addresses. We adapt the dataset for text classification by labeling the posts by the blogs' topics. Posts with the topic label "unknown" were removed. After the removal, the dataset consists of 430,269 posts with 39 unique labels. Preprocessing was kept to a minimal: non-printable ASCII characters, non-ASCII characters (e. g., Korean letters), and URL links were removed, otherwise the text remained unchanged.

4.2 Procedure and Implementation

The procedure of our experiments follows the pipeline as illustrated in Fig. 1. Below, we describe the details of each step. All experiments were conducted on a NVIDIA A100 HGX GPU with 40 GB of RAM.

Fine-Tuning. The experiments are based on the Hugging Face implementation of the BERT base uncased model [59]. For single-label classification, a custom classifier head is attached to the base model consisting of a Dropout and a Linear layer. In the *Full* and *Partial* setup we used the standard Adam optimizer, while in the *Differentially Private* fine-tuning, we changed it to the DPAdam optimizer from the Opacus library [61] with a microbatch size of 1.

Text Generation. For text generation, we first removed the classifier heads from the fine-tuned models and attached a pre-trained MLM head instead. We then used the sequential generation method described in Sect. 3, with the addition of beam search and nucleus sampling combined with a temperature parameter. During prompt creation, we sampled a string with a character length of 100 either from the Common Crawl dataset (naive prompting) or from the test set (informed prompting). We set the sequence length to 256 tokens. We removed the tokens of the prompt before saving the samples, i. e., if the prompt contained any entities, they are not considered for the evaluation. In total, we generated 20,000 text samples for each setup.

Named Entity Recognition. For collecting the named entities from the fine-tuning datasets, we employed the NER system of the spaCy library that utilizes a custom word embedding strategy, a transformer, and a transition-based approach to named entity parsing [26]. spaCy distinguishes between a total of 18 different named entity types. Out of these 18 entity types we selected seven (Person, Organization, Location, Geo-Political Event, Facility, Money, Cardinal), which have a high possibility to contain personal or privacy sensitive information. When creating the *Private* entity set described in Sect. 3.3, we cross-checked our fine-tuning entities with the pre-training datasets (the Book Corpus and Wikipedia datasets, available through the datasets library [31]) to discard the entities present both in fine-tuning and pre-training. The numbers of named entities per type in each of the three sets can be seen in Table 2.

4.3 Hyperparameter Optimization

Fine-Tuning. During fine-tuning, we carefully optimized the models on both datasets using manual tuning based on test accuracy. Dropout rates were fixed based on the default BERT base implementation of the huggingface library (0.1 for attention dropout and 0.3 for the classifier) [59]. For batch size, learning rate, and number of epochs a search space was defined based on previous works [13, 19]. Specifically, we chose the batch size from {8, 16, 32}, the learning rate

Table 2. Number of named entities found in the datasets sorted by type.

Named Entity Type	Enron			Blog Authorship		
	All	Private	Private 1-eidetic	All	Private	Private 1-eidetic
PERSON	10,712	7,717	4,844	209,434	137,892	113,599
ORG	9,933	7,178	5,001	168,068	107,480	90,594
LOC	316	175	125	10,562	4,902	4,049
GPE	1,551	739	490	37,691	17,781	13,196
FAC	367	230	174	12,824	7,137	6,349
MONEY	1,220	736	585	11,216	7,551	6,343
CARDINAL	2,918	1,924	1,386	24,075	13,020	10,810
Total	27,017	18,726	12,605	473,870	295,763	244,940

from {5e−3, 1e−3, 1e−4, 5e−5, 1e−5}, and the number of epoch from {3, 5, 10}. Across all setups we found that using a batch size of 32 leads to the best performance. On the Enron dataset the highest accuracy values were achieved when the number of epochs is set to 10, while the Blog Authorship dataset required 5 epochs to reach the highest values. In the *Full* setup a learning of 1e−5 was found to be the best performing on both datasets. In the *Partial* setup the best results were found with a 5e−5 learning rate for the Enron dataset and 1e−4 for the Blog Authorship dataset.

In the *DP* fine-tuning, the best results were achieved with a learning rate of 1e−3 on both datasets. In this setup two additional hyperparameters had to be optimized to achieve the highest possible accuracy while keeping the privacy budget ϵ single-digit. We set the per-example gradient clipping threshold to 10 based on a previous study using DP with BERT [62], and found the best values for the noise multiplier to be 0.5 for the Enron and 0.4 for the Blog Authorship dataset.

Text Generation. For text generation, we studied the effects of the different sampling parameters. We found the best results in terms of text diversity and coherence through manual tuning with the following value combinations: number of beams: 1, beam size: 30, nucleus sampling value: 0.8, temperature: 2.0, and n-gram repetition limit: 3.

4.4 Measures

To evaluate the performance on the downstream task, we use accuracy. In the *DP* setup, we measure privacy preservation with the privacy budget ϵ. In all models, the extent of unintended memorization of named entities found in the fine-tuning dataset is measured by counting their occurrences in generated samples and checking their k-eidetic value. A data point (or in our case an entity) is k-eidetic if it appears k times in the training corpus [8].

5 Results

5.1 Classification

Table 3 shows the singe-label text classification results for each setup. For both datasets, we observe a similar trend between the different fine-tuning setups: *Full* achieved the highest accuracy on the test set, *Partial* performed slightly worse, and the *DP* setup produced the worst results with a 16% point drop compared to the *Partial.* In general, the accuracy values are considerably higher on the Enron dataset. In the *DP* setup, the privacy budget ϵ is 9.79 for the Enron and $\epsilon = 7.38$ for Blog Authorship.

Table 3. Mean accuracy and standard deviation over five runs on the single-label text classification

Fine-tuning Setup	Enron	Blog Authorship
Full	86.83% (0.46)	51.69% (0.32)
Partial	85.95% (0.34)	49.58% (0.15)
DP	68.28% (0.88)	35.86% (0.12)

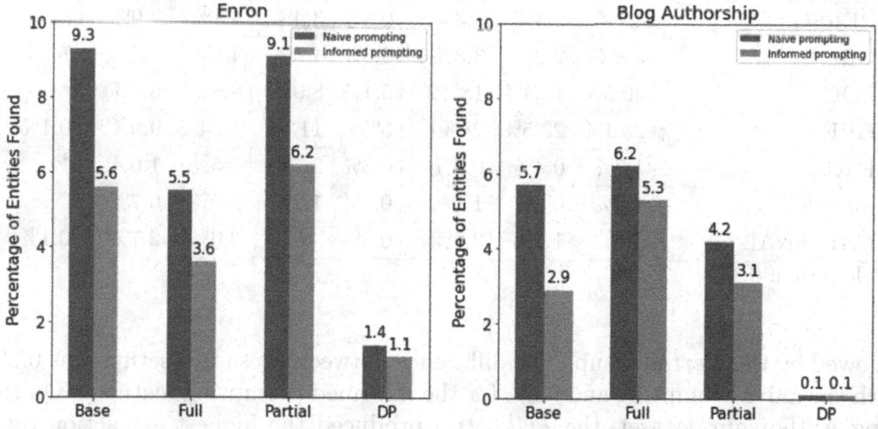

Fig. 3. The percentages of all entities successfully extracted from the models, compared by prompting methods.

5.2 Named Entity Memorization

For the named entity memorization experiments, we also included a pre-trained only BERT, i. e., without any fine-tuning, which we call the *Base* setup. Figure 3 shows our initial results on the *All* entity set. The highest extraction rate was 9.3% for the Enron and 6.2% for the Blog Authorship dataset. On the Enron dataset, the highest extraction rate was achieved on the *Base* setup. closely

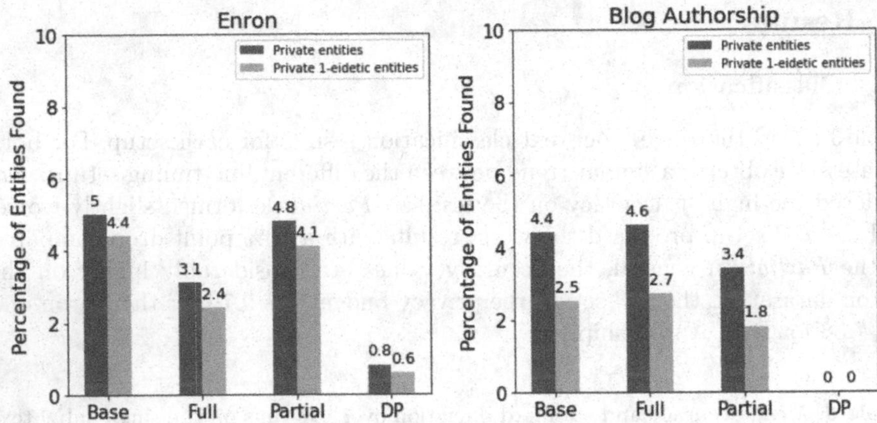

Fig. 4. The percentages of private entities and private 1-eidetic entities successfully extracted from the models with the use of naive prompting.

Table 4. Extraction ratio of entities from the *Private* set using naive prompting, grouped by entity types

Named Entity Type	Enron				Blog Authorship			
	Base	Full	Partial	DP	Base	Full	Partial	DP
PERSON	4.1%	2.3%	4.3%	0.8%	3.4%	4.1%	2.9%	*
ORG	3.8%	2.2%	3.3%	0.3%	4.5%	4.1%	3%	*
LOC	20.5%	15.4%	18.4%	5.1%	8.9%	8.6%	5.5%	*
GPE	28.1%	22.5%	28%	5%	11.5%	13.4%	9.9%	0.1%
FAC	1.7%	0.4%	0.8%	0.8%	2.7%	2.5%	1.6%	*
MONEY	1.5%	0.7%	1%	0.1%	1.2%	0.9%	0.7%	*
CARDINAL	4.8%	1.7%	3.9%	0.5%	4.4%	3.9%	2.7%	0.1%

* less than 0.1%

followed by the *Partial* setup. The difference between these two setups was 0.2% with the naive prompting and 0.6% for the informed prompting methods. On the Blog Authorship dataset, the *Full* setup produced the highest extraction rate, followed by the *Base* setup. Between these two setups, the naive prompting resulted in 0.5% and the informed prompting in a 2.4% difference. The *DP* setup produced the lowest extraction rates with 1.4% (naive prompting) and 1.1% (informed prompting) on the Enron and 0.1% in the Blog Authorship dataset (both, naive and informed prompting). Naive prompting consistently outperformed informed prompting in all fine-tuning setups on both datasets.

Figure 4 shows the comparison of the extraction rates between the private entities and the private 1-eidetic entities, using the naive prompting method. Compared to the results in Fig. 3, the extraction rates are consistently lower across all setups. The difference between *Base* and *Partial* on Enron, and *Base* and *Full* on Blog Authorship once again is negligible. Overall the memorization rate of private 1-eidetic entities is lower than the memorization rate of all private entities. But the difference is less than 1% point on the Enron and less than 2% points on the Blog Authorship dataset.

To further investigate the extracted private entities, we also measured the extraction ratio of each entity type in Table 4. The Location and Geo-Political Event types produced the highest percentages, while the Facility and Money types had results less than 3% across all setups and datasets. The extraction ratios on Blog Authorship are consistently lower in every entity type compared to Enron. The only exception can be seen in the Facility type, where the Blog Authorship results were 1 to 2% points higher.

6 Discussion

6.1 Key Insights

Prompting Methods. Our experiments show that the naive prompting method produces better results in all setups. Although for informed prompting the seed sequences will be more similar to the text sequences found in the fine-tuning data, this informed prompting likewise limits the possibilities of producing diverse outputs. Following Carlini et al. [8], we conclude that using random prompts sampled from a huge corpus unrelated to the training data yields better extraction results. This shows that adversaries do not need to have prior knowledge about the training data of the attacked model, a simple black-box approach is sufficient.

Named Entity Memorization in BERT. We extracted private named entities from the fine-tuned models at surprisingly low rates. In no setup, we extracted more than 10% of the private entities. Interestingly, our results further show that using a pre-trained *Base* model that has not been fine-tuned on the training set containing those extracted entities produces similar extraction ratios. Our assumption is that the small percentage of private entities that have been successfully extracted from both the *Base* and *Full* or *Partial* models have low level of complexity in terms of length and n-gram diversity. Therefore, they are more likely to be randomly generated by combining common subword tokens.

In order to better understand the reasons behind this observation, we conducted a more detailed analysis of the extracted entities. As can be seen from Table 4, distinct entity types have different probabilities to be extracted. From the seven types, we argue Location and Geo-Political Event are the least unique in their nature, therefore it is not suprising that the highest extraction rates have been achieved on them. The lower values in the Money and Cardinal types reinforce the findings that the subword tokenization in BERT is a suboptimal method to encode numerical values [56]. Overall our findings suggest that BERT could be rather resistant to training data extraction attacks unlike other large LMs such as GPT-2 [8]. This is most likely due to its smaller size as argued in [8]. It is also possible that auto-encoder transformers are generally less prone to these attacks compared to auto-regressive transformers as result of their different pre-training objectives.

Differentially Private Fine-Tuning. In all the experiments that used Differentially Private fine-tuning, the extraction rates of named entities were reduced by a large extent. Our samples have shown that the text quality in the *DP* setups was very low, both text coherence and text diversity decreased dramatically. Even though the performance on the downstream task was also considerably lower, we argue this trade-off between performance and privacy is still promising for future developments. Considering that the focus of our study was not on achieving state-of-the art performance for singe-label text classification, we only used the Adam variant of the original DPSGD algorithm [1]. We leave the use of other, more advanced DP algorithms like [62] to future work. One can expect that for tasks, where the ability of a model to generate text is irrelevant, the use of DP can be a viable solution to increase the privacy of the model.

6.2 Generalization

In our experiments, we intentionally used datasets of different characteristics. While the Enron dataset we used is small in size and is very cluttered due to its source (real world emails), the Blog Authorship Corpus is a public web corpus that contains a large amount of samples covering a broad range of domains with a higher text quality. Although, we only used single-label text classification (in which BERT is generally considered as state-of-the-art [19]) as a downstream task for fine-tuning, results should be similar on different downstream tasks since the memorization takes place in the encoding layers, irrespectively of the task-specific final layers of a model. Finally our conclusion about the memorization capabilities of the BERT base model is in line with the training data extraction study done on Clinical BERT, in which the authors were unable to reliably extract patient names from a specific BERT variant pre-trained on clinical data [30].

6.3 Threats to Validity

We acknowledge that the experimental datasets are limited to English. Although named entities are often unique to their respective language, we have no reason to believe that generating named entities would be significantly easier in other languages. For languages that have larger character sets (e. g., Chinese) or use long compound words (e. g., German), the probability of unintended memorization may even be smaller. Regarding the efficiency of the extraction of named entities, the results can be influenced by both the named entity recognition system and our text generation method. It is possible that some entities have been missed and some have been falsely identified. The missed entities are unlikely to influence the results since we still had a great amount of entities of differing k-eidetic values. Controlling for the falsely identified entities was a more difficult problem. Therefore, we decided to remove all entities with a character length of less than 4. Using a left-to-right sequential text generation method might also bias our results, as BERT uses context from both directions to predict a token during pre-training. This, we argue has more impact on text coherence rather than the ability to trigger a diverse output containing named entities. The latter was of higher importance to our study.

7 Conclusion

As the capabilities of large LMs increase, it is important that the privacy aspects of these models are also considered. We performed an investigation into the capabilities of BERT to memorize named entities. We ran experiments, in which we tried to extract private named entities from fine-tuned BERT models using three different fine-tuning methods and two prompting strategies. Overall, we could only extract a low percentage of named entities from BERT, and found that the pre-trained only model generates the same amount of entities as the fine-tuned models. We also employed a Differential Private fine-tuning method, which showed to be a promising privacy preserving method against training data extraction attacks with some trade-off on the downstream task performance. Although our results do not rule out the possibility to extract personal information from a fine-tuned BERT base model using more advanced methods, our findings suggest that doing so is at least not trivial. As for future work it would be interesting to re-run the experiments on other commonly used language models and to test the embedding layers of our BERT setups against membership inference attacks.

References

1. Abadi, M., et al.: Deep learning with differential privacy. In: Proceedings of the 2016 ACM SIGSAC conference on computer and communications security, pp. 308–318 (2016)
2. Acar, A., Aksu, H., Uluagac, A.S., Conti, M.: A survey on homomorphic encryption schemes: theory and implementation. ACM Comput. Surv. (CSUR) **51**(4), 1–35 (2018)
3. Alsentzer, E., et al.: Publicly available clinical bert embeddings. arXiv preprint arXiv:1904.03323 (2019)
4. Aono, Y., Hayashi, T., Wang, L., Moriai, S., et al.: Privacy-preserving deep learning via additively homomorphic encryption. IEEE Trans. Inf. Forensics Secur. **13**(5), 1333–1345 (2017)
5. Beltagy, I., Lo, K., Cohan, A.: SciBERT: a pretrained language model for scientific text. arXiv preprint arXiv:1903.10676 (2019)
6. Brown, T., et al.: Language models are few-shot learners. Adv. Neural. Inf. Process. Syst. **33**, 1877–1901 (2020)
7. Carlini, N., Liu, C., Erlingsson, Ú., Kos, J., Song, D.: The secret sharer: evaluating and testing unintended memorization in neural networks. In: 28th USENIX Security Symposium (USENIX Security 2019), pp. 267–284 (2019)
8. Carlini, N., et al.: Extracting training data from large language models. In: 30th USENIX Security Symposium (USENIX Security 2021), pp. 2633–2650 (2021)

9. Davody, A., Adelani, D.I., Kleinbauer, T., Klakow, D.: Robust differentially private training of deep neural networks. arXiv preprint arXiv:2006.10919 (2020)
10. De Cristofaro, E.: An overview of privacy in machine learning. arXiv preprint arXiv:2005.08679 (2020)
11. Devlin, J., Chang, M.W., Lee, K., Toutanova, K.: BERT: pre-training of deep bidirectional transformers for language understanding. arXiv preprint arXiv:1810.04805 (2018)
12. Diera, A., Lell, N., Garifullina, A., Scherp, A.: A study on extracting named entities from fine-tuned vs. differentially private fine-tuned BERT models. CoRR abs/2212.03749 (2022). https://doi.org/10.48550/arXiv.2212.03749
13. Dodge, J., Ilharco, G., Schwartz, R., Farhadi, A., Hajishirzi, H., Smith, N.: Fine-tuning pretrained language models: weight initializations, data orders, and early stopping. arXiv preprint arXiv:2002.06305 (2020)
14. Dupuy, C., Arava, R., Gupta, R., Rumshisky, A.: An efficient DP-SGD mechanism for large scale NLP models. arXiv preprint arXiv:2107.14586 (2021)
15. Dwork, C., Kenthapadi, K., McSherry, F., Mironov, I., Naor, M.: Our data, ourselves: privacy via distributed noise generation. In: Vaudenay, S. (ed.) EUROCRYPT 2006. LNCS, vol. 4004, pp. 486–503. Springer, Heidelberg (2006). https://doi.org/10.1007/11761679_29
16. Dwork, C., Roth, A., et al.: The algorithmic foundations of differential privacy. Found. Trends Theor. Comput. Sci. **9**(3–4), 211–407 (2014)
17. Fredrikson, M., Jha, S., Ristenpart, T.: Model inversion attacks that exploit confidence information and basic countermeasures. In: Proceedings of the 22nd ACM SIGSAC Conference on Computer and Communications Security, pp. 1322–1333 (2015)
18. Freitag, M., Al-Onaizan, Y.: Beam search strategies for neural machine translation. In: Proceedings of the First Workshop on Neural Machine Translation, pp. 56–60 (2017)
19. Galke, L., Scherp, A.: Bag-of-words vs. graph vs. sequence in text classification: questioning the necessity of text-graphs and the surprising strength of a wide MLP. In: Proceedings of the 60th Annual Meeting of the Association for Computational Linguistics (Volume 1: Long Papers), ACL 2022, 22–27 May 2022, pp. 4038–4051. Association for Computational Linguistics (2022)
20. Geman, S., Geman, D.: Stochastic relaxation, Gibbs distributions, and the Bayesian restoration of images. IEEE Trans. Pattern Anal. Mach. Intell. **6**, 721–741 (1984)
21. Ghazvininejad, M., Levy, O., Liu, Y., Zettlemoyer, L.: Mask-predict: parallel decoding of conditional masked language models. arXiv preprint arXiv:1904.09324 (2019)
22. Gilad-Bachrach, R., Dowlin, N., Laine, K., Lauter, K., Naehrig, M., Wernsing, J.: CryptoNets: applying neural networks to encrypted data with high throughput and accuracy. In: International Conference on Machine Learning, pp. 201–210. PMLR (2016)
23. Hassan, F., Domingo-Ferrer, J., Soria-Comas, J.: Anonymization of unstructured data via named-entity recognition. In: Torra, V., Narukawa, Y., Aguiló, I., González-Hidalgo, M. (eds.) MDAI 2018. LNCS (LNAI), vol. 11144, pp. 296–305. Springer, Cham (2018). https://doi.org/10.1007/978-3-030-00202-2_24
24. He, P., Liu, X., Gao, J., Chen, W.: DeBERTa: decoding-enhanced BERT with disentangled attention. arXiv preprint arXiv:2006.03654 (2020)
25. Holtzman, A., Buys, J., Du, L., Forbes, M., Choi, Y.: The curious case of neural text degeneration. arXiv preprint arXiv:1904.09751 (2019)

26. Honnibal, M., Montani, I., Van Landeghe, S., Boyd, A.: spaCy: industrial-strength natural language processing in python (2022). https://zenodo.org/record/121230
27. Howard, J., Ruder, S.: Universal language model fine-tuning for text classification. arXiv preprint arXiv:1801.06146 (2018)
28. Klimt, B., Yang, Y.: Introducing the Enron corpus. In: CEAS 2004 - First Conference on Email and Anti-Spam, 30–31 July 2004, Mountain View, California, USA (2004)
29. Lee, J., Tang, R., Lin, J.: What would Elsa do? Freezing layers during transformer fine-tuning. arXiv preprint arXiv:1911.03090 (2019)
30. Lehman, E., Jain, S., Pichotta, K., Goldberg, Y., Wallace, B.C.: Does BERT pre-trained on clinical notes reveal sensitive data? arXiv preprint arXiv:2104.07762 (2021)
31. Lhoest, Q., et al.: Datasets: a community library for natural language processing. In: Proceedings of the 2021 Conference on Empirical Methods in Natural Language Processing: System Demonstrations, pp. 175–184. Association for Computational Linguistics, Online and Punta Cana (2021)
32. Liu, B., Ding, M., Shaham, S., Rahayu, W., Farokhi, F., Lin, Z.: When machine learning meets privacy: a survey and outlook. ACM Comput. Surv. (CSUR) **54**(2), 1–36 (2021)
33. Liu, Y., et al.: RoBERTa: a robustly optimized BERT pretraining approach. arXiv preprint arXiv:1907.11692 (2019)
34. Liu, Z., Winata, G.I., Madotto, A., Fung, P.: Exploring fine-tuning techniques for pre-trained cross-lingual models via continual learning. arXiv preprint arXiv:2004.14218 (2020)
35. Mao, H.H.: A survey on self-supervised pre-training for sequential transfer learning in neural networks. arXiv preprint arXiv:2007.00800 (2020)
36. McMahan, H.B., Moore, E., Ramage, D., y Arcas, B.A.: Federated learning of deep networks using model averaging. arXiv preprint arXiv:1602.05629 (2016)
37. Mireshghallah, F., Taram, M., Vepakomma, P., Singh, A., Raskar, R., Esmaeilzadeh, H.: Privacy in deep learning: a survey. arXiv preprint arXiv:2004.12254 (2020)
38. Nadeau, D., Sekine, S.: A survey of named entity recognition and classification. Lingvisticae Invest. **30**(1), 3–26 (2007)
39. Oh, S.J., Schiele, B., Fritz, M.: Towards reverse-engineering black-box neural networks. In: Samek, W., Montavon, G., Vedaldi, A., Hansen, L.K., Müller, K.-R. (eds.) Explainable AI: Interpreting, Explaining and Visualizing Deep Learning. LNCS (LNAI), vol. 11700, pp. 121–144. Springer, Cham (2019). https://doi.org/10.1007/978-3-030-28954-6_7
40. Ouyang, L., et al.: Training language models to follow instructions with human feedback. arXiv preprint arXiv:2203.02155 (2022)
41. Parisot, M.P., Pejo, B., Spagnuelo, D.: Property inference attacks on convolutional neural networks: influence and implications of target model's complexity. arXiv preprint arXiv:2104.13061 (2021)
42. Raffel, C., et al.: Exploring the limits of transfer learning with a unified text-to-text transformer. arXiv preprint arXiv:1910.10683 (2019)
43. Rigaki, M., Garcia, S.: A survey of privacy attacks in machine learning. arXiv preprint arXiv:2007.07646 (2020)
44. Rogers, A., Kovaleva, O., Rumshisky, A.: A primer in BERTology: what we know about how BERT works. Trans. Assoc. Comput. Linguist. **8**, 842–866 (2020)

45. Rubinstein, B.I., Bartlett, P.L., Huang, L., Taft, N.: Learning in a large function space: privacy-preserving mechanisms for SVM learning. arXiv preprint arXiv:0911.5708 (2009)
46. Sanh, V., Debut, L., Chaumond, J., Wolf, T.: DistilBERT, a distilled version of BERT: smaller, faster, cheaper and lighter. arXiv preprint arXiv:1910.01108 (2019)
47. Schler, J., Koppel, M., Argamon, S., Pennebaker, J.: Effects of age and gender on blogging in proceedings of 2006 AAAI spring symposium on computational approaches for analyzing weblogs. In: Proceedings of 2006 AAAI Spring Symposium on Computational Approaches for Analyzing Weblogs (2006)
48. Sharir, O., Peleg, B., Shoham, Y.: The cost of training NLP models: a concise overview. arXiv preprint arXiv:2004.08900 (2020)
49. Shokri, R., Stronati, M., Song, C., Shmatikov, V.: Membership inference attacks against machine learning models. In: 2017 IEEE Symposium on Security and Privacy (SP), pp. 3–18. IEEE (2017)
50. Singhofer, F., Garifullina, A., Kern, M., Scherp, A.: A novel approach on the joint de-identification of textual and relational data with a modified Mondrian algorithm. In: DocEng 2021: ACM Symposium on Document Engineering 2021, 24–27 August 2021, pp. 14:1–14:10. ACM (2021)
51. Sun, W., Khan, H., Guenon des Mesnards, N., Rubino, M., Arkoudas, K.: Unfreeze with care: space-efficient fine-tuning of semantic parsing models. In: Proceedings of the ACM Web Conference 2022, pp. 999–1007 (2022)
52. Sweeney, L.: k-anonymity: a model for protecting privacy. Internat. J. Uncertain. Fuzziness Knowl.-Based Syst. **10**(05), 557–570 (2002)
53. Thakkar, O., Ramaswamy, S., Mathews, R., Beaufays, F.: Understanding unintended memorization in federated learning. arXiv preprint arXiv:2006.07490 (2020)
54. Tramèr, F., Zhang, F., Juels, A., Reiter, M.K., Ristenpart, T.: Stealing machine learning models via prediction APIs. In: 25th USENIX security symposium (USENIX Security 2016), pp. 601–618 (2016)
55. Vaswani, A., et al.: Attention is all you need. Adv. Neural Inf. Process. Syst. **30** (2017)
56. Wallace, E., Wang, Y., Li, S., Singh, S., Gardner, M.: Do NLP models know numbers? Probing numeracy in embeddings. In: Proceedings of the 2019 Conference on Empirical Methods in Natural Language Processing and the 9th International Joint Conference on Natural Language Processing (EMNLP-IJCNLP), pp. 5307–5315 (2019)
57. Wang, A., Cho, K.: BERT has a mouth, and it must speak: BERT as a Markov random field language model. arXiv preprint arXiv:1902.04094 (2019)
58. Wang, B., Gong, N.Z.: Stealing hyperparameters in machine learning. In: 2018 IEEE Symposium on Security and Privacy (SP), pp. 36–52. IEEE (2018)
59. Wolf, T., et al.: Transformers: state-of-the-art natural language processing. In: Proceedings of the 2020 Conference on Empirical Methods in Natural Language Processing: System Demonstrations, pp. 38–45 (2020)
60. Xu, Y., Ma, T., Tang, M., Tian, W.: A survey of privacy preserving data publishing using generalization and suppression. Appl. Math. Inf. Sci. **8**(3), 1103 (2014)
61. Yousefpour, A., et al.: Opacus: user-friendly differential privacy library in pytorch. arXiv preprint arXiv:2109.12298 (2021)
62. Yu, D., et al.: Differentially private fine-tuning of language models. arXiv preprint arXiv:2110.06500 (2021)
63. Zanella-Beguelin, S., et al.: Analyzing information leakage of updates to natural language models. In: Proceedings of the 2020 ACM SIGSAC Conference on Computer and Communications Security, pp. 363–375 (2020)

64. Zhang, H., Song, H., Li, S., Zhou, M., Song, D.: A survey of controllable text generation using transformer-based pre-trained language models. arXiv preprint arXiv:2201.05337 (2022)
65. Zhang, T., He, Z., Lee, R.B.: Privacy-preserving machine learning through data obfuscation. arXiv preprint arXiv:1807.01860 (2018)
66. Zhu, T., Ye, D., Wang, W., Zhou, W., Yu, P.: More than privacy: applying differential privacy in key areas of artificial intelligence. IEEE Trans. Knowl. Data Eng. **34**, 2824–2843 (2020)

Event and Entity Extraction from Generated Video Captions

Johannes Scherer[1], Deepayan Bhowmik[2], and Ansgar Scherp[1(✉)]

[1] Universität Ulm, Ulm, Germany
{johannes.scherer,ansgar.scherp}@uni-ulm.de
[2] Newcastle University, Newcastle upon Tyne, UK
deepayan.bhowmik@newcastle.ac.uk

Abstract. Annotation of multimedia data by humans is time-consuming and costly, while reliable automatic generation of semantic metadata is a major challenge. We propose a framework to extract semantic metadata solely from automatically generated video captions. As metadata, we consider entities, the entities' properties, relations between entities, and the video category. Our framework combines automatic video captioning models with natural language processing (NLP) methods. We use state-of-the-art dense video captioning models with masked transformer (MT) and parallel decoding (PVDC) to generate captions for videos of the ActivityNet Captions dataset. We analyze the output of the video captioning models using NLP methods. We evaluate the performance of our framework for each metadata type, while varying the amount of information the video captioning model provides. Our experiments show that it is possible to extract high-quality entities, their properties, and relations between entities. In terms of categorizing a video based on generated captions, the results can be improved. We observe that the quality of the extracted information is mainly influenced by the dense video captioning model's capability to locate events in the video and to generate the event captions.

An earlier version of this paper has been published on arXiv [20]. We provide the source code here:
https://github.com/josch14/semantic-metadata-extraction-from-videos.

Keywords: metadata extraction · vision models · natural language processing

1 Introduction

The annotation of multimedia with semantic metadata by humans is time-consuming and costly. Automatic extraction methods exist for different types of high-level metadata, but these methods usually have high error rates and therefore manual correction of the user is still required [19]. Thus, in contrast to

Published by Springer Nature Switzerland AG 2023
A. Holzinger et al. (Eds.): CD-MAKE 2023, LNCS 14065, pp. 280–300, 2023.
https://doi.org/10.1007/978-3-031-40837-3_17

the value of semantic metadata, especially when it can be generated automatically, the reliable automatic generation of semantic metadata is still a major challenge. For each semantic metadata type, one could use a different computer vision method to generate the data. For example, video object detection could be used to detect entities in a video, while video visual relation tagging methods find instances of relations between depicted entities. However, when using multiple methods, they need to be trained separately and the training is, especially for videos, computationally expensive.

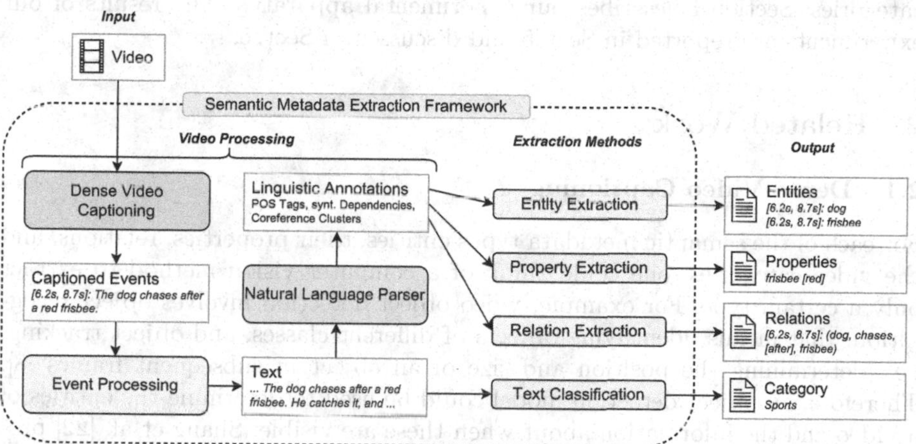

Fig. 1. Semantic metadata extraction and its key components: a dense video captioning model and a natural language parser

From this motivation, we propose a framework that generates semantic metadata from videos of not only one, but multiple types. Depending on the video application, there are various semantic metadata types of interest. We focus on four different of those types, namely the depicted *entities* and their *properties*, the observable *relations* between entities and the video *category*. Additionally, we consider semantic metadata on different levels, namely event-level, where temporal information is relevant, and video-level. Our framework combines several methods from the fields of computer vision and natural language processing (NLP) (see Fig. 1). For an input video, a dense video captioning (DVC) model generates a set of natural language sentences for multiple temporally localized video events, thus providing a richly annotated description of video semantics. We process the captioned events into text to make them accessible for different NLP methods. Text classification determines the category of a video, while the extraction methods for entities, properties, and relations rely on linguistic annotations of a language parser. In summary, our contributions are:

- A framework for extracting semantic metadata combining an automatic video captioning model with several NLP methods for entity detection, extraction of entity properties, relation extraction, and categorical text classification.

- We evaluate the capabilities of our framework using the ActivityNet Captions [10] dataset. We compare two state-of-the-art dense video captioning models with masked transformer (MT) [27] and parallel decoding (PVDC) [25].
- The quality of the extracted metadata mainly depends on the event localization in the video and the performance of the event caption generation.

Below, we discuss the related work. Section 3 introduces the methods used to extract semantic metadata in the form of entities, properties, relations, and categories. Section 4 describes our experimental apparatus. The results of our experiments are reported in Sect. 5 and discussed in Sect. 6.

2 Related Work

2.1 Dense Video Captioning

For each of the semantic metadata types entities, their properties, relations, and the video categories, one could think of a computer vision method to extract only a certain type. For example, video object detection involves object recognition, that means, identifying objects of different classes, and object tracking, i.e., determining the position and size of an object in subsequent frames [8]. Therefore, an object detection model could be used to determine the entities of a video and the information about when these are visible. Shang et al. [22] propose video visual relation tagging to detect relations between objects in videos. Here, relations are annotated to the whole video without the requirement of object localization. A relation is denoted by a triplet *(subject, predicate, object)*, where the predicate may be a transitive or intransitive verb, comparative, or spatial predicate. Further methods that could be used for the extraction of video semantics include video classification for determining the category of a video [14], and emotion recognition, which aims to classify videos into basic emotions [26]. However, it is not efficient to use one computer vision method at a time for the extraction of only one semantic metadata type.

Automatic video description involves understanding and detection of different types of information like background scene, humans, objects, human actions, and events like human-object interactions [1]. In such a way, automatic video description can be seen as a task that unites the mentioned computer vision tasks like object detection, visual relation tagging, and emotion recognition. Dense video captioning (DVC), as first introduced by Krishna et al. [10], generate *captioned events*, which not only involves the localization of multiple, potentially overlapping events in time, but also the generation of a natural language sentence description for each event. Because of the rich information DVC models provide, we utilize such model in our framework. We present two DVC models with masked transformer (MT) [27] and parallel decoding (PVDC) [25], which we use in our experiments, in detail in Sect. 3.1 (together with our framework).

2.2 Text Information Extraction and Classification

In our framework, semantic metadata is extracted from the captioned events generated by a DVC model. This includes the analysis of the events' textual descriptions, for which methods from Open Information Extraction (Open IE) can be employed [16]. Open IE is the task of generating a structured representation of the information extracted from a natural language text in the form of relational triples. A triple *(arg1, rel, arg2)* consists of a set of argument phrases and a phrase denoting a semantic relation between them [16]. Existing Open IE approaches make use of a set of patterns, which are either hand-crafted rules or automatically learned from labeled training data. Furthermore, both methodologies can be divided into two subcategories: approaches that use shallow syntactic analysis and approaches that utilize dependency parsing [18]. Fader et al. [6] proposed REVERB, which makes use of hand-crafted extraction rules. They restrict syntactic analysis to part-of-speech tagging and noun phrase chunking, resulting in an efficient extraction for high-confidence propositions. Relations are extracted in two major steps: first, relation phrases are identified that meet syntactic and lexical constraints. Then, for each relation phrase, a pair of noun phrase arguments is identified. Contrary to REVERB, ClausIE (clause-based Open IE) uses hand-crafted extraction rules based on a typed dependency structure [4]. It does not make use of any training data and does not require any postprocessing like filtering out low-precision extractions. First, a dependency parse of the sentence is computed. Then, using the dependency parse, a set of clauses is determined. The authors define seven clause types, where each clause consists of one subject, one verb and optionally of an indirect object, a direct object, a complement, and one or more adverbials. Finally, for each clause, one or more propositions are generated. Since dependency parsing is used, ClausIE is computationally more expensive compared to REVERB.

Algur et al. [2] argue that the proper category identification of a video is essential for efficient query-based video retrieval. This task is traditionally posed as a supervised classification of the features derived from a video. The features used for video classification can be of visual nature only, but if user-provided textual metadata (i.e., title, description, tags) is available, it can be used in a profitable way [3]. However, in our proposed framework the video category is predicted only with the textual information the DVC model provides. So although we address the video classification problem, we do this by utilizing existing work in the text classification area. Text classification models can be roughly divided into two categories [12]. First, traditional statistics-based models such as k-Nearest Neighbors and Support Vector Machines require manual feature engineering. Second, deep learning models consist of artificial neural networks to automatically learn high-level features for better results in text understanding. For example, TextCNN [9] is a text classification method using text-induced word-document cooccurence graph and graph learning. We use the pre-trained BERT (Bidirectional Encoder Representations from Transformers) [5] model, which set the state-of-the-art for text classification [7].

3 Semantic Metadata Extraction from Videos

In our framework, the extraction of semantic metadata is based only on the captioned events generated by the DVC model. As a result, certain metadata types such as emotions are difficult to extract. This depends mainly on how detailed the DVC model is able to describe video semantics. Considering the capabilities of current video description models, we focus on four semantic metadata types that we aim to extract from a video: the depicted **entities** (i.e., persons, objects, locations) and their **properties**, observable visual **relations** between entities, and the video **category**, see Fig. 2. We distinguish between *event-level* and *video-level* semantic metadata depending on whether semantic metadata is assigned to a specific time interval or not. For example, assume that at some point in a video there is a *cat* visible. On video-level, the corresponding entity item only stores the information that there is a cat occurring in the video. On event-level, the metadata item does not only store the name of the entity, but also a time interval in which the entity is visible.

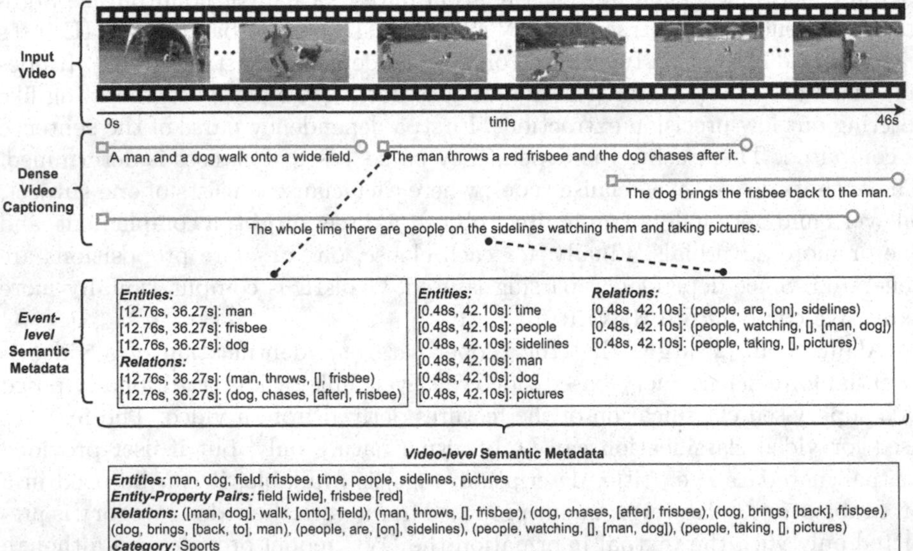

Fig. 2. Our framework extracts semantic metadata in the form of entities, properties of entities, relations between entities, and the video category from automatically generated captioned events. We distinguish event-level and video-level semantic metadata, depending on whether semantic metadata is assigned to a specific time interval or not. Image adapted from [13].

Revisiting Fig. 1, it can be seen that our framework consists of several methods. For an input video, a DVC model generates captioned events (Sect. 3.1), which are then processed into text to make them accessible for different methods (Sect. 3.2). The natural language parser produces linguistic annotations, namely part-of-speech (POS) tags, a dependency parse, and coreference clusters (Sect. 3.3). The extraction methods for the semantic metadata types entities (Sect. 3.4), properties (Sect. 3.5), and relations (Sect. 3.6) use these linguistic annotations, while the text classification method determines the category of a video using only the generated captioned events (Sect. 3.7). The lexical database WordNet [15] is used at various points to ensure that extracted semantic metadata consists of linguistically correct English nouns, verbs, adjectives, and adverbs.

3.1 Dense Video Captioning (DVC)

From an input video, DVC models generate a set of captioned events. Each captioned event consists of the event itself, a temporal segment which potentially overlaps with segments of other captioned events, and a natural language sentence that captions the event. While introducing the task of DVC, Krishna et al. [10] proposed a model which consists of a proposal module for event localization, and a separate captioning module, an attention-based Long Short-Term Memory network for context-aware caption generation. Zhou et al. [27] argue that the model of Krishna et al. is not able to take advantage of language to benefit the event proposal module. To this end, they proposed an end-to-end DVC model with masked transformer (MT) that is able to simultaneously produce event proposals and event descriptions. Like many methods that tackle the DVC task, Zhou et al.'s model consists of three components. The video encoder, composed of multiple self-attention layers, extracts visual features from video frames. The proposal decoder takes the features from the encoder and produces event proposals, i.e., temporal segments. The captioning decoder takes input from the visual encoder and the proposal decoder to caption each event.

Wang et al. [25] state that methods like the model of Zhou et al. follow a two-stage "localize-then-describe" scheme, which heavily relies on hand-crafted components. In contrast to the usual structure of DVC models, they proposed a simpler framework for end-to-end DVC with parallel decoding (PDVC). Their model directly decodes extracted frame features into a captioned event set by applying two parallel prediction heads: localization head and captioning head. They propose an event counter, which is stacked on top of the decoder to predict the number of final events. The authors claim that PVDC is able to precisely segment the video into a number of events, avoiding to miss semantic information as well as avoiding replicated caption generation.

3.2 Event Processing

The event processing module processes the captioned events generated by the preceding DVC model into text in order to make semantic information accessible to the natural language parser and the text classification method. In detail, the sentences of the captioned events are sorted in ascending order of the start times of the corresponding events. Afterwards, the sentences are concatenated, resulting in a single text per video, and forwarded to the language parser and text classification method, respectively. By not passing the sentences separately to the language parser, this enables it to use coreference resolution (see Sect. 3.3). When extracting entities and relations on event-level, we annotate each entity and relation with the temporal segment of the captioned event whose sentence contains the name of the entity or the words of the relation, resp. (see Fig. 3).

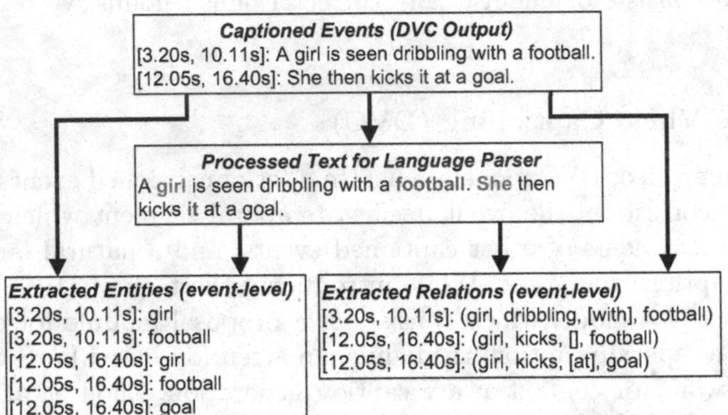

Fig. 3. Captioned events are processed into text and made accessible to the language parser and text classifier.

3.3 Language Processing

The extraction of entities, the entities' properties, and relations is done through syntactic analysis based on the linguistic annotations generated by the natural language parser. It is required to provide POS tags of tokens, a dependency parse, and coreference clusters. The POS tag is a label assigned to the token to indicate its part of speech. The dependency parse consists of a set of directed syntactic relations between the tokens of the sentence. Coreference clusters aim to find all language expressions that refer to the same entity in a text.

We use a CNN-based model from spaCy[1]. The model generates the desired linguistic annotations in a pipe-lined manner: First, the tokenizer segments the input text into tokens. Afterwards, the tagger and dependency parser assign POS tags and dependency labels to tokens, respectively. Finally, coreference clusters are determined using spaCy's NeuralCoref extension. For an input sentence, the dependency parse produced by spaCy models is a tree where the head of a sentence, which is usually a verb, has no dependency. Every other token of the dependency tree has a dependency label that indicates its syntactic relation to its *parent*. The *children* of a token are all its immediate syntactic dependents, i.e., the tokens of the dependency tree for which it is the parent.

[1] spaCy is available at: https://github.com/explosion/spaCy.

3.4 Entity Extraction

The entity extraction method determines which (video-level) and when (event-level) entities like persons, objects, and locations are visible in the video. A video-level **entity** only consists of a *name* describing the entity. On event-level, the entity additionally consists of a *temporal segment* containing the information about when the entity is seen. Entities are extracted by determining (compound) nouns with the POS tags and dependency labels of tokens, and coreference clusters, which are all provided by the language parser.

On event-level, before assigning a temporal segment to each name in order to obtain the entities, each pronoun is analysed whether it refers to a noun or not (see Fig. 3). First, the tokens of the text are filtered for pronouns. Using the computed coreference clusters, for each pronoun it is checked whether it refers to a noun that has been determined in the previous step or not. If this is the case, then the pronoun is replaced with the corresponding noun, i.e., the name of a new entity. Finally, the event-level entities are built by assigning temporal segments to the names of the entities. Here, for an entity, the temporal segment is the segment of the corresponding sentence in which its name (or the pronoun that was previously replaced) occurs.

3.5 Property Extraction

The property extraction method determines properties of entities such as their color, size, and shape. The method is only used to extract video-level information in order to collect information for an entity from different captioned events. An **entity-property pair** is a tuple consisting of an *entity* (i.e., its name) and a *property*, which further describes the entity. For each (video-level) entity, extracted in the previous section, properties are determined as follows. The candidate properties for an entity are the children of the corresponding token (or tokens for compound nouns) that are marked with the dependency labels. We use WordNet to recognize the token or its lemma as an adjective. Such candidate property is considered as a property of the entity. An entity-property pair is formed with the name of the entity and the detected property. In such way, the method results in a set of video-level entity-property pairs. There is no restriction on the tag a property token needs to have. Therefore, properties may be marked by the language parser as `ADJ`ective (e.g., *round* ball), `VERB` (e.g., *provoking* film), or other tags.

3.6 Relation Extraction

As seen for visual relation tagging (Sect. 2.1) or Open IE (Sect. 2.2), relations are usually formulated as triples. In contrast, we define relations as follows. A video-level **relation** is a 4-tuple of the form *(subjects, verb, modifiers, objects)*. With a *temporal segment*, an event-level relation has an additional element containing the information about when the relation is observed. The first element *subjects* is a list containing the names of the relation's acting entities.

Objects, also a list of names of entities, contains the entities that are the receiver of the action. Note that usually subjects and objects contain only a single entity. In some cases, such as in the sentence *"A boy and a girl are seen playing football"*., multiple entities are the actors in a relation: *([boy, girl], play, [], football)*. With the above definition, we are able to capture relations with different types of verbs: single-word verbs and multi-word verbs, i.e., prepositional verbs (verb+preposition), phrasal verbs (verb+particle), and phrasal-prepositional verbs. In both cases, *verb* contains the verb. *Modifiers* is an empty list for single-word verbs. For multi-word verbs, however, *modifiers* contains the verb's particles and prepositions. Both, particles and prepositions, provide information about how the verb and the object are related to each other. For example, the relation *(girl, catches, [up, with], kids)* for the sentence *"The girl catches up with the other kids"*. is better understood than the relation *(girl, catches, kids)*.

As for entity extraction, the relation extraction method utilizes POS tags and dependency labels of tokens and coreference clusters. In fact, the entity extraction method is used here in order to determine valid subjects and objects for relations. The relation extraction method proceeds in three steps: search for candidate verbs, search for candidate tuples consisting of a subject and a verb, and search for corresponding objects and modifiers for each verb in a candidate tuple. In brief, these steps are conducted by analyzing the dependency tree, exploiting the POS tags, and exploiting WordNet.

The video-level relation extraction is finished here. Event-level relations are built by assigning temporal segments to the extracted relations. Here, for a determined relation, the event is the temporal segment of the corresponding sentence from which the relation was extracted.

3.7 Text Classification

Although our proposed framework performs video classification, it predicts the category of a video only with the textual information its DVC model provides. The motivation is to see how far a text classifier on generated video captions can correctly classify a video. As text classifier, we adopt BERT [5].

4 Experimental Apparatus

4.1 Datasets

We introduce the datasets to evaluate the DVC models and the different tasks of our metadata extraction framework.

Dense Video Captioning. We use the large-scale benchmark dataset ActivityNet Captions [10] to train and evaluate the dense video captioning models. It consists of 20k YouTube videos of various human activities split up into train/-val/test sets of 0.5/0.25/0.25. Each video is annotated with captioned events, each consisting of a descriptive sentence and a specific temporal segment to which the description refers. On average, each video is annotated with 3.65 temporally-localized sentences. Each captioned event on average covers 36 s and is composed of 13.5 words. Temporal segments in the same video can overlap in time, which enables DVC models to learn complex events and relations.

Entity, Property, and Relation Extraction. The information that ActivityNet Captions provides for each video is limited to captioned events, i.e., pairs of temporal segments and sentences. To be able to evaluate the entity, property, and relation extraction methods, we need information about depicted entities, properties of entities, and relations of videos. For this purpose, we utilize the gold standard captioned events of ActivityNet Captions' validation videos. We extract entities, their properties, and relations from the captioned events, and treat the results as gold standard for semantic metadata extraction. We generate five different datasets for videos in the ActivityNet Captions validation set: each one for event-level entities, video-level entities, entity-property pairs, event-level relations, and video-level relations. Using these datasets, we evaluate the framework's ability to extract the semantic metadata and compare it to metadata extracted from the captioned events generated by the DVC models.

Text Classification. We build a new dataset to train and evaluate our text classification method. Here, we take advantage of the fact that ActivityNet Captions consists of YouTube videos. For each video of the ActivityNet Captions' train and validation sets, we query the corresponding category from the YouTube Data API. We were able to query the category of 12,579 ActivityNet Captions videos (on March 20, 2022). We split the videos in train/val/test of 0.6/0.2/0.2, while ensuring that the category distribution is the same for all splits.

4.2 Procedure

Dense Video Captioning. We train MT and PDVC using the ActivityNet Captions dataset. For action recognition, both models adopt the same action recognition network, a pre-trained temporal segment network [24], to extract frame-level features. To ensure consistency, we evaluate both models on the ActivityNet Captions validation set (using both annotation files) and compare the performances with those reported in the corresponding works.

For our proposed semantic metadata extraction methods, the number of captioned events forwarded to each method is an important parameter. With increasing number of captioned events forwarded to an extraction method, it can extract more semantic information. In this work, we denote the number of captioned events that are generated by a DVC model and forwarded to a specific method as $|E|$. MT and PDVC, the dense video captioning models of our choices, internally calculate confidence scores for their generated captioned events. If the number of considered captioned events is limited, then the generated captioned events with the highest confident scores are used. For high values of $|E|$, DVC models tend to produce identical sentences for different temporal segments. However, our property extraction, and video-level entity and relation extraction methods rely mainly on the textual information that the captioned events provide. Therefore, captioned events with duplicate sentences do not provide further semantic information. Because of that, we only forward events with distinct captions to these methods. We denote the number of distinct captioned events that are forwarded to a specific method as $|dist(E)|$. If two events share the exact same caption, then the event with the higher confidence score is forwarded. We finally evaluate MT and PDVC with $|E|$ set to 10, 25, 50, and 100, and $|dist(E)|$ set to 1, 3, 10, and 25.

Entity, Property, and Relation Extraction. For both trained dense video captioning models, MT and PDVC, we extract event-level entities and relations from captioned events that they generate for ActivityNet Captions validation videos with $|E|$ set to 10, 25, 50, and 100. Video-level entities and relations, and entity-property pairs are extracted with $|dist(E)|$ set to 1, 3, 10, and 25. We then evaluate the entity, property, and relation extraction methods by comparing the extracted semantic metadata with our generated gold standards for entities, entity-property pairs, and relations.

For the evaluation of video-level entity extraction, we introduce the entity frequency threshold f. We extract video-level entities from ActivityNet Captions' train and validation videos. The frequency of a video-level entity is the number of different videos in which it occurs. We evaluate the generated video-level entities with f set to 0, 10, 25, and 50, meaning that the gold standard only contains those entities which have a frequency higher than the frequency threshold. Thus, a larger f means that the DVC models are given a higher chance to learn and reproduce the entities in the videos.

Text Classification. We train and evaluate the text classification model of our framework in three different settings: using the captioned events generated by the MT and PDVC model, respectively, and using the gold standard captioned events provided by ActivityNet Captions. This allows us to analyze how useful the automatically generated captioned events are for text classification compared to gold standard captioned events. For each data input, we set $|dist(E)|$ to 10.

4.3 Hyperparameter Optimization

For the training of the DVC models, we use largely the same parameters that were used to train the models in their original works (refer to MT [27] and PDVC [25] or their latest codebases). For MT, we have to switch to a batch size of 84 and a learning rate of 0.06 to make sure that the model training converges. In the default configuration, PDVC generates a maximum of 10 events per video. We change the corresponding parameter such that 100 captioned events are generated to ensure that there is enough information for our metadata extraction methods to work on. For both, MT and PDVC, we use the models' best states w.r.t. their METEOR score performances (see Sect. 4.4). For MT, this was achieved after 34 epochs, and for PDVC after 13 epochs.

For the entity, property, and relation extraction methods, no hyperparameter optimization is necessary. In our text classification model, we use an uncased BERT model. Each input text is truncated to 128 tokens. For training, we use cross-entropy loss and the Adam optimizer with 10% warmup steps. We use class weights to help the model learn on the imbalanced data of our classification dataset. While using captioned events from ActivityNet Captions as training data, we perform grid search over dropout rates in $\{0.0, 0.1, \ldots, 0.5\}$, maximum number of training epochs in $\{1, 2, \ldots, 5\}$, learning rates between $5e$-6 and $1e$-4, and a batch sizes in $\{1, 2, 4, 8\}$. The lowest validation loss was achieved after epoch 2 while using dropout of 0.1, 2 maximum training epochs, learning rate of $2e$-5, and batch size of 4. Using this parameter configuration, we train the model with each data input separately. We repeat the training three times and select the model that achieved the lowest final validation loss.

4.4 Measures and Metrics

Dense Video Captioning. To evaluate the dense video captioning models MT and PDVC, we use the official evaluation toolkit provided by ActivityNet Captions Challenge 2018[2] that measures the capability to localize and describe events in videos. For evaluation of dense video captioning, we report METEOR [11] and BLEU [17] scores. Using temporal Intersection over Union (tIoU) thresholds at {0.3, 0.5, 0.7, 0.9} for captioned events, we report recall and precision of their temporal segments and METEOR and BLEU scores of their sentences. Given a tIoU threshold, if the event proposal has a segment overlapping larger than the threshold with any gold standard segment, the metric score is computed for the generated sentence and the corresponding gold standard sentence. Otherwise, the metric score is set to 0. The scores are then averaged across all event proposals and finally averaged across all tIoU thresholds.

Entity, Property, and Relation Extraction. For property extraction and video-level entity and relation extraction, we measure micro-averaged precision, recall, and their harmonic mean, the F1 score. Similarly to the evaluation of DVC models, when evaluating entity and relation extraction on event-level, we also take the quality of the predictions' temporal segments into account. Here, we compute micro-averaged precision and recall across all videos using tIoU thresholds at {0.3, 0.5, 0.7, 0.9}, and then average the results across all thresholds. Additionally, we report the F1 score of averaged precision and recall values.

For the evaluation of entities, entity-property pairs, and relations, we use WordNet to compare words in terms of whether they are synonyms of each other or not. This is fair as the variety of extracted semantic metadata is large. For example, *"stand up"* is considered a correct prediction for the verb *"get up"*. However, this means that for a video the number of predictions that are treated

[2] See: https://github.com/ranjaykrishna/densevid_eval/.

as correct (the set TP_p) and the number of gold standard targets (denoted as TP_g) is not necessarily the same. With other words, the gold standard is enriched with synonyms. For example, $|TP_p| = 2$ and $|TP_g| = 1$ when accepting *person* and *individual* as synonymously correct predictions for $TP_g = \{person\}$. To ensure the validity of precision (proportion of correct predictions) and recall (proportion of correctly predicted targets), we use TP_p for calculating precision and TP_g for recall, respectively.

All entities, properties, and relations are compared using their word lemmas. On video-level, this means that potential duplicates of entities and relations are removed, i.e., they are unique. For example, $\{man, men\}$ results in $\{man\}$.

Text Classification. We report weighted and macro-averages of precision, recall, and F1 score across all categories.

5 Results

5.1 Dense Video Captioning

We introduced the parameters $|E|$ and $|dist(E)|$, which are primarily used in our framework to control the amount of information that is forwarded from the DVC model to the semantic metadata extraction methods. Using these parameters, we are also able to compare the DVC models in a fair way, meaning that their performances are evaluated while generating an equal number of captured events. Table 1 shows the event localization and dense video captioning performances of MT and PDVC with respect to $|E|$ and $|dist(E)|$, the number of (distinct) captioned events that are generated by each DVC model and used for the evaluation. For event localization, PDVC constantly achieves better recall than MT, while MT constantly outperforms PDVC in terms of precision. For both models, higher values of $|E|$ and $|dist(E)|$ results in better recall and worse precision performance in event localization. Looking at the results for dense video captioning, we can state that the METEOR performances of both models degrade with increasing number of captioned events, both distinct and non-distinct, except for when increasing $|dist(E)|$ from 1 to 3. When increasing $|E|$ from 10 to 100, the METEOR performance of PDVC drops by 3.71 points in total, while the METEOR performance of MT drops by 1.21 points. This is a consequence of the declining precision for event localization for higher $|E|$, which affects the PDVC model more heavily than the MT model.

5.2 Entity Extraction

Table 2 shows precision, recall, and F1 score performances of our framework for **video-level entity** extraction. We evaluate the extracted video-level entities with entity frequency threshold f set to 0, 10, 25, and 50. One observation is that, with increasing $|dist(E)|$, precision decreases and recall improves for both models and all thresholds. For both DVC models and all thresholds, the best F1

Table 1. Event localization and dense video captioning results of MT and PDVC for different numbers of generated captioned events on the ActivityNet Captions validation set. We report recall and precision of temporal segments, METEOR (M), and BLUE@N (B@N) of generated captioned events.

Event Localization & Dense Video Captioning						
DVC Model	$\|E\|$	avg. Recall	avg. Precision	2018 eval. toolkit B@3	B@4	M
MT	10	43.61	48.96	2.18	1.02	5.89
	25	56.55	46.15	2.33	1.14	5.74
	50	67.70	41.60	2.31	1.13	5.35
	100	76.33	34.78	2.12	1.04	4.68
PDVC	10	61.88	45.41	3.10	1.59	6.34
	25	73.81	36.89	2.54	1.23	5.17
	50	78.76	25.75	1.92	0.90	3.88
	100	82.24	15.63	1.36	0.63	2.63

(a) Results for varying numbers of generated captioned events $|(E)|$.

Event Localization & Dense Video Captioning						
DVC Model	$\|dist(E)\|$	avg. Recall	avg. Precision	2018 eval. toolkit B@3	B@4	M
MT	1	22.04	50.99	1.55	0.68	5.37
	3	32.72	49.65	1.97	0.90	5.76
	10	50.08	44.75	2.22	1.05	5.58
	25	63.67	36.23	2.05	0.97	4.80
PDVC	1	20.05	49.63	2.52	1.30	5.86
	3	40.28	48.26	2.98	1.54	6.46
	10	62.38	44.10	2.85	1.41	6.01
	25	71.62	31.69	2.14	1.00	4.54

(b) Results for varying numbers of generated distinct captioned events $|dist(E)|$.

score performance is achieved with $|dist(E)| = 10$. This indicates a limited level of semantic information that any further generated captioned events provide. For higher entity frequency thresholds, for both DVC models and all $|dist(E)|$, our framework is able to predict video-level entities with improved recall, at the cost of only slightly lower precision. When our framework is using the PDVC model, it achieves better precision and F1 scores for all $|dist(E)|$. The framework achieves better recall performances when using the MT model for $|dist(E)|$ set to 10 and 25. Overall, for video-level entity extraction, our framework achieves its highest F1 scores when using the PDVC model with $|dist(E)|$ set to 10. Here, the achieved F1 scores range from 31.27 for $f = 0$ to 34.21 for $f = 50$.

Table 3 shows the results of **event-level entity** extraction of our framework. Depending on the framework's used DVC model and the number of generated captioned events $|E|$, we report precision and recall for different temporal Intersection over Union (tIoU) thresholds. F1 score is calculated using the averages

Table 2. Video-level entity extraction using dense video captioning models MT and PDVC. We report precision, recall, and F1 score for different numbers of generated distinct captioned events $|dist(E)|$ and entity frequency thresholds f.

Video-level Entity Extraction													
DVC Model	$\|dist(E)\|$	Precision(@f)				Recall(@f)				F1(@f)			
		0	10	25	50	0	10	25	50	0	10	25	50
MT	1	39.94	39.88	39.66	38.91	15.38	16.49	17.57	18.99	22.21	23.33	24.35	25.52
	3	33.91	33.83	33.61	32.78	23.44	25.12	26.72	28.70	27.72	28.83	29.77	30.60
	10	26.44	26.37	26.16	25.43	32.59	34.90	37.06	39.67	29.20	30.04	30.67	30.99
	25	20.76	20.69	20.49	19.82	40.69	43.55	46.16	49.15	27.49	28.05	28.38	28.25
PDVC	1	45.13	45.05	44.91	44.43	15.44	16.56	17.68	19.28	23.01	24.22	25.37	26.89
	3	39.01	38.95	38.83	38.28	23.33	25.03	26.72	29.04	29.20	30.48	31.66	33.03
	10	31.77	31.71	31.60	31.03	30.78	33.01	35.21	38.11	31.27	32.35	33.30	34.21
	25	26.30	26.25	26.14	25.61	36.18	38.79	41.33	44.66	30.46	31.31	32.02	32.55

Table 3. Event-level entity extraction results. Reported are precision and recall for different numbers of generated distinct captioned events $|E|$ and tIoU thresholds. The F1 scores are the averages of precision and recall across all thresholds.

Event-level Entity Extraction												
DVC Model	$\|E\|$	Precision(@tIoU)					Recall(@tIoU)					F1
		0.3	0.5	0.7	0.9	Avg	0.3	0.5	0.7	0.9	Avg	
MT	10	28.97	20.08	8.55	1.26	14.72	21.11	15.31	9.59	2.94	12.24	13.37
	25	28.12	18.44	7.40	1.03	13.75	26.48	20.66	14.29	5.58	16.75	15.10
	50	26.30	15.93	6.10	0.83	12.29	31.15	25.32	18.69	8.47	20.91	15.48
	100	23.31	12.70	4.52	0.59	10.28	36.35	30.05	22.89	11.04	25.08	14.58
PDVC	10	30.47	19.79	9.11	2.53	15.47	26.36	21.13	14.52	5.87	16.97	16.19
	25	27.16	15.83	6.41	1.53	12.73	31.14	26.34	20.01	8.21	21.42	15.97
	50	20.75	10.49	3.84	0.86	8.98	33.66	28.89	22.40	8.75	23.42	12.98
	100	13.76	6.36	2.22	0.48	5.71	35.35	30.65	24.00	9.02	24.75	9.28

of precision and recall. Note that for a prediction to be correct for an event-level entity, a condition is that its temporal segment overlaps with the gold standard entity's temporal segment larger than the tIoU threshold. Therefore, precision and recall decreases at higher tIoU thresholds. Regardless of the DVC model used, the framework's precision decreases for higher $|E|$, while at the same time it benefits in recall performance. When using the PDVC model, the framework's precision drops more (1.13 on average) when increasing $|E|$ from 10 to 100 as compared to when it is using the MT model (0.57 on average). Note that we made the same observation when we evaluated the event localization and dense video captioning performances of the DVC models. For event localization, the PDVC model could not convert the large drops in precision for higher $|E|$ into better recall performance. Thus, on the one hand, this results in worse dense video captioning performance. On the other hand, when our framework is using the PDVC model for event-level entity extraction, this results in degrading F1 score performance for higher $|E|$. Still, our framework achieves its highest F1 score with 16.19 when it uses the PDVC model with $|E| = 10$. On the other hand, when using the MT model, the highest achieved F1 score is 15.48 for $|E| = 50$.

5.3 Property Extraction

The results of the extraction of entity-property pairs with our framework are shown in Table 4. In general, increasing $|dist(E)|$ leads to drops in precision, however, our framework benefits from much improved recall. Consequently, our framework achieves its highest F1 scores for $|dist(E)| = 25$. In contrast to video-level entity extraction where highest F1 scores were achieved for $|dist(E)| = 10$, we can observe that increasing $|dist(E)|$ from 10 to 25 leads to even better property extraction performance with respect to the F1 score, indicating that the DVC models still provide meaningful semantic information about the properties of entities when many captioned events are generated. For all $|dist(E)|$, our framework achieves better recall and F1 scores when using the MT model for captioned events generation and better precision when using the PDVC model. The highest achieved precision is 9.08 when using PDVC with $|dist(E)| = 1$. Using the MT model with $|dist(E)| = 25$ results in the framework's highest achieved recall (8.86) and F1 score (4.94).

Table 4. Results for the extraction of entity-property pairs.

Property Extraction				
DVC Model	$\|dist(E)\|$	Prec.	Rec.	F1
MT	1	6.48	0.82	1.45
	3	6.64	1.85	2.90
	10	5.66	3.54	4.36
	25	4.53	5.43	4.94
PDVC	1	9.08	0.72	1.34
	3	8.76	1.61	2.72
	10	6.96	2.68	3.87
	25	4.79	3.76	4.21

5.4 Relation Extraction

Table 5 shows the results of video-level relation extraction with our framework. In general, using the PDVC model leads to better precision performance except for $|dist(E)| = 1$, while using the MT model leads to better recall performance except for $|dist(E)| = 3$. For both DVC models, our framework achieves its highest F1 scores for $|dist(E)| = 10$. This suggests that any larger number of generated captioned events can provide more semantic information for relations only at a higher cost of precision, the same observation as we made for video-level entity extraction. However, for property extraction, the highest F1 scores were achieved for $|dist(E)| = 25$. The highest achieved F1 score of our framework for video-level relation extraction is 5.02 while using the PDVC model with $|dist(E)| = 10$.

The results for event-level relation extraction are shown in Table 6. Very similar observations can be made as for event-level entity extraction. The framework's precision decreases for higher $|E|$ while benefiting in recall performance. When using the PDVC model, the framework's precision performance suffers more for higher $|E|$ compared to when it is using the MT model. As observed before for event-level entity extraction, this is not converted into much higher recall, resulting in a degradation of the framework's F1 score performance. Therefore, for $|E|$ set to 50 and 100, i.e., high numbers of generated captioned events, our framework achieves its highest F1 scores when using the MT model, while for $|E|$ set to 10 and 25 the highest F1 scores are achieved when using the PDVC model. The event-level relation extraction achieves its best performance with respect to F1 score when it uses the PDVC model with $|E| = 10$.

Table 5. Results for video-level relation extraction.

Video-level Relation *Extraction*				
DVC Model	$\|dist(E)\|$	Prec.	Rec.	F1
MT	1	5.76	2.07	3.04
	3	4.88	4.07	4.44
	10	3.64	6.61	4.70
	25	2.89	8.86	4.35
PDVC	1	5.64	2.06	3.02
	3	5.02	4.18	4.56
	10	4.09	6.50	5.02
	25	3.47	8.36	4.91

Table 6. Experimental results for event-level relation extraction.

Event-level Relation *Extraction*												
DVC Model	$\|E\|$	Precision(@tIoU)					Recall(@tIoU)					F1
		0.3	0.5	0.7	0.9	Avg	0.3	0.5	0.7	0.9	Avg	
MT	10	3.76	2.52	1.08	0.14	1.87	3.81	2.81	1.73	0.42	2.20	2.02
	25	3.61	2.28	0.93	0.12	1.74	4.89	3.90	2.56	0.87	3.05	2.22
	50	3.35	1.99	0.77	0.10	1.55	5.87	4.78	3.29	1.32	3.81	2.20
	100	2.96	1.59	0.56	0.07	1.30	7.06	5.60	3.91	1.63	4.55	2.02
PDVC	10	3.64	2.33	1.14	0.32	1.86	4.87	3.87	2.63	1.06	3.11	2.33
	25	3.33	1.92	0.80	0.20	1.56	6.01	4.95	3.49	1.37	3.95	2.24
	50	2.66	1.33	0.50	0.11	1.15	6.53	5.37	3.81	1.41	4.28	1.81
	100	1.75	0.82	0.30	0.07	0.73	6.75	5.53	3.91	1.41	4.40	1.25

5.5 Text Classification

Finally, Table 7 shows the weighted and macro-averages of precision, recall, and F1 score performances of the framework's text classification method, while trained and evaluated in three different settings: using the captioned events generated by the MT or PDVC model, respectively, and using the captioned events provided by ActivityNet Captions. The most noteworthy observation is that the classification performance of our framework, when using captioned events generated by MT and PDVC, is not far from the classification performance that is achieved when using gold standard captioned events of ActivityNet Captions. Therefore, we can state that the DVC models generate specific semantic information for videos of different categories at a similar level as the captioned events of the gold standard provide. This is important for the text classifier to categorize videos successfully.

When using automatically generated captioned events for video classification, our framework achieves its best performances for all metrics, both weighted and macro-averaged, when using the PDVC model. Here, for weighted precision, recall, and F1 score, our framework performs around 2 points better as compared to when using the MT model. When using PDVC, our framework achieves an overall accuracy (i.e., weighted recall) of 50.22, which is only 0.59 points lower than the accuracy achieved when classifying videos using the gold standard captioned events of ActivityNet Captions. Taking the category imbalance in our dataset into account, we observe that the achieved macro-averages of precision, recall and F1 scores are much lower as compared to their weighted-averages.

Table 7. Results for classification of video captions in the settings: (i) captioned events generated by MT, (ii) PDVC, and (iii) gold standard captioned events from ActivityNet Captions. For MT and PDVC, $|dist(E)| = 10$ is used.

Text Classification				
Averaging	Captioned events input	Prec.	Rec.	F1
Weighted	MT	43.72	48.24	44.80
	PDVC	45.75	50.22	46.96
	ActivityNet Captions	46.97	50.81	48.23
Macro	MT	27.51	30.45	28.08
	PDVC	34.31	32.30	30.88
	ActivityNet Captions	33.42	34.74	33.19

6 Discussion

6.1 Key Results

The experiments show that our proposed framework is able to automatically generate multiple types of semantic metadata in a meaningful way. We have to keep in mind that in our framework semantic metadata is extracted only from captioned events that are automatically generated by its DVC model, i.e., a model that is designed and trained for a different task. To extract higher quality semantic information for each semantic metadata type, a dedicated computer vision method could be used, such as video object detection and video visual relation tagging. Here, however, our framework trades quality for effectiveness as it only requires training for one computer vision model, the DVC model. We must also bear in mind that the variety of entities, entity-property pairs, and relations that occur in our gold standards, and that are extracted by our framework, is large, in particular when compared to related computer vision tasks. For example, the VidOR (Video Object Relation) dataset [21,23], which is used to train models for visual relation detection from videos, contains annotations of 80 categories of objects and 50 categories of relation predicates. In our gold standard for relations, however, 2,493 different entities acted either as subject (905 entities) or object (2,299 entities), while 905 different verbs occur in the relations. Also, for video-level entity extraction, we observed that for higher entity frequency thresholds f, the framework's precision decreases only slightly, while recall performance improves greatly. This observation is not surprising, as the number of different entities in our gold standard for entities, which is based on ActivityNet Captions, is large, and many entities occur only a few times. This leads to the conclusion that the DVC models are only able to learn entities when there is enough training data, i.e., the models see them sufficiently enough during training. Regarding the video classification results (based on the generated captions), our framework could not reliably predict the video category. This is because the captioned events generated by the DVC models do not contain sufficiently specific semantic information for videos of different categories. Here, a visual-based video classifier is preferred over a purely text-based approach.

6.2 Threats to Validity and Future Work

We generated a gold standard using the captioned events of ActivityNet Captions validation videos. As described in Sect. 4.1, we use our proposed entity, property, and relation extraction methods on the processed captioned events, and treat the results as gold standards for semantic metadata. This requires that the extraction methods work sufficiently well, i.e., they are able to extract semantic metadata using the linguistic annotations provided by the framework's language parser. In order to validate this hypothesis, we annotated a subset of 25 ActivityNet Captions videos with the video-level entities, entity-property pairs, and video-level relations that we expect the methods to extract from captioned events. In total, we made annotations of 110 captioned events in these 25 videos. Table 8 shows the precision and recall performances of our entity, property, and relation extraction methods on these manually annotated videos. The results show that our entity and property extraction methods are certainly reliable. In some cases, however, our entity extraction method wrongly determines entities. For example, for the sentence *"the camera pans around the field"*, spaCy wrongly classifies *camera*, *pans* and *field* as nouns, while *pans* actually acts as verb in the sentence.

Since *pan* is listed in WordNet as a noun, *pans* is still determined as an entity. The relation extraction method is able to determine only around 70% of the relations. This is due to the complexity of relation extraction.

Table 8. Evaluation of the entity, property, and relation extraction methods on captioned events from 25 manually annotated videos.

Semantic Metadata Type	Prec.	Rec.
Entities (video-level)	94.21	98.39
Properties	92.98	91.38
Relations (video-level)	78.36	70.62

7 Conclusion

We presented a framework for metadata extraction of various types from generated video captions. The metadata quality mainly depends on two factors: The event localization and video captioning performance of the dense video captioning model, and the number of captioned events forwarded from the dense video captioning model to the semantic metadata extraction methods. This opens the path for future research on integrated models for semantic metadata extraction.

References

1. Aafaq, N., Mian, A., Liu, W., Gilani, S.Z., Shah, M.: Video description: a survey of methods, datasets, and evaluation metrics. ACM Comput. Surv. **52**(6) (2019)
2. Algur, S., Bhat, P.: Metadata construction model for web videos: a domain specific approach. IJECS **3** (2014)
3. Algur, S.P., Bhat, P.: Web video mining: metadata predictive analysis using classification techniques. IJ Inf. Technol. Comput. Sci. **2** (2016)
4. Del Corro, L., Gemulla, R.: Clausie: clause-based open information extraction. In: World Wide Web. ACM (2013)
5. Devlin, J., Chang, M., Lee, K., Toutanova, K.: BERT: pre-training of deep bidirectional transformers for language understanding. CoRR abs/1810.04805 (2018)
6. Fader, A., Soderland, S., Etzioni, O.: Identifying relations for open information extraction. In: EMNLP. ACL (2011)
7. Galke, L., Scherp, A.: Bag-of-words vs. graph vs. sequence in text classification: questioning the necessity of text-graphs and the surprising strength of a wide MLP. In: Association for Computational Linguistics. ACL (2022)
8. Jiao, L.: New generation deep learning for video object detection: a survey. IEEE Trans. Neural Netw. Learn. Syst. (2021). https://doi.org/10.1109/TNNLS.2021.3053249
9. Kim, Y.: Convolutional neural networks for sentence classification. In: EMNLP. ACL (2014)
10. Krishna, R., Hata, K., Ren, F., Fei-Fei, L., Niebles, J.C.: Dense-captioning events in videos. In: ICCV (2017). https://doi.org/10.1109/ICCV.2017.83

11. Lavie, A., Agarwal, A.: METEOR: an automatic metric for MT evaluation with high levels of correlation with human judgments. In: Workshop on Statistical Machine Translation. ACL (2007)
12. Li, Q., et al.: A survey on text classification: from shallow to deep learning. CoRR abs/2008.00364 (2020)
13. Li, Y., Yao, T., Pan, Y., Chao, H., Mei, T.: Jointly localizing and describing events for dense video captioning. In: 2018 IEEE/CVF Conference on Computer Vision and Pattern Recognition (2018). https://doi.org/10.1109/CVPR.2018.00782
14. Long, X., Gan, C., de Melo, G., Wu, J., Liu, X., Wen, S.: Attention clusters: purely attention based local feature integration for video classification. In: CVPR (2018). https://doi.org/10.1109/CVPR.2018.00817
15. Miller, G.A.: WordNet: a lexical database for English. Commun. ACM **38**(11) (1995)
16. Niklaus, C., Cetto, M., Freitas, A., Handschuh, S.: A survey on open information extraction. In: International Conference Computer Linguistics. ACL (2018)
17. Papineni, K., Roukos, S., Ward, T., Zhu, W.J.: Bleu: a method for automatic evaluation of machine translation. In: ACL. ACL (2002)
18. Sarhan, I.; Spruit, M.; Atzmueller, M.; Duivesteijn, W.: Uncovering algorithmic approaches in open information extraction: a literature review. In: 30th Benelux Conference on Artificial Intelligence (2018). https://dspace.library.uu.nl/handle/1874/374300
19. Sarvas, R., Herrarte, E., Wilhelm, A., Davis, M.: Metadata creation system for mobile images. In: MobiSys. ACM (2004)
20. Scherer, J., Scherp, A., Bhowmik, D.: Semantic metadata extraction from dense video captioning. CoRR abs/2211.02982 (2022). https://doi.org/10.48550/arXiv.2211.02982
21. Shang, X., Di, D., Xiao, J., Cao, Y., Yang, X., Chua, T.S.: Annotating objects and relations in user-generated videos. In: Multimedia Retrieval. ACM (2019)
22. Shang, X., Ren, T., Guo, J., Zhang, H., Chua, T.S.: Video visual relation detection. In: Multimedia. ACM (2017)
23. Thomee, B., et al.: YFCC100M: the new data in multimedia research. Commun. ACM **59**(2) (2016)
24. Wang, L., et al.: Temporal segment networks: towards good practices for deep action recognition. In: Leibe, B., Matas, J., Sebe, N., Welling, M. (eds.) ECCV 2016. LNCS, vol. 9912, pp. 20–36. Springer, Cham (2016). https://doi.org/10.1007/978-3-319-46484-8_2
25. Wang, T., Zhang, R., Lu, Z., Zheng, F., Cheng, R., Luo, P.: End-to-end dense video captioning with parallel decoding. ICCV (2021)
26. Zhou, H., et al.: Exploring emotion features and fusion strategies for audio-video emotion recognition. In: 2019 International Conference on Multimodal Interaction (2019). https://doi.org/10.1145/3340555.3355713
27. Zhou, L., Zhou, Y., Corso, J.J., Socher, R., Xiong, C.: End-to-end dense video captioning with masked transformer. In: CVPR (2018). https://doi.org/10.1109/CVPR.2018.00911

Fine-Tuning Language Models for Scientific Writing Support

Justin Mücke, Daria Waldow, Luise Metzger, Philipp Schauz, Marcel Hoffman, Nicolas Lell, and Ansgar Scherp(✉)

Universität Ulm, Ulm, Germany
{justin.muecke,daria.waldow,luise.metzger,philipp.schauz, marcel.hoffman,nicolas.lell,ansgar.scherp}@uni-ulm.de

Abstract. We support scientific writers in determining whether a written sentence is scientific, to which section it belongs, and suggest paraphrasings to improve the sentence. Firstly, we propose a regression model trained on a corpus of scientific sentences extracted from peer-reviewed scientific papers and non-scientific text to assign a score that indicates the scientificness of a sentence. We investigate the effect of equations and citations on this score to test the model for potential biases. Secondly, we create a mapping of section titles to a standard paper layout in AI and machine learning to classify a sentence to its most likely section. We study the impact of context, i. e., surrounding sentences, on the section classification performance. Finally, we propose a paraphraser, which suggests an alternative for a given sentence that includes word substitutions, additions to the sentence, and structural changes to improve the writing style. We train various large language models on sentences extracted from arXiv papers that were peer reviewed and published at A*, A, B, and C ranked conferences. On the scientificness task, all models achieve an MSE smaller than 2%. For the section classification, BERT outperforms WideMLP and SciBERT in most cases. We demonstrate that using context enhances the classification of a sentence, achieving up to a 90% F1-score. Although the paraphrasing models make comparatively few alterations, they produce output sentences close to the gold standard. Large fine-tuned models such as T5 Large perform best in experiments considering various measures of difference between input sentence and gold standard.

Code is provided here: https://github.com/JustinMuecke/SciSen.

Keywords: Scientific Writing · Language Models · Paraphrasing

1 Introduction

Scientific writing is a complex task with many resources helping researchers and students write better text [2,41]. A good structure and language facilitate the readers' understanding of the relevant content. Sentences in scientific papers can

A. Holzinger et al. (Eds.): CD-MAKE 2023, LNCS 14065, pp. 301–318, 2023.
https://doi.org/10.1007/978-3-031-40837-3_18

be expected to follow a certain scientific style, which is distinct from colloquial texts. A typical structure of research papers with methodological and empirical contributions such as in AI and machine learning is the section sequence of an introduction, related work, methods, results, discussion, conclusion, and optionally an appendix [36]. Although these sections might vary based on writing styles and problem-specific content (e. g., in machine learning literature, the methods section is often separated into method and experimental apparatus), readers expect to find certain pieces of information in certain sections. Placing content into sections contrary to a reader's expectation makes it more difficult to find said information. We investigate whether this structural clarity is better reflected in published papers of higher quality (i. e., CORE database rankings[1]) compared to less prestigious publications. Besides structural clarity, finding the best phrasing is a challenge, since a sentence with the same meaning can be phrased in many different ways. While there are already solutions for related sub-tasks of sentence paraphrasing [3,13,14,18,20,26,39], these are not specific to the domain of scientific papers. Other tools like Grammarly[2] and ChatGPT[3] are limited to online use only and do not guarantee any data protection. We propose a simple training procedure for paraphrasers to perform insertions, deletions, and modifications on the input text and apply it to state-of-the-art paraphrasers on scientific text. In summary, our contributions are:

(i) Scientificness score: We train regression models to discern non-scientific from scientific sentences by determining a sentence's scientificness.
(ii) Section classification: We train multi-label classifiers to indicate to which sections a sentence belongs. Additionally, we investigate the impact of the context length and scientific quality (i. e., CORE conference rank) of the input sentence on the classification performance.
(iii) Sentence paraphrasing: We fine-tune the language models BART [19] and T5 v1.1 [29] small, base, and large. We evaluate the sentence paraphrasing on these models as well as using Pegasus [42] and GPT-2 [28].

The paper is structured as follows: We summarize the related work in Sect. 2. The experimental apparatus is described in Sect. 3. The results are reported in Sect. 4 and discussed in Sect. 5, before we conclude.

2 Related Work

We discuss the literature on language models and their capabilities on our tasks, i. e., scoring, multi-label classification, and paraphrasing. We provide a brief overview of existing commercial tools for writing assistance to further demonstrate the relevance of this area of research.

---

[1] http://portal.core.edu.au/conf-ranks/.
[2] https://app.grammarly.com/.
[3] https://openai.com/blog/chatgpt.

2.1 Pre-trained Encoder Language Models

Encoder-only language models learn representations for each token of an input sequence. BERT [7] is a encoder-only language model pre-trained using masked language modelling (MLM) and next sentence prediction (NSP). The pre-trained model can be fine-tuned on various downstream tasks [7]. There are many variations of BERT [1,5,19,21] with SciBERT [1] being the most relevant to us. It is pre-trained on a corpus of scientific papers from bio-medicine and computer science to increase its performance in those domains [12]. Due to the high computational cost for pre-training, ELECTRA [5] aims to increase pre-training efficiency. ELECTRA uses two neural networks, a generator which is discarded after training and a discriminator. The generator plausibly substitutes masked tokens from an input sentence. The discriminator has to distinguish between tokens from the original input sequence and tokens generated by the generator. This way, each token of the input sequence contributes to the loss of the discriminator, instead of only the masked tokens as in BERT.

2.2 Pre-trained Decoder Language Models

Decoder language models [19,28,29] are designed for text generation. They take textual input and generate a new output sequentially token by token. Auto-regressive decoders use already generated tokens to generate the following tokens [19,28,29].

We use four language models for paraphrasing, namely BART [19], Pegasus [42], T5 v1.1 [29], and GPT-2 [28]. BART [19] is a general-purpose sequence-to-sequence model that adds a left-to-right auto-regressive decoder to BERT. Pegasus [42] is a sequence-to-sequence language model trained by gap-sentence generation, which is comparable to MLM, but masks whole sentences instead of words. We use a variant of Pegasus fine-tuned on paraphrasing [30]. The encoder-decoder model T5 [29] introduces the concept of task instructions such as translation, classification, and summarization as part of the prompts. These instructions are provided to T5 while being fine-tuned on multiple tasks at the same time. We use a version of T5 that is yet not fine-tuned on multiple tasks, i. e., does not provide a token-based task execution. Instead, we fine-tune our T5 (in version 1.1) on the task of paraphrasing as this is the only task we want to perform. Thus, we omit using task-specific prefixes during fine-tuning. GPT-2 [28] is a decoder-only model and is trained on the next token prediction objective. It generates text in an auto-regressive manner to continue the prompt.

2.3 Text Classification

A classification task can be either single-label or multi-label. In multi-label classification, a classified object can be associated with none, one, or more classes. BERT-based architectures achieve state-of-the-art results in many tasks, including single-label and multi-label text classification [9,10]. For scientific texts, SciBERT has been used to perform citation intent classification [24] and classification of paper titles and abstracts to research disciplines [11]. SciBERT and

BioBERT outperformed BERT on texts from STEM domains, but were outperformed on texts on language or history [11]. The performance of these models can be influenced by characteristics of the input data, e. g., adding document context can improve task performance [22]. Galke et al. [9] showed that WideMLP [10], a Multi-Layer Perceptron (MLP) model with a wide hidden layer, is a strong baseline for text classification in both the single-label and multi-label scenarios. However, BERT achieved state-of-the-art results in text classification for various datasets. To the best of our knowledge, there has been no attempt to classify sentences of scientific text according to their corresponding section.

2.4 Sentence Transformation and Paraphrasing

A common task performing sentence transformations is Neural Machine Translation (NMT) [6]. Many approaches for machine translation require large amounts of training data [13,26,39], with transformers achieving state-of-the-art performance [39]. Besides translation, a text can be transformed by paraphrasing, which changes an existing sentence while preserving its meaning [17,23,37]. Many paraphrasing approaches are limited to word-level changes [18,20]. Rudnichenko et al. [32] propose a system that paraphrases individual sentences, including changes to the word order. These sentence transformation methods are all supervised, i. e., the training datasets have a parallel corpus containing two versions of each sentence. Such datasets are expensive to create. To tackle this challenge, unsupervised paraphrasing approaches create training data by inserting, replacing, or deleting words from a sentence [3,14,18,20]. Other approaches create multiple alternative sentences [13,18] and apply evaluation methods on each suggestion. A special case of paraphrasing is text style transfer (TST) which aims to change the style of a text to imitate a specific writing style [15,16].

2.5 Tools to Improve Writing Quality

There are various tools to assess writing quality, which target spelling mistakes, grammar errors, long sentences, and suggest paraphrases. Specifically, Writefull[4] is a tool for scientific writing. It allows sentence paraphrasing and is trained on scientific text, which sets it apart from tools for general English language. QuillBot[5] provides paraphrasing for general writing. LanguageTool[6] and Grammarly[7] provide general spelling and grammar improvements. However, even if a sentence is grammatically, orthographically, and semantically correct, it could still be non-scientific in style. Recent developments suggest that these tasks can be tackled by tools like ChatGPT (based on InstructGPT [25]). However, it cannot be used offline, is expensive to run, and does not guarantee data protection. Thus, we train large language models ourselves to perform paraphrasing tasks on a local infrastructure.

[4] https://www.writefull.com/.
[5] https://quillbot.com/.
[6] https://languagetool.org/.
[7] https://www.grammarly.com/.

3 Experimental Apparatus

In this section, we describe the datasets, the preprocessing, the procedure for each task, the hyperparameter search, and the evaluation measures.

3.1 Datasets

We use papers published on arXiv until May 2022 with LaTeX available that were accepted at A*, A, B, and C ranked conferences of the Australian CORE2021 database. To map papers with their respective conferences, we use the Papers With Code database[8]. Since we extract the structure of the papers, we drop all papers that are not using any `\section{...}` command in LaTeX. Overall, we have a total of 26,201 papers, of which 21,774 are from A*, 3,665 from A, 530 from B, and 232 from C-ranked conferences.

For the scientificness score task, we complement our arXiv text with non-scientific sentences from Reddit comments[9], sci-fi stories[10], and subsets of different Twitter datasets [27,38]. For the section task, we can use our arXiv dataset as-is. Finally, for paraphrasing, we create two parallel datasets by reducing the quality of the sentences, e. g., replacing words with colloquial synonyms. The first dataset is Pegasus-DS, which is created by changing sentences using Pegasus fine-tuned for paraphrasing [42]. The second dataset IDM-DS is created by randomly inserting, deleting, and modifying up to half of the tokens of each sentence based on MLM using BERT [7]. To evaluate the paraphraser, we additionally use Grammarly's Yahoo Answers Formality Corpus (GYAFC) [31] for testing. GYAFC contains informal and formal sentences with four human-written paraphrases. We use 1,332 sentences from the category family and relationships, as the dataset provides output sentences from other models in this category. The statistics of the datasets are summarized in Table 1.

3.2 Preprocessing

Citations and references were replaced by a `<reference>`-token. In the case of `\citeauthor`, `\citet`, etc., which produces author names in the LaTeX output, we insert a random name[11] to preserve the structure of the sentence. Math syntax was replaced by `<equation>`-tokens. For the section classifier, we remove all sections with titles that cannot be mapped to one of our predefined categories, i. e., the classes our models are trained on. These classes are "introduction", "related work", "method", "experiment", "result", "discussion", and "conclusion". Section titles extracted from the papers that fall into more than one category are mapped to all of the categories they consist of. For example, the paper

[8] https://production-media.paperswithcode.com/about/papers-with-abstracts.json.gz.

[9] https://files.pushshift.io/reddit/comments/.

[10] https://www.kaggle.com/datasets/jannesklaas/scifi-stories-text-corpus.

[11] https://www.kaggle.com/datasets/jojo1000/facebook-last-names-with-count.

Table 1. The number of sentences in the datasets and the sentences removed by applying filters. The filters remove sentences with non-ASCII characters, minimum length threshold, maximum length threshold, and if they contained a non-capitalized first character or did not end with a punctuation.

Dataset name	Number	Filter					Remaining
		ASCII	Short	Long	First	Last	
arXiv	5,283,451	51,201	61,705	1,905	197,081	12,696	4,958,863
w. section ID	2,864,755	27,357	32,110	790	110,467	6,667	2,687,364
Books	1,763,465	0	149,215	1,006	10,673	0	1,613,244
Reddit	279,288	11,774	51,582	340	5,638	0	217,225
Twitter	268,419	233,272	241	9	0	0	35,108

section entitled Introduction and Background is mapped internally to the two classes "introduction" and "related work". Our corpus includes machine learning papers containing `[MASK]` as a word. Since the insert, delete, and modify (IDM) process recognizes `[MASK]` as a special input token, we removed the brackets in the IDM-DS dataset.

We split the input at end-of-sentence punctuation symbols `.`, `?`, and `!` to obtain sentences. As documented in Table 1, we drop sentences containing non-ASCII characters to ensure that classification tasks are not trivial due to emojis or similar characters. We limit the length of extracted sentences to be at least 4 and at most 100 words. The upper limit was set as five times the average sentence-length in non-fiction writing as well as five times the highest recommended sentence length in English writing [33]. We filter sentences that do not follow basic orthography, i. e., that do not start with a capital letter or end with end-of-sentence punctuation.

3.3 Procedure

Scientificness Score. We then fine-tune BERT base [7] and SciBERT [1] and train a Bag-of-Words WideMLP [9] with one hidden layer from scratch to predict the scientificness score. This is interpreted as a regression score, where we assign a score of 0.9 to scientific sentences and 0.1 to non-scientific sentences during training. We evaluate whether the conference rank of the paper affects the models' scores.

Furthermore, we investigate the effect of using the `<equation>` and `<reference>` tokens by separately evaluating scientific sentences with and without such tokens. We also add these tokens to 100,000 randomly sampled sentences from the Books dataset and compare the scores of the modified (i. e., with tokens ingested) and original sentences.

Section Classification. We use BERT base [7], SciBERT [1], and a Bag-of-Words WideMLP [9,10] with one hidden layer. Since a sentence might have multiple section labels, we train the models as multi-label classifiers. We examine the influence of the amount of context provided as input to the model by varying the context length in training and testing. The input contexts provided to the models are a single-sentence, two sentences, and three sentences (up to BERT's maximum input length of 512 tokens). Two-sentence input contains the sentence of interest plus its predecessor, and three-sentence input contains the sentence of interest plus its predecessor and successor. Additionally, we examine the influence of the conference rank on classification performance, i. e., we separately evaluate sentences from conferences ranked as A*, A, and B and C combined. Thus, papers from B and C are treated as one bucket, since the number of C papers (232 publications) is small.

Sentence Paraphrasing. The training of the paraphrasing models is based only on text from A* and A conference papers to ensure high-quality training data. The models are fine-tuned on Pegasus-DS and IDM-DS to reconstruct the original scientific sentence from the corrupted version. We fine-tune models based on T5 v1.1 in the variants small, base, and large, and BART base. We use GPT-2 with the prompt prefix "In scientific language," and include identity as a baseline. For all models, we apply beam search with a width of 5 to generate paraphrases and select the one with the highest probability.

The metrics are computed on the test split of each dataset. The test splits are divided into buckets which reflect the amount of changes made compared to the gold standard relative to the sentence length. The changes range from 0% to 50% in 10% steps resulting in six buckets, where, for example, 40% means that 6 words are changed in a 15 words long sentence. For IDM-DS, the number of changes is known from creating the dataset. Thus, we control the amount of changes in the sentences, but it may happen in an unlikely case that a sequence of operations could undo an earlier change on a sentence, e. g., an insert followed by a delete later on. For Pegasus-DS, we use the word error rate [40] (WER) between the original and corrupted sentence to measure the amount of changes. WER is a word-level version of the edit distance representing the number of substitutions, deletions, and insertions divided by the original sequence length.

Additionally, we evaluate the performance of our models on the GYAFC dataset to assess their capabilities to transform text from informal to formal writing. We compare our models to the results of the GYAFC paper [31], which includes a non-scientific paraphraser, a rule-based approach, and a NMT-based model combined with rules (denoted as NMT combined). We also compare to the results of two text-style transfer models, DualRL and DAST-C [15].

3.4 Hyperparameter Optimization

For all tasks and datasets, we use random 70:20:10 train, validate, and test split. We tune hyperparameters on a 10% subset of the train and validation data.

Scientificness Score. For the scientificness score, we fine-tune BERT, SciBERT, and WideMLP using AdamW. We test the learning rates $1 \cdot 10^{-5}$, $3 \cdot 10^{-5}$, $5 \cdot 10^{-5}$, the dropout rates 0.1, 0.3, and 0.5, and the values 0.05, 0.01, and 0.001 for the weight decay. We train BERT and SciBERT for five epochs and WideMLP for ten epochs, since the loss stopped decreasing there. We use a batch size of 8 as this was the highest one to fit on our GPU. SciBERT performed best with a learning rate of $1 \cdot 10^{-5}$, 0.3 dropout rate, and 0.1 weight decay. BERT performed best using a learning rate of $1 \cdot 10^{-5}$, 0.1 dropout rate, and 0.5 weight decay. WideMLP performed best with a learning rate of 0.05, 0.3 dropout rate, and 0.05 weight decay.

Section Classification. We use Adam for fine-tuning BERT and SciBERT for multi-label classification. We set a maximum of 15 epochs with early stopping if the validation loss did not decrease for two epochs. Since the models stopped improving after 1 to 3 epochs, we did not tune the number of epochs further. We train with a batch size of 32, which was the maximum that reliably fit on our GPU. We experimented with learning rates of $1 \cdot 10^{-5}, 3 \cdot 10^{-5}, 5 \cdot 10^{-5}$ and with λ thresholds of 0.5, 0.3, 0.2, and 0.1, i. e., the threshold above which a label is assigned in multi-label classification. The best-performing parameters were found to be a learning rate of $1 \cdot 10^{-5}$ and $\lambda = 0.2$ for all transformer models. For WideMLP, we use AdamW and train for 100 epochs with a learning rate of 10^{-1} for all datasets, following Galke et al. [9]. After testing the λ thresholds, we achieved the best results with $\lambda = 0.2$.

Sentence Paraphrasing. We use AdamW to fine-tune T5 and BART. Performance stopped improving after three to five epochs, so we set the number of epochs to 5. As we did not observe a performance impact of changing the batch size, we use the highest batch size that fits on our GPU, which was between 20 to 200 depending on the model. We use a learning rate of $2 \cdot 10^{-5}$ for the experiments. We test the values of 0.05, 0.01, and 0.001 for weight decay. Different metrics favored different values, so we use 0.001 as weight decay because the models perform consistently well for all metrics using this value.

3.5 Measures

Since the scientificness score is a regression task, we evaluate it using mean squared error (MSE). For the section classification, we use sample-based F1, following Galke et al. [9]. For the sentence paraphrasing, BLEU, METEOR, and BERTScore measure the difference to the gold standard and self-BLEU measures the difference to the input. BLEU calculates n-gram similarity with $n = 4$ and is the standard metric for paraphrasing [14,15,18,20,31]. METEOR is similar to BLEU but includes synonym matching to better match human judgements [14]. BERTScore [43] measures semantic changes by calculating the cosine similarity of sentence embeddings of two sentences [3,15,18]. We use SciBERT [1] to generate these embeddings, since we apply the score on scientific text. The self-BLEU calculates a BLEU score between the input sentence and output sentence and is a measure of the amount of changes done by each model.

4 Results

For the scientificness score task, we achieved an MSE of 0.181% for the fine-tuned BERT, 0.213% for fine-tuned SciBERT, and 0.049% for the best performing WideMLP. The results of our study of the effect of `<equation>` or `<reference>` tokens on the models are presented in Table 2. For scientific text, the score is roughly the same for sentences with and without such tokens. However, the standard deviation with tokens is three orders of magnitude lower for sentences with the tokens. Adding such tokens to non-scientific text pushes the score towards more scientificness and also increases the standard deviation.

Table 2. Scientificness score of sentences grouped by conference rank. Left: Only sentences without `<equation>` or `<reference>` tokens. Right: Only sentences with such tokens are evaluated. We also report scores for non-scientific sentences (NSC) and modified-NSC (m-NSC), where the equation and reference tokens were artificially inserted at random.

Model	MSE	Without equation and reference tokens					With equation and reference tokens				
		A*	A	B	C	NSC	A*	A	B	C	m-NSC
BERT	Avg	.8993	.8985	.8984	.8918	.1054	.9016	.9016	.9016	.9016	.8142
	SD	.0392	.0449	.0450	.0804	.0786	.0001	.0001	.0001	.0001	.2334
SciBERT	Avg.	.9004	.8988	.8992	.8892	.1034	.9032	.9032	.9032	.9031	.8819
	SD	.0438	.0548	.0514	.0977	.0778	.0000	.0000	.0000	.0000	.1096
WideMLP	Avg	.8914	.8880	.8889	.8645	.1388	.8913	.8878	.8885	.8648	.5168
	SD	.0522	.0617	.0611	.0997	.1197	.0523	.0615	.0606	.0988	.1387

For the section classification, the best sample-based F1-score was achieved by BERT trained on a three-sentence input taken from conferences ranked A*. See Table 3 for detailed results. Table 4 shows the results for the context length experiment. In this setting, the BERT model trained on two and evaluated on three sentences achieved the best performance.

For sentence paraphrasing, the results in Table 5 show that T5 large performed best on the fine-tuning datasets. On the IDM-DS, sentences are changed more (self-BLEU) than on the Pegasus-DS, and at the same time the changed sentences are closer to the gold standard (BLEU). On the GYAFC dataset, see results in Table 6, T5 base has the highest BLEU score. Overall, the fine-tuning on IDM-DS performed better than Pegasus-DS as the BLEU score is higher, but at the cost of a higher self-BLEU.

The BLEU and METEOR scores improve with larger model sizes, i. e., the generated sentences are closer to the original sentences when larger models are

Table 3. Sample-based F1-score (in %) on section classification. Model trained on all data with different context sizes and evaluated per conference level. 1-sentence input uses the current sentence only, 2-sentence additionally considers the previous, and 3-sentence additionally the previous and next sentence.

Input	Model	all	A*	A	B/C
1-sentence	BERT	68.37	68.97	64.69	64.60
	SciBERT	68.68	69.30	64.96	64.42
	WideMLP	40.97	41.42	38.20	38.61
2-sentences	BERT	79.40	79.77	77.05	77.04
	SciBERT	79.16	79.58	76.62	76.20
	WideMLP	60.36	61.26	54.73	54.58
3-sentences	BERT	**90.10**	**90.26**	**89.13**	**88.86**
	SciBERT	88.87	89.05	87.67	87.54
	WideMLP	67.43	68.37	61.30	62.05

Table 4. Sample-based F1-score (in %) of the section classification task from papers of all ranks. Context (Train) indicates the context during training, while Context (Eval) refers to the context for evaluation. 1-sentence input uses the current sentence only, 2-sentence additionally considers the previous, and 3-sentence additionally the previous and next sentence.

Context (Train)	Model	Context (Eval)		
		1-sentence	2-sentences	3-sentences
1-sentence	BERT	68.37	78.35	81.72
	SciBERT	68.68	78.81	81.81
	WideMLP	40.97	42.46	43.16
2-sentences	BERT	73.96	79.40	**90.30**
	SciBERT	73.05	79.16	89.39
	WideMLP	51.44	60.36	64.82
3-sentences	BERT	72.68	90.04	90.10
	SciBERT	71.48	88.35	88.87
	WideMLP	49.51	62.53	67.43

used. The BERTScore also improves for larger models, showing that the sentences' semantics is preserved better. Increasing the model size decreases self-BLEU, i. e., larger models change the input sentences to a higher degree. On the Pegasus-DS the difference of BLEU and METEOR is quite large, while being quite small on the IDM-DS. This means that on IDM-DS, the models have a higher chance of replacing words with the correct synonyms, while the paraphrasing output on the Pegasus-DS remains to have larger differences to the gold standard. The models fine-tuned on Pegasus-DS have a higher self-BLEU than the IDM-DS models. Therefore, the models' inputs are closer to their outputs, i.

e., these models make fewer changes on average. Table 6 shows that our models have higher BLEU and self-BLEU scores on the GYAFC dataset, i. e., our models make fewer changes to the input sentences and still produce outputs close to the gold standard.

Table 5. Results (in %) for Pegasus-DS (left) / IDM-DS (right) divided into buckets based on Word Error Rate (WER) and change rate (CR), respectively. All models are fine-tuned on the respective dataset. Identity returns the input.

WER/CR	Model	BLEU ↑	METEOR ↑	BERT ↑	sBLEU
0	identity	69.56/74.62	78.03/83.91	**87.79**/97.41	100.00/100.00
	T5 small	69.01/84.91	76.91/87.66	86.94/90.53	93.50/**58.17**
	T5 base	69.00/85.95	77.23/88.44	86.74/90.81	88.35/76.00
	T5 large	**69.57**/86.85	78.07/89.03	86.86/**98.50**	**85.27**/74.93
	BART base	69.24/**87.71**	**78.35/91.23**	87.44/91.56	86.11/75.44
10%	identity	48.25/67.16	63.05/81.53	82.84/96.47	100.00/100.00
	T5 small	48.47/80.03	62.56/86.17	81.91/89.82	90.15/**55.24**
	T5 base	48.83/81.64	63.45/87.07	81.54/90.24	80.73/71.09
	T5 large	**49.57**/83.13	64.78/87.78	81.81/**98.20**	**75.73**/69.52
	BART base	49.42/**83.72**	**65.31/90.09**	**82.88**/91.08	79.15/70.23
20%	identity	36.28/58.16	55.28/78.71	79.71/94.23	100.00/100.00
	T5 small	36.95/71.73	55.08/83.47	78.65/88.26	87.13/**49.86**
	T5 base	37.59/74.09	56.29/84.60	78.42/88.91	75.31/66.27
	T5 large	**38.60**/76.40	57.60/85.58	78.83/**97.54**	69.40/63.87
	BART base	38.27/**76.41**	**58.04/87.76**	**80.06**/89.87	**74.22**/64.86
30%	identity	27.32/52.38	50.00/77.05	76.87/92.40	100.00/100.00
	T5 small	28.34/65.34	50.15/81.58	76.03/87.01	85.01/45.87
	T5 base	29.50/68.41	51.55/82.87	76.15/87.87	71.31//63.72
	T5 large	**30.59**/**/71.25**	52.75/83.96	76.63/**97.04**	**64.45/60.80**
	BART base	30.18/70.91	**53.16/86.06**	**77.92**/88.87	69.84/62.12
40%	identity	22.02/46.58	47.06/75.46	74.86/90.43	100.00/100.00
	T5 small	23.29/59.10	47.88/79.65	74.83/85.67	84.62/**41.76**
	T5 base	25.16/62.61	49.62/81.04	75.28/86.71	67.98/60.95
	T5 large	**26.43/65.89**	50.93/82.25	75.87/**96.47**	**60.26**/57.39
	BART base	25.82/65.18	**51.24/84.30**	**76.98**/87.76	65.62/59.16
50%	identity	**36.28**/41.28	55.28/73.82	**79.71**/88.38	100.00/100.00
	T5 small	29.03/52.89	55.13/77.69	75.94/84.19	80.81/**37.63**
	T5 base	32.65/56.76	58.52/79.22	77.28/85.41	63.32/58.79
	T5 large	34.67/**60.40**	60.46/80.51	78.19/**95.88**	**56.28**/54.73
	BART base	34.19/59.51	**61.15/82.57**	79.21/86.54	60.17/56.79

Table 6. Results (in %) of our models on the GYAFC dataset. All models are evaluated with the same implementation of the metrics for either our own models ("own", i. e., we trained the models), on the models' output provided by the original papers (marked as "output" in the provenance column), or model weights (indicated by "weights"). The best scores per metric are marked in bold.

Model	Fine-tuning	BLEU↑	METEOR↑	BERTScore↑	sBLEU↓	Provenance
Original Informal	–	55.01	20.25	94.00	100.00	output [31]
Rule-based	–	49.49	17.20	94.39	57.92	output [31]
NMT Combined	GYAFC	52.50	17.23	**94.93**	47.86	output [31]
DualRL	GYAFC	39.75	16.93	92.38	45.99	output [15]
DAST-C	GYAFC	36.14	18.52	90.99	47.81	output [15]
Pegasus	–	49.72	16.80	86.33	35.98	weights [30]
IDM	–	49.52	17.93	92.76	80.02	own
GPT-2	–	1.48	17.30	78.84	**1.38**	own
T5 small	Pegasus-DS	48.73	22.04	84.35	65.23	own
T5 base	Pegasus-DS	50.04	**22.22**	66.30	70.30	own
T5 large	Pegasus-DS	49.91	21.56	85.65	74.65	own
BART base	Pegasus-DS	54.56	20.75	67.13	86.42	own
T5 small	IDM-DS	58.47	20.75	88.66	86.65	own
T5 base	IDM-DS	**57.21**	20.63	67.62	90.60	own
T5 large	IDM-DS	55.23	20.45	87.89	91.28	own
BART base	IDM-DS	54.48	20.42	67.40	93.24	own

5 Discussion

5.1 Key Results

Scientificness Score. All models score sentences from scientific papers at a value of around 0.9 (see Table 2) with the highest scores provided by SciBERT. For all models, the mean output decreases, and the standard deviation increases with decreasing conference rank. This suggests that lower-ranked conferences contain, on average, fewer scientific sentences. Therefore, low-ranked conferences have a broader range of sentence quality and include more sentences with a lower scientificness score.

The experiment on the influence of `<equation>` and `<reference>` tokens (see Table 2) shows that transformer models rank sentences containing such tokens higher than sentences without such tokens. This indicates that the models have learned to connect sentences containing these tokens with higher scientificness. The low standard deviation indicates a more stable prediction of the scientificness score for sentences containing these tokens.

We performed an additional experiment to analyze the influence of the `<equation>` and `<reference>` tokens on non-scientific sentences. We modified the non-scientific sentences by inserting the specific tokens. As shown in Fig. 1, SciBERT now scores most non-scientific sentences containing such a token with

0.9, while BERT still keeps a small amount of sentences with scores in the non-scientific range. For WideMLP, we can see that the influence of the tokens is much smaller. The mean score here is 0.52, which is lower then in BERT (0.81) and SciBERT (0.88). Therefore, WideMLP relies less on these tokens, which makes it more suitable for non-scientific text containing equations.

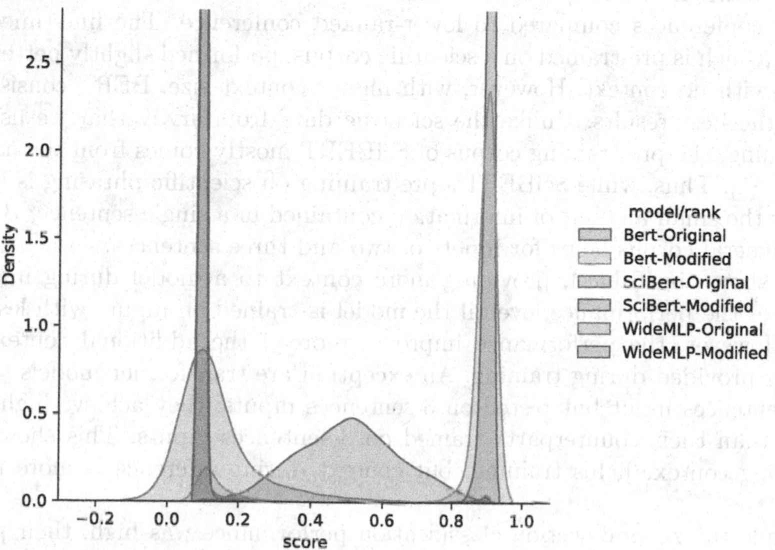

Fig. 1. Scorings for non-scientific sentences with no modifications (original) and the same sentences with `<equation>` and `<reference>` tokens being randomly inserted (modified).

Section Classification. For section classification, the WideMLP baseline is consistently outperformed by transformer-based models. This might be a result of the lack of sequence information which the Bag-of-Words approach neglects. Also, we use pre-trained transformer models but train WideMLP from scratch, which means that the transformer models start with some understanding of language already.

We observe that the classification performance increases with more context provided (from 1-sentence to 3-sentences). For example, the BERT model classifies the one-sentence input

> "*Then a weighted sum of attention is carried out to get an attended attention over the document for the final predictions.*"

as possibly fitting into an introduction, related work, or methods. However, enhanced by the surrounding sentences to the following input

> "We present a novel neural architecture, called attention-over-attention reader, to tackle the cloze-style reading comprehension task. *Then a*

> *weighted sum of attention is carried out to get an attended attention over the document for the final predictions.* Among several public datasets, our model could give consistent and significant improvements over various state-of-the-art systems by a large margin."

the sentence is correctly placed in the conclusion.

As shown in Table 3, the performance is better for sentences from higher-ranked conferences compared to lower-ranked conference. The fine-tuned SciBERT, which is pre-trained on a scientific corpus, performed slightly better than BERT with no context. However, with higher context size, BERT consistently yields the best results. Unlike the scientific data from arXiv that we used for fine-tuning, the pre-training corpus of SciBERT mostly comes from the medical domain [1]. Thus, while SciBERT's pre-training on scientific phrasing is beneficial for the small amount of information contained in a single sentence, BERT's more general corpus helps for inputs of two and three sentences.

As shown in Table 4, providing more context to a model during inference improves the performance, even if the model is trained on inputs with less context. However, the performance improves more if the additional context was already provided during training. An exception are transformer models trained on 2-sentences input but tested on 3-sentences inputs: they achieve higher F1-scores than their counterparts trained on 3-sentences inputs. This shows that providing context helps training, but context during inference is more important.

While the general section classification performance was high, their predictions include label combinations one would not expect to find in a scientific paper. For example, sentences were assigned no label, more than two labels, or sections that would not typically contain similar sentences (e. g., "introduction" and "experiment"). In pre-experiments, applying individual thresholds per class or limiting the number of assigned labels to one or two labels per sentence affected <1.21% of outputs and improved the sample-based F1-score only by <0.01% for the best model. Therefore, the influence can be neglected.

Sentence Paraphrasing. Our task differs from general paraphrasing, since we focus specifically on scientific sentence improvement, where we expect that the input is already quite good and only few changes are necessary. However, most baseline models have higher BERTScores, which means that these paraphrasers can still keep the semantics of the input, which makes them better general-purpose paraphrasers. We observe that fine-tuning on IDM-DS gives a 7% larger BLEU score than fine-tuning on Pegasus-DS. GPT-2 with our custom prompt has a low self-BLEU but high BLEU score, which means that it changes the input a lot and that the output is different from the gold standard. The low performance of GPT-2 may be attributed to the lack of fine-tuning the model. Finally, we observe that the amount of changes in the sentences increases with a higher corruption level. This means that more sentences are changed when the dissimilarity to the original scientific sentence increases.

5.2 Threats to Validity

We provide a method for distinguishing whether a sentence is scientific or not. The selection and labeling of the non-scientific datasets may pose a limitation, which could be improved by using a wider range of non-scientific datasets and more fine-grained scientificness scores. We carefully investigated the influence of `<equation>` and `<reference>` tokens. Although the experiments showed that the tokens increase the scientificness of a sentence, this is not an issue, since references and citations are in fact indicators of high scientificness. For the sentence paraphrasing, the output sentence can be equal to the input sentence. An unchanged sentence can be a problem for general paraphrasing, where the model should provide a variety of different suggestions. However, this is not an issue for us, since the input sentence can be already (quite) scientific.

5.3 Ethical Considerations

The development of AI systems in fields like scientific writing needs consideration of ethical and social impact. Common problems of recent language models are authorship and hallucinations [44]. Our models do not present these concerns. The only models we trained that generate text are the paraphrasers, which aim to maintain the meaning of the input sentence without introducing any new information, whether real or fake. If one were to deploy our models for interactive writing support, users should check the suggested paraphrases and not blindly integrate them into their text. This applies to all writing support tools, even simple non-AI variants like Overleaf's dictionary that at times may suggest wrong replacements for technical terms or unknown words. As with other language models, there is a possibility of extracting training samples [4, 8]. However, the pre-training checkpoints of our generative models are publicly available and we fine-tuned them on public papers only, which should not contain sensitive private information. In contrast to proprietary language models that may entail high costs, both in training as well as operation, and thus makes some inaccessible, our models are open-source and accessible to anyone.

6 Conclusion

While scientific writing remains a complex task, machine learning methods can be leveraged to be of assistance. We show that transformer models achieve the best results in computing a score of scientificness for a sentence, classifying a sentence to a section within the structure of a scientific paper, and paraphrasing scientific sentences. SciBERT, which is pre-trained on a scientific corpus comprising mostly papers from the broad biomedical domain [1], does not outperform the general-purpose BERT model [7] on tasks for scientific texts from the computer science domain. We also showed that transformer models profit from context during training and evaluation, with providing more context during evaluation being more important than providing it during training.

There are also other datasets such as unarXiv [34] that we considered using. Due to the lack of providing section information with the text parapgrahs, we created our own section extraction and mapping approach. In March 2023, an updated unarXive 2022 dataset [35] was released that provides structured full text, i. e., per paragraph the section title, section type, content type, etc. It would be interesting to repeat the experiments with this dataset that was not available yet at the time of writing.

Acknowledgement. This work is co-funded under the 2LIKE project by the German Federal Ministry of Education and Research (BMBF) and the Ministry of Science, Research and the Arts Baden-Württemberg within the funding line Artificial Intelligence in Higher Education. We thank C. Schindler, D. Podjavorsek, and S. Birkholz for an early version of the section headings synonym dictionary.

References

1. Beltagy, I., Lo, K., Cohan, A.: Scibert: a pretrained language model for scientific text. In: Proceedings of EMNLP-IJCNLP 2019. ACL (2019). https://doi.org/10.18653/v1/D19-1371
2. Bottomley, J.: Academic Writing for International Students of Science. Routledge, Abingdon (2022)
3. Cao, Y., Wan, X.: DivGAN: towards diverse paraphrase generation via diversified generative adversarial network. In: Findings of the Association for Computational Linguistics: EMNLP 2020. ACL (2020). https://doi.org/10.18653/v1/2020.findings-emnlp.218
4. Carlini, N., et al.: Extracting training data from large language models. In: Bailey, M., Greenstadt, R. (eds.) 30th USENIX Security Symposium, USENIX Security 2021(August), pp. 11–13, 2021. pp. 2633–2650. USENIX Association (2021). https://www.usenix.org/conference/usenixsecurity21/presentation/carlini-extracting
5. Clark, K., Luong, M., Le, Q.V., Manning, C.D.: ELECTRA: pre-training text encoders as discriminators rather than generators. In: ICLR 2020. OpenReview.net (2020)
6. Dabre, R., Chu, C., Kunchukuttan, A.: A survey of multilingual neural machine translation. ACM Comput. Surv. **53**(5): 1–38 (2020). Article No. 99. https://doi.org/10.1145/3406095
7. Devlin, J., Chang, M., Lee, K., Toutanova, K.: BERT: pre-training of deep bidirectional transformers for language understanding. In: NAACL-HLT 2019. ACL (2019). https://doi.org/10.18653/v1/n19-1423
8. Diera, A., Lell, N., Garifullina, A., Scherp, A.: A study on extracting named entities from fine-tuned vs. differentially private fine-tuned BERT models. CoRR abs/2212.03749 (2022). https://doi.org/10.48550/arXiv.2212.03749
9. Galke, L., et al.: Are we really making much progress? bag-of-words vs. sequence vs. graph vs. hierarchy for single- and multi-label text classification. CoRR (2022). https://doi.org/10.48550/arXiv.2204.03954
10. Galke, L., Scherp, A.: Bag-of-words vs. graph vs. sequence in text classification: questioning the necessity of text-graphs and the surprising strength of a wide MLP. In: ACL 2022. ACL (2022). https://doi.org/10.18653/v1/2022.acl-long.279

11. Garcia-Silva, A., Gomez-Perez, J.M.: Classifying scientific publications with BERT - is self-attention a feature selection method? In: Hiemstra, D., Moens, M.-F., Mothe, J., Perego, R., Potthast, M., Sebastiani, F. (eds.) ECIR 2021. LNCS, vol. 12656, pp. 161–175. Springer, Cham (2021). https://doi.org/10.1007/978-3-030-72113-8_11
12. Gu, N., Gao, Y., Hahnloser, R.H.R.: Local citation recommendation with hierarchical-attention text encoder and SciBERT-based reranking. In: Hagen, M., et al. (eds.) ECIR 2022. LNCS, vol. 13185, pp. 274–288. Springer, Cham (2022). https://doi.org/10.1007/978-3-030-99736-6_19
13. Gupta, A., Agarwal, A., Singh, P., Rai, P.: A deep generative framework for paraphrase generation. In: McIlraith, S.A., Weinberger, K.Q. (eds.) Proceedings of the Thirty-Second AAAI Conference on Artificial Intelligence, (AAAI-18), the 30th innovative Applications of Artificial Intelligence (IAAI-18), and the 8th AAAI Symposium on Educational Advances in Artificial Intelligence (EAAI-18), New Orleans, Louisiana, USA, 2–7 February 2018, pp. 5149–5156. AAAI Press (2018). https://www.aaai.org/ocs/index.php/AAAI/AAAI18/paper/view/16353
14. Hegde, C.V., Patil, S.: Unsupervised paraphrase generation using pre-trained language models. CoRR (2020)
15. Hu, Z., Lee, R.K., Aggarwal, C.C., Zhang, A.: Text style transfer: a review and experimental evaluation. SIGKDD Explor. (2022). https://doi.org/10.1145/3544903.3544906
16. Jin, D., Jin, Z., Hu, Z., Vechtomova, O., Mihalcea, R.: Deep learning for text style transfer: a survey. COLING 2022 (2022). https://doi.org/10.1162/coli_a_00426
17. Knight, K., Marcu, D.: Statistics-based summarization - step one: sentence compression. In: Proceedings of the Seventeenth National Conference on Artificial Intelligence and Twelfth Conference on on Innovative Applications of Artificial Intelligence, 2000. AAAI Press/The MIT Press (2000)
18. Kumar, D., Mou, L., Golab, L., Vechtomova, O.: Iterative edit-based unsupervised sentence simplification. In: ACL 2020. ACL (2020). https://doi.org/10.18653/v1/2020.acl-main.707
19. Lewis, M., et al.: BART: denoising sequence-to-sequence pre-training for natural language generation, translation, and comprehension. In: ACL 2020. ACL (2020). https://doi.org/10.18653/v1/2020.acl-main.703
20. Liu, X., Mou, L., Meng, F., Zhou, H., Zhou, J., Song, S.: Unsupervised paraphrasing by simulated annealing. In: ACL 2020. ACL (2020). https://doi.org/10.18653/v1/2020.acl-main.28
21. Liu, Y., et al.: RoBERTa: a robustly optimized BERT pretraining approach. CoRR (2019)
22. Luoma, J., Pyysalo, S.: Exploring cross-sentence contexts for named entity recognition with BERT. In: Proceedings of the 28th International Conference on Computational Linguistics, COLING 2020, Barcelona, Spain (Online), 8–13 December 2020. International Committee on Computational Linguistics (2020). https://doi.org/10.18653/v1/2020.coling-main.78
23. McKeown, K.R.: Paraphrasing questions using given and new information. Am. J. Comput. Linguist. **9**(1), 1–10 (1983)
24. Motrichenko, D., Nedumov, Y., Skorniakov, K.: Bag of tricks for citation intent classification via SciBERT. In: 2021 Ivannikov Ispras Open Conference (ISPRAS) (2021). https://doi.org/10.1109/ISPRAS53967.2021.00022
25. Ouyang, L., et al.: Training language models to follow instructions with human feedback. CoRR (2022). https://doi.org/10.48550/arXiv.2203.02155

26. Prakash, A., et al.: Neural paraphrase generation with stacked residual LSTM networks. In: COLING 2016. ACL (2016)
27. Preda, G.: COVID19 Tweets (2020). https://doi.org/10.34740/KAGGLE/DSV/1451513
28. Radford, A., Wu, J., Child, R., Luan, D., Amodei, D., Sutskever, I., et al.: Language models are unsupervised multitask learners. OpenAI blog (2019)
29. Raffel, C., et al.: Exploring the limits of transfer learning with a unified text-to-text transformer. J. Mach. Learn. Res. **21**(1), 5485–5551 (2020). Article No. 140
30. Rajauria, A.: Pegasus paraphraser. https://huggingface.co/tuner007/pegasus_paraphrase. Accessed November 2022
31. Rao, S., Tetreault, J.R.: Dear sir or madam, may i introduce the GYAFC dataset: corpus, benchmarks and metrics for formality style transfer. In: NAACL-HLT 2018. ACL (2018). https://doi.org/10.18653/v1/n18-1012
32. Rudnichenko, N., Vychuzhanin, V., Shibaeva, N., Antoshchuk, S., Petrov, I.: Intellectual information system for supporting text data rephrasing processes based on deep learning. In: Proceedings of the 2nd International Workshop on Intelligent Information Technologies & Systems of Information Security with CEUR-WS. CEUR-WS.org (2021)
33. Rudnicka, K.: Variation of sentence length across time and genre. Diachronic corpora, genre, and language change (2018)
34. Saier, T., Färber, M.: unarXive: a large scholarly data set with publications' full-text, annotated in-text citations, and links to metadata. Scientometrics **125**(3), 3085–3108 (2020)
35. Saier, T., Krause, J., Färber, M.: unarXive 2022: All arXiv Publications Pre-Processed for NLP, Including Structured Full-Text and Citation Network. CoRR abs/2303.14957 (2023)
36. Schultz, D.M.: The structure of a scientific paper. Am. Meteorol. Soc. (2009). https://doi.org/10.1007/978-1-935704-03-4_4
37. Shah, P., et al.: Building a conversational agent overnight with dialogue self-play. CoRR (2018)
38. Wando, B.: Ukraine Conflict Twitter Dataset (2022). https://doi.org/10.34740/KAGGLE/DSV/4787803
39. Wang, S., Gupta, R., Chang, N., Baldridge, J.: A task in a suit and a tie: paraphrase generation with semantic augmentation. In: AAAI, IAAI, EAAI 2019, AAAI Press (2019). https://doi.org/10.1609/aaai.v33i01.33017176
40. Woodard, J., Nelson, J.: An information theoretic measure of speech recognition performance (1982)
41. Wymann, C.: Checkliste Schreibprozess : Ihr Weg zum guten Text: Punkt für Punkt. Verlag Barbara Budrich (2018)
42. Zhang, J., Zhao, Y., Saleh, M., Liu, P.J.: PEGASUS: pre-training with extracted gap-sentences forabstractive summarization. In: ICML 2020. PMLR (2020)
43. Zhang, T., Kishore, V., Wu, F., Weinberger, K.Q., Artzi, Y.: Bertscore: evaluating text generation with BERT. In: ICLR 2020. OpenReview.net (2020)
44. Zhou, J., Müller, H., Holzinger, A., Chen, F.: Ethical ChatGPT: Concerns, Challenges, and Commandments. CoRR abs/2305.10646 (2023)

Author Index

Published by Springer Nature Switzerland AG 2023
A. Holzinger et al. (Eds.): CD-MAKE 2023, LNCS 14065, pp. 319–320, 2023.
https://doi.org/10.1007/978-3-031-40837-3

Printed in the United States
by Baker & Taylor Publisher Services